Advanced Research and Applications of Deep Learning and Neural Network in Image Recognition

Advanced Research and Applications of Deep Learning and Neural Network in Image Recognition

Editors

Ganggang Dong
Yuanxin Ye
Zhongling Huang

Basel • Beijing • Wuhan • Barcelona • Belgrade • Novi Sad • Cluj • Manchester

Editors
Ganggang Dong
Xidian University
Xi'an
China

Yuanxin Ye
Southwest Jiaotong
University
Chengdu
China

Zhongling Huang
Northwestern Polytechnical
University
Xi'an
China

Editorial Office
MDPI
St. Alban-Anlage 66
4052 Basel, Switzerland

This is a reprint of articles from the Special Issue published online in the open access journal *Electronics* (ISSN 2079-9292) (available at: https://www.mdpi.com/journal/electronics/special_issues/57UONM477H).

For citation purposes, cite each article independently as indicated on the article page online and as indicated below:

Lastname, A.A.; Lastname, B.B. Article Title. *Journal Name* **Year**, *Volume Number*, Page Range.

ISBN 978-3-7258-0423-8 (Hbk)
ISBN 978-3-7258-0424-5 (PDF)
doi.org/10.3390/books978-3-7258-0424-5

Contents

About the Editors

Ganggang Dong

Ganggang Dong received an M.S. and Ph.D. in Information and Communication Engineering from the National University of Defense Technology, Changsha, China, in 2012 and 2016, respectively. His research interests include but are not limited to deep learning, SAR imaging, radar target detection and recognition, cognitive radio, and radar image interpretation. He has authored over 40 scientific papers in peer-reviewed journals and conferences, including *IEEE TIP*, *IEEE TCYB*, *IEEE TGRS*, *IEEE TIM*, and *Pattern Recognition*. Dr. Dong is currently an associate professor at Xidian University. He has accumulated over 1410 citations on Google Scholar. He was awarded the 2017 Excellent Doctoral Thesis from the Chinese Institute of Electronics and was a reviewer for various top-tier journals on remote sensing and image processing.

Yuanxin Ye

Yuanxin Ye (Member, IEEE) received a B.S. degree in Remote Sensing Science and Technology from Southwest Jiaotong University, Chengdu, China, in 2008 and a Ph.D. degree in Photogrammetry and Remote Sensing from Wuhan University, Wuhan, China, in 2013., He is a Professor at the Faculty of Geosciences and Environmental Engineering department at Southwest Jiaotong University. His research interests include remote sensing image processing, image registration, change detection, and object detection. Dr. Ye received the ISPRS Prize for Best Papers by Young Authors at the 23rd International Society for Photogrammetry and Remote Sensing Congress, Prague, Czech Republic, in 2016 and the Best Youth Oral Paper Award at ISPRS Geospatial Week 2017, Wuhan, in 2017.

Zhongling Huang

Zhongling Huang received a B.Sc. degree from Beijing Normal University in 2015 and a Ph.D. from the University of Chinese Academy of Sciences (UCAS) and the Aerospace Information Research Institute, Chinese Academy of Sciences, in 2020. She was a Visiting Scholar with the German Aerospace Center from 2018 to 2019. She is an Associate Professor at the Brain and Artificial Intelligence Laboratory (BRAIN LAB), School of Automation, Northwestern Polytechnical University, Xi'an, China. Her research interests include explainable deep learning for synthetic aperture radar (XAI4SAR), SAR image interpretation, deep learning, and remote sensing data mining.

Editorial

Guest Editorial: Foreword to the Special Issue on Advanced Research and Applications of Deep Learning and Neural Network in Image Recognition

Ganggang Dong [1,*], Yuanxin Ye [2] and Zhongling Huang [3]

[1] National Key Laboratory of Radar Signal Processing, Xidian University, Xi'an 710071, China
[2] Faculty of Geosciences and Environmental Engineering, Southwest Jiaotong University, Chengdu 611756, China; yeyuanxin@home.swjtu.edu.cn
[3] School of Automation, Northwestern Polytechnical University, Xi'an 710072, China; huangzhongling@nwpu.edu.cn
* Correspondence: dongganggang@nudt.edu.cn; Tel.: +86-029-88201246

Citation: Dong, G.; Ye, Y.; Huang, Z. Guest Editorial: Foreword to the Special Issue on Advanced Research and Applications of Deep Learning and Neural Network in Image Recognition. *Electronics* **2024**, *13*, 557. https://doi.org/10.3390/electronics13030557

Received: 21 January 2024
Accepted: 26 January 2024
Published: 30 January 2024

Over the last two decades, the realm of image recognition has undergone a remarkable transformation, characterized by an astonishing pace of advancement. Object detection performance, once stagnant at around 30-percentage in mean average precision (mAP), has now impressively soared to 90-percentage and beyond, particularly in benchmarks like PASCAL VOC. Equally remarkable is the surpassing of human capabilities by contemporary learning algorithms, achieving unprecedented accuracy in image classification tasks such as those found in the ImageNet dataset. These strides in image classification signal a new era with profound implications for practical applications, spanning areas like video surveillance, autonomous driving, intelligent healthcare, remote sensing image interpretation, and artificial intelligence.

At the core of this revolutionary progress in image recognition lie deep learning algorithms, whose success is rooted in two critical factors: the availability of extensive training datasets and the computational prowess of state-of-the-art platforms. Notably, deep neural networks consistently outperform meticulously crafted manual image features across a spectrum of image tasks. However, amidst the resounding success achieved by deep learning in image recognition, numerous challenges persist, emphasizing the necessity for further exploration and innovation.

This Special Issue goes beyond mere acknowledgment of these challenges; it is devoted to showcasing novel solutions poised to overcome these hurdles. By delving into the intricacies of these issues, our aim is to contribute to the ongoing discourse and advancement in the ever-evolving field of image recognition. This collection of research endeavors not only commemorates achievements thus far but also charts the course for the next frontier in the continuous evolution of image recognition technologies.

For this Special Issue, which was open to all researchers, 12 papers have undergone a meticulous review process and have been ultimately accepted for publication. These papers cover a broad spectrum of vision task topics as follows.

- Point cloud deep learning. The efficacy of 3D vision surpasses that of 2D vision due to its enhanced spatial and depth perception capabilities. Point cloud representation, commonly utilized in relevant applications, preserves the original geometric information in three-dimensional space without discretization. However, the inherent disorder of point clouds poses challenges for their integration into deep learning frameworks. In 2017, deep convolutional neural networks were used to address the sparsity and disorder issues associated with point clouds, forming the famous models, PointNet and PointNet++. Various kinds of improved models were presented subsequently, leading to the expanded application of point clouds in areas such as 3D object detection (Contributor 1), pose estimation (Contributor 2), and more (Contributor 3).

- Pixelwise semantic segmentation. Semantic segmentation is a fundamental task in computer vision field. It refers to classify each pixel in an image, enabling segmentation into semantically distinct regions. In 2015, the full convolutional network (FCN) was presented, in which the task of semantic segmentation were revolutionized by employing transposed convolutions. The fully connected layers were replaced by the convolutional layers to achieve end-to-end segmentation. The presented fashion have been expanded across various fields. Weng et al. enhanced the DeepLabV3+ model, addressing railway track extraction errors with morphological algorithm optimization (Contributor 4). Zheng et al. applied multi-scale semantic segmentation to fire smoke, incorporating global information (Contributor 5). These advancements showcase the widespread impact of semantic segmentation in diverse fields.

- Zeroshot learning. The common supervised learning tasks were often struggled with the limited or the unlabeled examples. This challenge was tackled by zero-shot learning, in which the transferable representations can be exploited. The discriminative and semantic-relevant features can be therefore encompassed into the learned representations. In some researchers, the semantic-relevant representations were emphasized through visual-semantic alignment. Likewise, the discrimination techniques for broader generalization were focused on in the other studies. Recently, the shared representations between these sub-tasks were targeted. In this special issue, Wang et al. introduces a novel partially-shared multi-task representation method preserving complementary knowledge, namely PS-GZSL (Contributor 6). The emerging algorithms like federated learning and contrast learning are used too to offer new solutions to zero-sample learning (Contributor 7).

- Model optimization. A significant challenge in the current advancement of deep learning lies in the extensive computation and parameters involved. It is infeasible to deploy the resource-intensive convolutional neural networks (CNNs) on computing devices with limitations, such as embedded systems and mobile devices (Contributor 8). To tackle these issues, considerable research efforts have been dedicated to compression techniques, including channel pruning, low-rank decomposition, and weight quantization. In this special issue, a new trick via dynamic pruning and layer fusion is presented to optimize the deep model (Contributor 9). Through the incorporation of knowledge distillation and short–long fine-tuning, the redundant layers with minimal accuracy loss can be eliminated. The primary objective is to reduce memory access more significantly than reducing computational complexity.

- Multimodal applications. In the real-world visual tasks, a multi-dimensional framework is involved, encompassing spatial, temporal, and modal dimensions. Spatially, tasks may be spanned from image-level and region-level to pixel-level assignments. Temporally, the challenges are then extended beyond the static images to include the processing of time-series videos. In terms of modalities, the inputs and the outputs can be displayed in a variety of fashions, such as images, text, videos, or other types like body poses (Contributor 10) and depth maps (Contributor 11). So, it refers to another important research fields, the data engineering (Contributor 12). Given the diverse range of application scenarios (Contributor 13), it is challenging to achieve the universality for model design. Consequently, the future development trajectory of deep visual systems will focus on constructing more versatile models capable of accommodating a wide array of input and output types, effectively addressing the varied demands arising from different scenarios.

In the end, we would like to express our gratitude to these authors who have dedicated their efforts to in-depth research in the field of computer vision. Their contributions are of significant importance in addressing current challenges. Additionally, we extend our thanks to all the reviewers for their time, dedication, and valuable insights during the evaluation process. This helps ensure the selection of high-quality papers in accordance with standards.

Funding: This work was supported by the National Natural Science Foundation of China under Grant 61971324, 61525105, and the foundation of National Key Laboratory of Radar Signal Processing under Grant JKW202310.

Conflicts of Interest: The authors declare no conflicts of interest.

List of Contributions

1. Zhang, L.; Meng, H.; Yan, Y.; Xu, X. Transformer-Based Global PointPillars 3D Object Detection Method. *Electronics* **2023**, *12*, 3092.
2. Wang, Q.; Lei, H.; Qian, W. Siamese PointNet: 3D Head Pose Estimation with Local Feature Descriptor. *Electronics* **2023**, *12*, 1194.
3. Wang, Q.; Qian, W. Z.; Lei, H.; Chen, L. Siamese Neural PointNet: 3D Face Verification under Pose Interference and Partial Occlusion. *Electronics* **2023**, *12*, 620.
4. Weng, Y.; Li, Z.; Chen, X.; He, J.; Liu, F.; Huang, X.; Yang, H. A Railway Track Extraction Method Based on Improved DeepLabV3+. *Electronics* **2023**, *12*, 3500.
5. Zheng, Y.; Wang, Z.; Xu, B.; Niu, Y. Multi-Scale Semantic Segmentation for Fire Smoke Image Based on Global Information and U-Net. *Electronics* **2022**, *11*, 2718
6. Wang, G.; Tang, S. Generalized Zero-Shot Image Classification via Partially-Shared Multi-Task Representation Learning. *Electronics* **2023**, *12*, 2085.
7. Yu, L.; Huang, J. Cyclic Federated Learning Method Based on Distribution Information Sharing and Knowledge Distillation for Medical Data. *Electronics* **2022**, *11*, 4039.
8. Zhao, M.; Li, M.; Peng, S.L.; Li, J. A Novel Deep Learning Model Compression Algorithm. *Electronics* **2022**, *11*, 1066.
9. Li, Q.; Li, H.; Meng, L. Deep Learning Architecture Improvement Based on Dynamic Pruning and Layer Fusion. *Electronics* **2023**, *12*, 1208.
10. Gong, Z.; Zhang, Y.; Lu, D.; Wu, T. Vision-Based Quadruped Pose Estimation and Gait Parameter Extraction Method. *Electronics* **2022**, *11*, 3702.
11. Han, H.; Liu, J.; Wang, W.; Gao, C.; Shi, J. An Improved CNN for Polarization Direction Measurement. *Electronics* **2023**, *12*, 3723.
12. Liu, S.; Zhang, L.; Liu, W.; Hu, J.; Gong, H.; Zhou, X.; Gong, D. RERB: A Dataset for Residential Area Extraction with Regularized Boundary in Remote Sensing Imagery for Mapping Application. *Electronics* **2022**, *11*, 2790.
13. Chen, X.; Feng, X.; Li, Y. An Image Unmixing and Stitching Deep Learning Algorithm for In-Screen Fingerprint Recognition Application. *Electronics* **2023**, *12*, 3768.

Short Biography of Authors

Ganggang Dong received the M.S. and Ph.D. degrees in information and communication engineering from National University of Defense Technology, Changsha, China, in 2012 and 2016, respectively. His research interests include, but not limited to deep learning, SAR imaging, radar target detection and recognition, cognitive radio, radar image interpretation. He has authored more than 40 scientific papers in peer-reviewed journals and conferences, including IEEE TIP, IEEE TCYB, IEEE TGRS, IEEE TIM, and Pattern Recognition. Dr. Dong is currently an associate professor with Xidian University. He received more than 1410 citations in Google Scholar. He was awarded the 2017 Excellent Doctoral Thesis of the Chinese Institute of Electronics. He served as a reviewer for some top-tier journals on remote sensing and image processing.

 *electronics*

Article

Multi-Scale Semantic Segmentation for Fire Smoke Image Based on Global Information and U-Net

Yuanpan Zheng, Zhenyu Wang *, Boyang Xu and Yiqing Niu

College of Computer and Communication Engineering, Zhengzhou University of Light Industry, Zhengzhou 450002, China

* Correspondence: wang_zhenyu0801@163.com

Abstract: Smoke is translucent and irregular, resulting in a very complex mix between background and smoke. Thin or small smoke is visually inconspicuous, and its boundary is often blurred. Therefore, it is a very difficult task to completely segment smoke from images. To solve the above issues, a multi-scale semantic segmentation for fire smoke based on global information and U-Net is proposed. This algorithm uses multi-scale residual group attention (MRGA) combined with U-Net to extract multi-scale smoke features, and enhance the perception of small-scale smoke. The encoder Transformer was used to extract global information, and improve accuracy for thin smoke at the edge of images. Finally, the proposed algorithm was tested on smoke dataset, and achieves 91.83% *mIoU*. Compared with existing segmentation algorithms, *mIoU* is improved by 2.87%, and *mPA* is improved by 3.42%. Thus, it is a segmentation algorithm for fire smoke with higher accuracy.

Keywords: multi-scale residual group attention; U-Net; smoke segmentation; global information

Citation: Zheng, Y.; Wang, Z.; Xu, B.; Niu, Y. Multi-Scale Semantic Segmentation for Fire Smoke Image Based on Global Information and U-Net. *Electronics* **2022**, *11*, 2718. https://doi.org/10.3390/electronics11172718

Academic Editor: George A. Papakostas

Received: 31 July 2022
Accepted: 24 August 2022
Published: 30 August 2022

Publisher's Note: MDPI stays neutral with regard to jurisdictional claims in published maps and institutional affiliations.

1. Introduction

Occurrence of fire not only threatens the safety of human life and property, but also damages the natural environment. According to the World Fire Statistics Center (WFSC), in 2019, there were 3.1 million fires in 34 countries around the world. About 19,000 civilians died in fires, and 68,000 were injured in fires. Moreover, as of 15 August 2020, California fires have destroyed 1.2 million acres of forest and burned a total of 1.4 million acres of fields for the year. Therefore, the segmentation of fire smoke images plays an important role in providing preconditions for fire smoke detection and early warning.

Semantic segmentation is one of the main tasks in the field of computer vision applications. It classifies each pixel in an image, so that the image can be segmented into regions with different semantics. In 2015, Long et al. [1] proposed the segmentation network FCN, which brought semantic segmentation into a new era. The network achieved end-to-end semantic segmentation using transposed convolutions to replace fully connected layers in the network. In the same year, Badrinarayanan et al. [2] proposed SegNet, which uses index information to perform up-sampling in a decoder to save computing power overhead, but this leads to the loss of more feature information. By using spatial pyramid pooling, PSPNet [3] adds local and global information to the feature map, which improves the semantic understanding ability of the network. In order to enhance feature information, Fu et al. [4] proposed DANet, in the form of a dual-path attention channel, to capture global feature dependencies simply and effectively. With the development of semantic segmentation, the semantic relationship between different images is gradually excavated by researchers. For example, Zhou et al. [5] used a graph model to build the semantic dependencies between a set of images, and the graph dropout was used to avoid ignoring isolated objects. Wang et al. [6] proposed a pixel-wise contrastive learning, which used the cross-image pixel-to-pixel relation to learn a well-structured pixel semantic embedding space to replace the traditional image-wise-based training paradigm. Recently, Zhou et al. [7] proposed a novel non-parametric segmentation method based on non-learnable prototypes, and achieved excellent performance.

There are two main types of algorithms for smoke segmentation: one is based on traditional image processing. According to color, texture, and motion characteristics of smoke, algorithms such as thresholding, clustering, and moving object detection are used to segment smoke regions. Appana et al. [8] converted smoke images from RGB color space to HSV color space, and performed thresholding on its saturation and brightness to achieve segmentation of smoke regions. Zhao et al. [9] used fuzzy c-means algorithm (FCM) to segment fire smoke areas based on pixel color information. Reference [10] proposed a multi-scale segmentation algorithm for smoke using a wavelet module based on smoke texture, and experiments show that it is better than traditional edge segmentation algorithms. Peng et al. [11] used background difference method to segregate smoke regions from videos based on a Gaussian mixture model (GMM), which is very effective for slow-moving smoke. Wu et al. [12] found that the optical flow of blue channel can effectively reflect motion characteristics of smoke, and proposed an algorithm based on dense optical flow to segregate smoke. Such methods effectively overcome the interference of smoke-like objects, but there will be a lot of interfering objects in the segmentation result, when some non-smog moving objects exist.

Some other smoke segmentation is performed based on deep learning methods. Compared with traditional image processing algorithms, convolutional neural networks can automatically learn deep-level pixel information in smoke images, which not only reduces the complexity of feature extraction, but also has higher anti-interference capability. Salman et al. [13] used DeepLab v3+ with an encoder and decoder to segment smoke, and the accuracy is significantly improved by 3%. Yuan et al. [14] proposed a dual-classification-assisted gated recurrent network (CGRNet) for smoke segmentation. The results are significantly better than existing algorithms, and satisfactory results are achieved on thin smoke. Considering the visibility of haze in the sky, Taanya et al. [15] adopted the dark channel pre-processing method to reduce the amount of haze in images, and combined dense optical flow with mask R-cnn to improve the anti-interference capability of the segmentation algorithm. Zhu et al. [16] proposed a 3D CNN with an encoder–decoder based on the motion features of smoke, which effectively mitigated the interference of moving objects.

Algorithms for smoke segmentation based on deep learning show significant advantages, but there are still many shortcomings: on the one hand, due to the variable size of smoke, the accuracy of existing algorithms for segmentation of small smoke is not high. On the other hand, the translucent character of thin smoke at the edges results in missed segmentation and wrong segmentation

Therefore, to solve above issues, this paper proposes a high-precision segmentation algorithm for smoke based on multi-scale features and global information of smoke. The main work is as follows: (1) a multi-scale residual group attention was proposed to extract and enhance smoke features of different scales, which improved the accuracy of segmentation in small smoke; (2) the self-attention of Transformer was used to generate global features of smoke and fuse them with multi-scale features to reduce the probability of missed segmentation and wrong segmentation; (3) the effectiveness of the method proposed in this paper was verified by ablation experiments and comparative experiments.

2. Related Work

2.1. Neural Attention

An attention mechanism is similar to the way that humans observe objects. More attention is given to locally important information by weighting the feature maps. It is very beneficial to enhance feature representation and reduce noise.

The most representative of existing channel attention is SE-Net [17], as shown in Figure 1. SE-Net generates attention maps by global average pooling, and applies them to input features. The purpose is to suppress useless channels, and enhance beneficial channels by building correlations between channels.

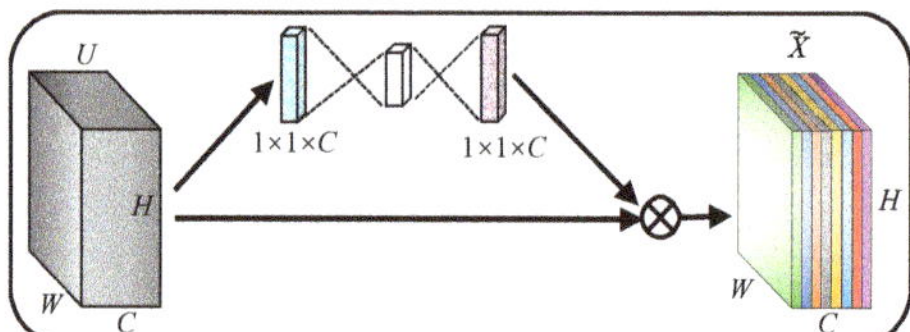

Figure 1. Channel attention (CA).

Spatial attention enhances the effective information in the feature map by assigning higher weights to pixels. Spatial attention in CBAM [18] is one of the most representative models. Spatial attention performs maximum pooling and average pooling along the channel. Then, the weight matrix is generated by convolution calculation. Finally, the weight matrix is multiplied with each channel of the feature map to obtain the attentional features, as shown in Figure 2.

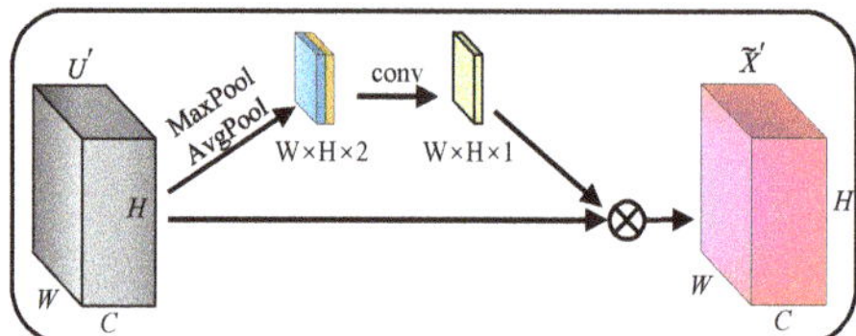

Figure 2. Spatial attention (SA).

2.2. Transformer

Transformer was first proposed by Vaswani [19] for machine translation and established state-of-the-art technology in many NLP tasks. To make Transformers also applicable for computer vision tasks, several modifications were made. For instance, Alexey et al. [20] proposed Vision Transformer (ViT), and achieved state-of-the-art results in ImageNet classification by directly applying Transformers with global self-attention to full-sized images. Swin Transformer [21] uses sliding windows to improve the computational speed of Transformer, and achieved excellent results in various vision tasks

3. Materials and Methods

3.1. Smoke–U-Net

U-Net [22], proposed by Ronneberger, is composed of an encoder and decoder. It is widely used in medical image segmentation. Smoke–U-Net is improved based on U-Net and Transformer, as shown in Figure 3. It mainly contains two parts. The first part is MRGA–U-Net at the bottom of Figure 3, which is improved based on U-Net. MRGA–U-Net replaces the convolution after up-sampling and down-sampling in each layer of U-Net with MRGA. On the one hand, MRGA is used to enhance the representation of smoke features, to make up for the loss of feature information. On the other hand, it can also enhance the perception of a model to multi-scale smoke. The second part is the GFM at the top of Figure 3, which is based on the encoder of Transformer. The input of GFM is an original RGB smoke image and the output is a feature map $\gamma \in \mathbb{R}^{H \times W \times 1}$, which represents the global information. The global feature γ is fused to MRGA–U-Net by matrix multiplication. Then, the fused features are sent to 1×1 convolution and Softmax to obtain the result of segmentation.

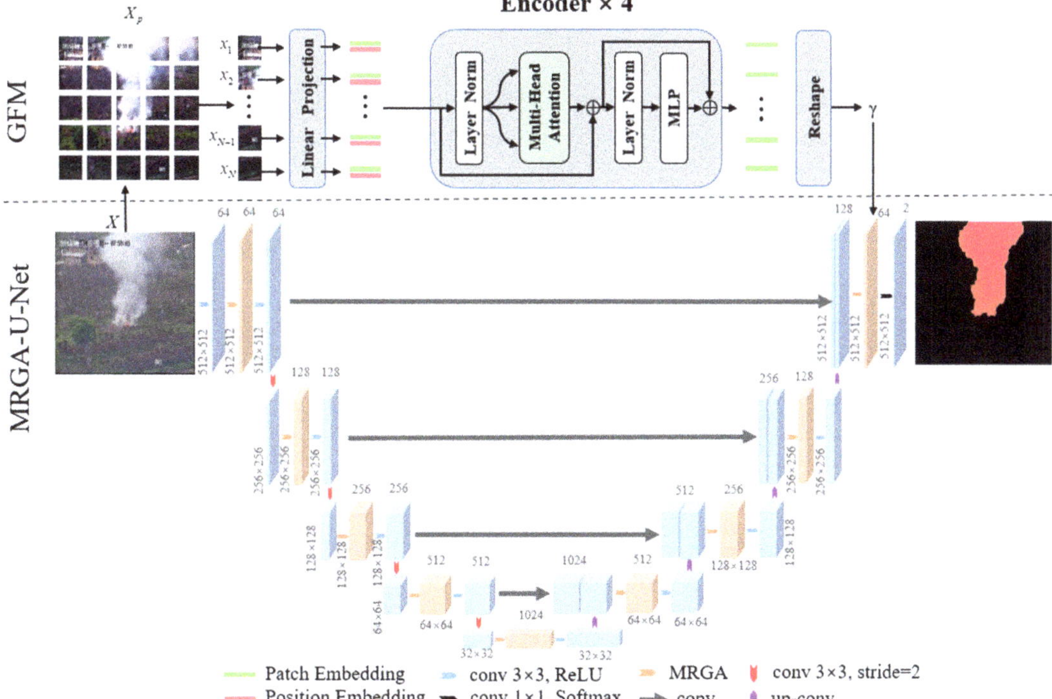

Figure 3. Smoke–U-Net algorithm.

3.2. Multi-Scale Residual Group Attention (MRGA)

Colors, textures, and edges are the main features of smoke. However, in algorithms of smoke segmentation with convolutional neural networks, on the one hand, down-sampling reduces or even removes the feature of small-sized smoke, and the features are difficult to recover during up-sampling, which causes more difficult segmentation on small smoke. To cope with the weakening and loss of information, MRGA uses multi-scale convolution kernels to retain more features without increasing parameters. On the other hand, the higher fusion of small smoke with the background, and the more noise caused by similar pixels, leads to errors in the classification of pixels. To enhance the features of smoke, the MRGA uses channel attention and spatial attention to enhance the representation of features, as shown in Figure 4.

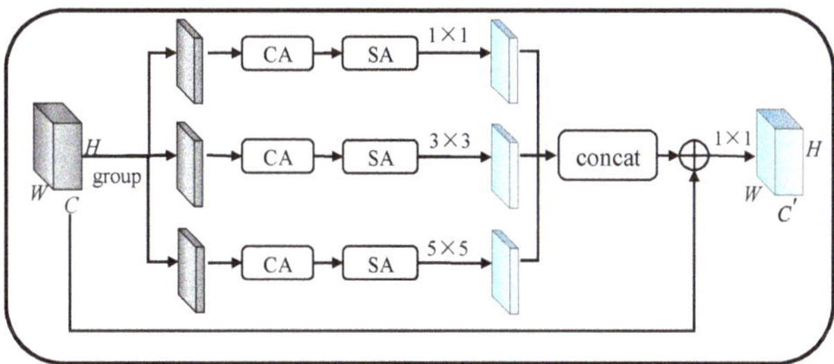

Figure 4. Multi-scale residual group attention block (MRGA).

MRGA first groups the feature maps, and then extracts multi-scale features of smoke through convolution kernels of different scales. Then, 1×1 convolution is used to extract small smoke features, 5×5 convolution is used to extract larger smoke features, and 3×3 convolution is used to extract regular size features. The extracted multi-scale features are then passed through channel attention and spatial attention to enhance channel features and spatial features, and output a refined feature map. Finally, smoke features at different scales are concatenated. MRGA directly propagates rich, low-frequency information by residual connections, which speeds up the training of the network and alleviates gradient degradation. At the same time, in order to ensure the exchange of information between different grouping features, and enhance the expressiveness of features, MRGA uses 1×1 convolution to rearrange output features.

3.3. Global Features Module (GFM)

To solve inaccurate segmentation due to thinness of smoke at the edge, GFM uses the encoder of Transformer to capture global semantic information of the image. We divided the input image $X \in \mathbb{R}^{H \times W \times C}$ into non-overlapping patches $X_p = \{x_1, \ldots, x_N\}$, where $x_N \in \mathbb{R}^{N \times C \times p \times p}$, $p \times p$ denotes the dimension of each patch and $N = \frac{HW}{p^2}$ is the length of image sequence. Through Linear Projection, we flattened each patch into a K dimensional vector I, where $K = C \times p \times p$. To maintain the spatial information of each patch, we added a *Positional Embedding* I_{pos} to the *Patch Embedding I*, in order to preserve positional information T. Finally, we use the encoder of Transformer to encode T and output the image-level global information $\gamma \in \mathbb{R}^{1 \times H \times W}$ by *Reshape*.

$$T = I + I_{pos} \tag{1}$$

3.4. Loss Function

Segmentation results depends not only on network structure, but also on the choice of loss function. Dice loss function is proposed for the problem of low segmentation accuracy due to small target. It is derived from binary classification, which essentially measures the overlap of two regions. In this paper, a loss function combining binary cross entropy (*BCE*) and Dice loss function was used to alleviate the inaccurate prediction effect caused by the difference in the proportion of foreground and background, as shown in Equations (2)–(4).

$$BCE = -\frac{1}{N} \sum_{i=1}^{N} y_i \log(p_i) + (1 - y_i) \log(1 - p_i) \tag{2}$$

$$Dice = 1 - \frac{2 \sum_{i=1}^{N} y_i p_i + \varepsilon}{\sum_{i=1}^{N} y_i + \sum_{i=1}^{N} p_i + \varepsilon} \tag{3}$$

$$BDLoss = 0.6 \times BCE + 0.4 \times Dice \tag{4}$$

where y represents the true label, p represents the predicted result, and N represents the set of all pixels. In order to prevent the numerator and denominator of *Dice* from being extremely small during the training process, which may exceed the computer's storage range for the float number, a minimum value ε is set, and ε takes 1×10^{-5} in the experiment.

4. Results

4.1. Experimental Platform

Our experiment used PyTorch framework to build the proposed model. Training and testing were performed on a computer with an Intel Core i7-9700 CPU and $4 \times$ NVIDIA Tesla V100 GPU/32G.

Parameters were updated by Adam optimizer with a learning rate of 0.001, a momentum of 0.99, and a decay factor of 0.9 for the learning rate. Batch size was 16.

4.2. Dataset

The smoke dataset comes from The State Key Laboratory of Fire Science, University of Science and Technology of China. The url is http://smoke.ustc.edu.cn/datasets.htm (accessed on 7 July 2021). It is a public dataset that only has raw smoke images, and does not contain label images for model training. Therefore, we organized and annotated 2300 smoke images ($3 \times 512 \times 512$) with labelme software to obtain the corresponding binary label maps. It not only includes multiple scenes such as forests, fields, indoors, playgrounds, urban buildings, and roads, but also includes large, medium, and small smoke of different scales. Some samples are shown in Figure 5. During training, the dataset was divided into training set, validation set, and test set, according to the ratio of 7:2:1.

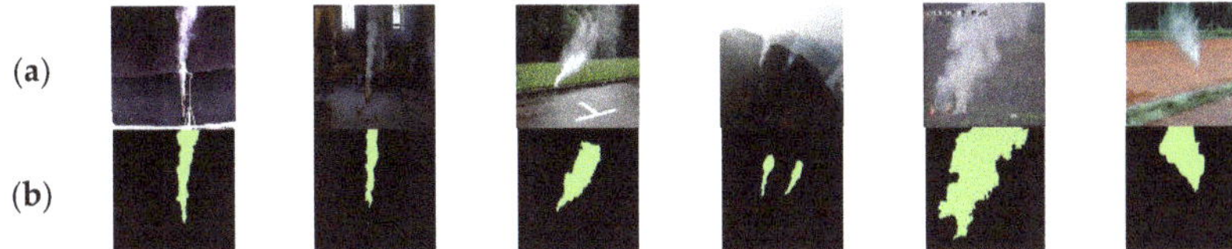

Figure 5. (**a**) samples of smoke images; (**b**) labels of smoke images.

4.3. Evaluation Index

Our experiment adopted mean intersection over union (*mIoU*), mean pixel accuracy (*mPA*), and the frames per second (FPS) as evaluation indicators.

mIoU indicates the average of overlap between the segmentation result and real label. It is commonly used to evaluate the accuracy of algorithms, and its calculation formula is as follows:

$$mIoU = \frac{1}{k+1} \sum_{i=0}^{k} \frac{p_{ii}}{\sum_{j=0}^{k} p_{ij} + \sum_{j=0}^{k} p_{ji} - p_{ii}} \tag{5}$$

where k is the number of classes of objects in the foreground, and p_{ij} is the number of pixels belonging to class i that are classified as class j.

mPA, shown in Equation (6), indicates the proportion of correctly classified pixels to all pixels:

$$mPA = \frac{1}{k+1} \sum_{i=0}^{k} \frac{p_{ii}}{\sum_{j=0}^{k} p_{ij}} \tag{6}$$

4.4. Results of Train and Validation

In this experiment, the method proposed in this paper was trained and verified for 145 epoch in a smoke dataset, and then the *mIoU* curve and loss curve were drawn, as shown in Figure 6.

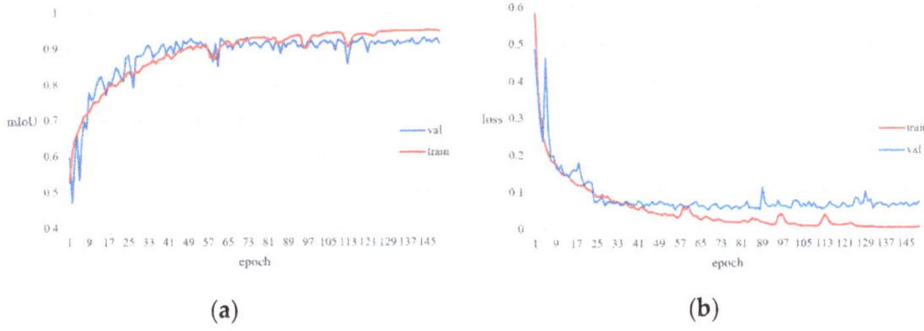

Figure 6. (**a**) Mean value of *mIoU* at convergence of train and validation is 93.51% and 91.82%, respectively; (**b**) mean value of DBLoss at convergence of train and validation is 0.033 and 0.078%, respectively.

5. Discussion

5.1. Ablation Experiment

In order to validate the effectiveness of MRGA and GFM, we performed ablation experiments to compare and analyze the impact of each improvement on the algorithm. The results are shown in Table 1.

Table 1. Results of ablation experiments. $\sqrt{}$ means the current module is included in the method.

Method	MRGA	GFM	*mIoU*	*mPA*	FPS
U-Net			86.81	91.84	22.36
GFM–U-Net		$\sqrt{}$	89.40	94.32	18.95
MRGA–U-Net	$\sqrt{}$		89.12	94.18	24.55
Smoke–U-Net	$\sqrt{}$	$\sqrt{}$	91.82	96.62	20.17

The results show that the model with MRGA has a 2.31% improvement in *mIoU* and a 2.34% improvement in *mPA* compared with the model without MRGA. It illustrates that MRGA can effectively reduce the negative effect caused by information loss and improve the accuracy of segmentation. Compared with the model without GFM, the *mIoU* and *mPA* of the model with GFM improve by 2.59% and 2.48%, respectively. Meanwhile, we can see that GFM slightly outperforms MRGA on *mIoU* and *mPA*, however, GFM has a lower FPS. This is related to the fact that GFM has more parameters.

5.2. Comparative Experiment of Multi-Scale Segmentation

To evaluate the effectiveness of Smoke–U-Net, four representative algorithms of segmentation, including SegNet, PSPNet, DeepLab v3+ [23], and U-Net, were selected for experiments on smoke datasets. Firstly, the validation set was divided into three subsets of large, medium, and small, according to the size of the smoke. Then, each algorithm was tested on the subsets to compare the capability of segmentation of smoke of different sizes. The results of different algorithms are shown in Table 2. Compared with other algorithms, Smoke-U-Net significantly improves the accuracy of segmentation for multi-scale smoke, and the average *mIoU* reaches 91.82%. From these results, it is shown that Smoke-U-Net achieves state-of-the-art results on all three subsets. In particular, in the results of small smoke, the *mIoU* of Smoke–U-Net increases by 3.11% and 4.54%, compared with DeepLab v3+ and U-Net, respectively.

Table 2. Comparison of *mIoU* for segmentation at multi-scale smoke and bold font indicates best grade.

Method	Large	Medium	Small
SegNet	85.34%	82. 89%	78.52%
PSPNet	86.53%	82.30%	80.18%
U-Net	89.72%	86.92%	83.82%
DeepLab v3+	91.03%	88.61%	85.25%
Smoke–U-Net	**94.97%**	**92.15%**	**88.36%**

Figure 7 shows the results of segmentation of different sized smoke by each algorithm. Combined with the definition of *mIoU*, the performance of each method for segmentation can be visualized from the degree of overlap between the red markers and smoke. It can be seen that Smoke–U-Net effectively reduces the wrong segmentation for small smoke and the missed segmentation for thin smoke at the edge. In summary, for different size smoke, the results of Smoke–U-Net are more complete, and the boundary is clearer, compared with the other algorithms.

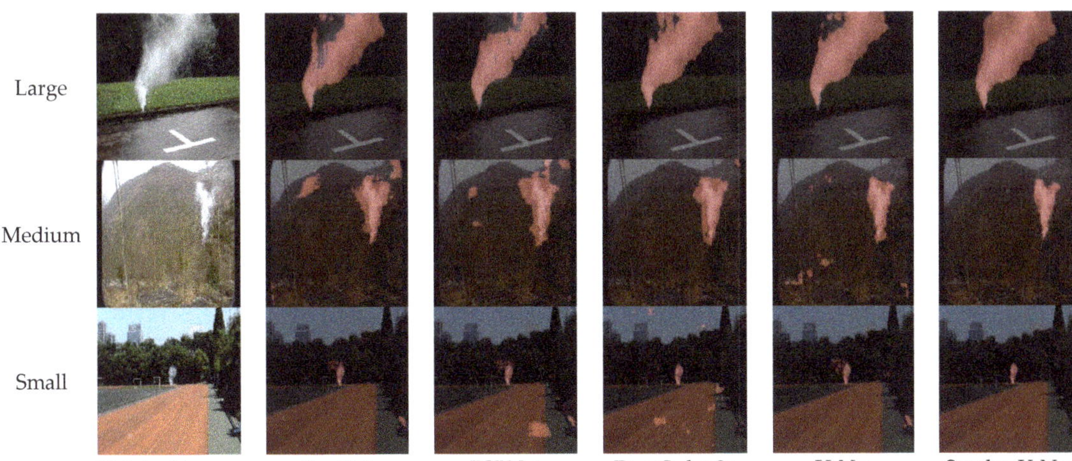

Figure 7. Comparison of segmentation results on multi-scale smoke.

5.3. Comprehensive Experiments

As shown in Table 3, eight semantic segmentation models, including Smoke–U-Net, are selected for comparison of comprehensive performance. FCN and SegNet are representatives of the semantic segmentation models with an encoder–decoder. PSPNet and DeepLab v3+ are representatives of semantic segmentation models that perform feature fusion on the basis of an encoder–decoder. DANet is a representative of semantic segmentation models that utilize the parallel attention mechanism to capture semantic features. Vision transformer is a typical representative of self-attention. Comprehensive experiments are performed on all smoke datasets.

Table 3. Comprehensive performance of each algorithm and bold font indicates best grade.

Model	*mIoU*/%	*mPA*%	FPS
FCN	80.26	85.95	18.28
SegNet	82.32	88.67	20.65
PSPNet	85.81	90.07	16.47
DeepLab v3+	88.96	93.20	19.01
U-Net	86.01	91.84	**22.36**
DANet	88.47	92.65	16.05
Vision Transformer	88.24	92.32	10.30
Smoke–U-Net	**91.83**	**96.62**	20.17

Experiments show that Smoke–U-Net achieves the highest scores on *mIoU* and *mPA*, reaching 91.83% and 96.62%, respectively. It benefits from the accuracy of feature extraction for small-scale smoke by MRGA, and the fusion of global smoke information by GFM. With the increase in sampling times, the receptive field of traditional convolution becomes larger and larger, which causes the small target feature to be larger than its original size when mapped back to original image. MRGA uses 1×1 convolution with a smaller receptive field to pass and enhance small features, which makes the feature extraction of small targets more accurate. At the same time, GFM constructs the global information of smoke concentration changes, which improves the recognition ability of thin smoke at the edge of images. Therefore, the higher score achieved by Smoke–U-Net is expected. In terms of frame rate, SegNet and U-Net are two simple encoder–decoder structures, and have significant advantage in computation. In contrast, the encoder of GFM has more parameters and computation. Therefore, Smoke–U-Net is second only to U-Net and SegNet in speed. However, compared to the increase in mIOU (2.70%) and *mPA* (2.44%), the slightly decrease in speed is acceptable. Vision transformer is more complex, and has a large amount of parameters in the encoder and decoder, so it achieves a lower FPS. Figure 8 shows the results of the comprehensive experiment. Taking into account the speed and accuracy of Smoke–U-Net, it is more applicable to the actual requirements for smoke segmentation than other models.

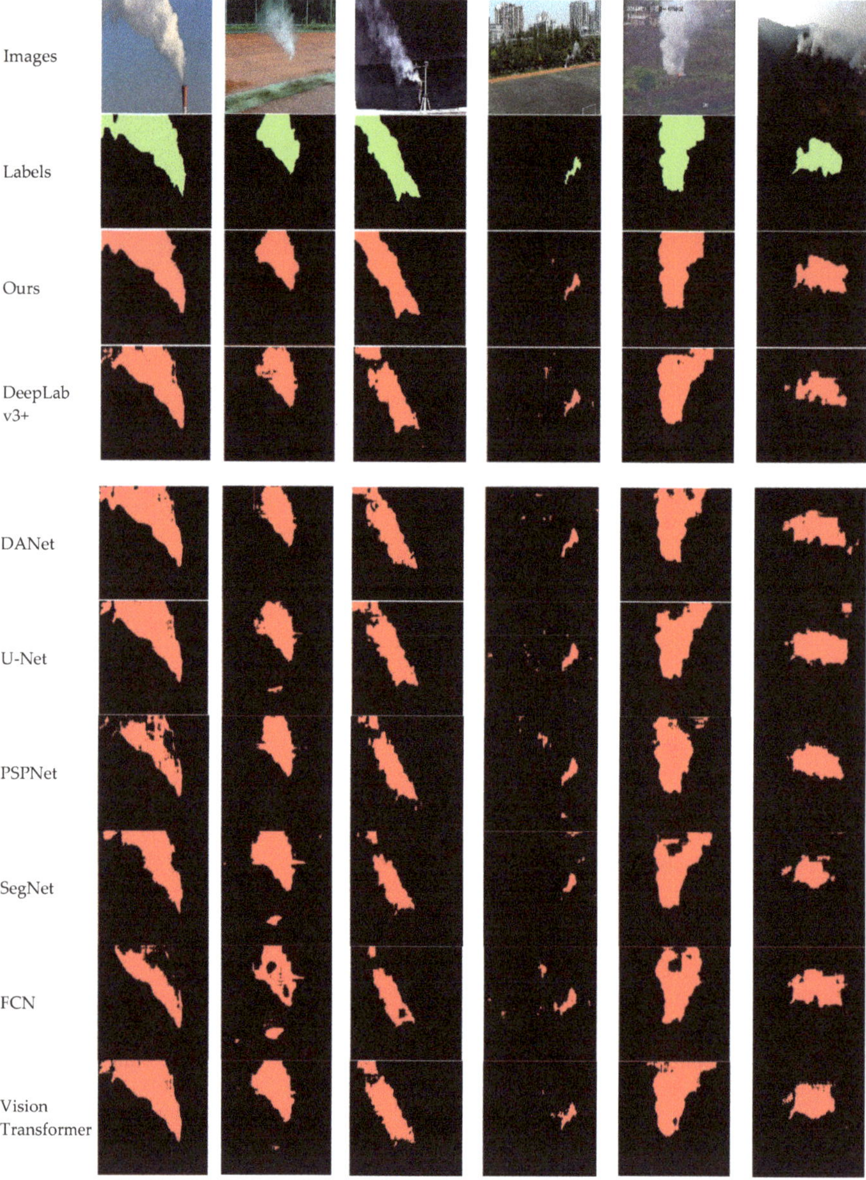

Figure 8. Comparison of comprehensive performance of the algorithms (Chinese characters in the picture have no meaning).

The method proposed in this paper is effective in multi-scale smoke segmentation. However, the method is not always highly accurate, as shown in Figure 9. When blurred background is almost integrated with the low-concentration smoke, or there are some solid-color interfering objects in the background, sometimes wrong segmentation occurs. Therefore, we still have a lot of work to do in terms of smoke segmentation.

Figure 9. Some failure cases of the method in this paper (Chinese characters in the picture have no meaning).

6. Conclusions

In this paper, we analyzed and discussed the current difficulties of smoke segmentation from the demand of detection of fire smoke. To solve these difficulties, an algorithm called Smoke–U-Net was proposed. This algorithm improves U-Net, and proposes a multi-scale residual group attention module, which not only reduces the loss of semantic information, but also enhances the smoke features through channel attention and spatial attention. In addition, the encoder Transformer was used to extract the global information of smoke, and establish the relationship of smoke concentration, so as to improve the segmentation accuracy of thin smoke. Finally, the effectiveness of Smoke–U-Net for smoke segmentation at multiple scales was experimentally verified. It achieves 91.82% $mIoU$ and 96.62% mPA.

Author Contributions: Writing—original draft, Y.Z. and Z.W.; writing—review and editing, Y.Z. and Z.W.; Data curation, B.X.; project administration, Y.N.; software, B.X. and Y.N. All authors have read and agreed to the published version of the manuscript.

Funding: This study was financially supported by the Science and Technology Key Project of Henan Province (Grant No. 202102210180).

Data Availability Statement: The data presented in this study are available on request from the corresponding author. The data are not publicly available, as the research group's fire smoke semantic segmentation related research is still being carried on, the later work will also rely on the current dataset.

Conflicts of Interest: The authors declare no conflict of interest.

References

1. Long, J.; Shelhamer, E.; Darrell, T. Fully convolutional network for semantic segmentation. In Proceedings of the IEEE International Conference on Computer Vision and Pattern Recognition, Boston, MA, USA, 7–12 June 2015; IEEE: Piscataway, NJ, USA, 2015; pp. 3431–3440.
2. Badrinarayanan, V.; Kendall, A.; Cipolla, R. SegNet: A deep convolutional encoder-decoder architecture for image segmentation. *Pattern Anal. Mach. Intell.* **2017**, *39*, 2481–2495. [CrossRef] [PubMed]
3. Zhao, H.; Shi, J.; Qi, X.; Wang, X.; Jia, J. Pyramid scene parsing network. In Proceedings of the IEEE Conference on Computer Vision and Pattern Recognition, Honolulu, HI, USA, 21–26 July 2017; IEEE Computer Society Press: Los Alamitos, CA, USA, 2017; pp. 2881–2890.
4. Fu, J.; Liu, J.; Tian, H.; Li, Y.; Bao, Y.; Fang, Z.; Lu, H. Dual attention network for scene segmentation. In Proceedings of the IEEE/CVF Conference on Computer Vision and Pattern Recognition, Long Beach, CA, USA, 5–20 June 2019; IEEE Computer Society Press: Los Alamitos, CA, USA, 2019; pp. 3141–3149.
5. Zhou, T.; Li, L.; Li, X.; Feng, C.M.; Li, J.; Shao, L. Group-wise learning for weakly supervised semantic segmentation. *IEEE Trans. Image Processing* **2021**, *31*, 799–811. [CrossRef] [PubMed]
6. Wang, W.; Zhou, T.; Yu, F.; Dai, J.; Konukoglu, E.; Van Gool, L. Exploring cross-image pixel contrast for semantic segmentation. In Proceedings of the IEEE/CVF International Conference on Computer Vision, Montreal, QC, Canada, 10–17 October 2021.
7. Zhou, T.; Wang, W.; Konukoglu, E.; Van Gool, L. Rethinking Semantic Segmentation: A Prototype View. *arXiv* **2022**, arXiv:2203.15102v2.
8. Appana, D.K.; Islam, R.; Khan, S.A.; Kim, J.M. A video-based smoke detection using smoke flow pattern and spatial-temporal energy analyses for alarm systems. *Inf. Sci.* **2017**, *418*, 91–101. [CrossRef]

9.	Zhao, Z.M.; Zhu, Z.L.; Liu, M. Fuzzy c-means clustering method for image segmentation insensitive to class size. *Laser Optoelectron. Prog.* **2020**, *57*, 56–65.

10.	Wang, R.; Yao, A.; Yang, R. Application of multi-scale image edge detection based on wavelet transform modulus maxima in smoke image. *Foreign Electron. Meas. Technol.* **2020**, *39*, 63–67.

11.	Peng, Y.; Wang, Y. Real-time forest smoke detection using hand-designed features and deep learning. *Comput. Electron. Agric.* **2019**, *167*, 105029. [CrossRef]

12.	Wu, Y.; Chen, M.; Wo, Y.; Han, G. Video smoke detection base on dense optical flow and convolutional neural network. *Multimed. Tools Appl.* **2021**, *80*, 35887–35901. [CrossRef]

13.	Khan, S.; Muhammad, K.; Hussain, T.; Del Ser, J.; Cuzzolin, F.; Bhattacharyya, S.; Akhtar, Z.; de Albuquerque, V.H. DeepSmoke: Deep learning model for smoke detection and segmentation in outdoor environments. *Expert Syst. Appl.* **2021**, *182*, 115125. [CrossRef]

14.	Yuan, F.; Zhang, L.; Xia, X.; Huang, Q.; Li, X. A gated recurrent network with dual classification assistance for smoke semantic segmentation. *IEEE Trans. Image Processing* **2021**, *30*, 4409–4422. [CrossRef] [PubMed]

15.	Gupta, T.; Liu, H.; Bhanu, B. Early Wildfire Smoke Detection in Videos. In Proceedings of the 25th International Conference on Pattern Recognition, Milan, Italy, 10–15 January 2021; IEEE: Piscataway, NJ, USA, 2021; pp. 8523–8530.

16.	Zhu, G.; Chen, Z.; Liu, C.; Rong, X.; He, W. 3D video semantic segmentation for wildfire smoke. *Mach. Vis. Appl.* **2020**, *31*, 50. [CrossRef]

17.	Hu, J.; Shen, L.; Sun, G. Squeeze-and-Excitation Networks. In Proceedings of the 31st IEEE/CVF Conference on Computer Vision and Pattern Recognition (CVPR), Salt Lake City, UT, USA, 18–23 June 2018; IEEE: Piscataway, NJ, USA, 2018; p. 00745.

18.	Woo, S.; Park, J.; Lee, J.Y.; Kweon, I.S. CBAM: Convolutional block attention module. In Proceedings of the 15th European Conference on Computer Vision (ECCV), Munich, Germany, 8–14 September 2018; Springer: Cham, Switzerland, 2018; pp. 8–14.

19.	Vaswani, A.; Shazeer, N.; Parmar, N.; Uszkoreit, J.; Jones, L.; Gomez, A.N.; Kaiser, Ł.; Polosukhin, I. Attention is All You Need. In Proceedings of the 31st International Conference on Neural Information Processing Systems, Long Beach, CA, USA, 4–9 December 2017; Curran Associates Inc.: Redhook, NY, USA, 2017; pp. 6000–6010.

20.	Dosovitskiy, A.; Beyer, L.; Kolesnikov, A.; Weissenborn, D.; Zhai, X.; Unterthiner, T.; Dehghani, M.; Minderer, M.; Heigold, G.; Gelly, S.; et al. An Image Is Worth 16 × 16 Words: Transformers For Image Recognition At Scale. *arXiv* **2020**, arXiv:2010.11929.

21.	Liu, Z.; Lin, Y.; Cao, Y.; Hu, H.; Wei, Y.; Zhang, Z.; Lin, S.; Guo, B. Swin Transformer: Hierarchical Vision Transformer using Shifted Windows. *arXiv* **2021**, arXiv:2103.14030.

22.	Ronneberger, O.; Fischer, P.; Brox, T. U-Net: Convolutional Networks for Biomedical Image Segmentation. In Proceedings of the Medical Image Computing and Computer-Assisted Intervention, Munich, Germany, 5–9 October 2015; pp. 234–241.

23.	Chen, L.C.; Zhu, Y.; Papandreou, G.; Schroff, F.; Adam, H. Encoder-decoder with atrous separable convolution for semantic image segmentation. In Proceedings of the European Conference on Computer Vision, Munich, Germany, 8–14 September 2018; Springer: Berlin/Heidelberg, Germany, 2018; pp. 801–818.

 *electronics*

Article

RERB: A Dataset for Residential Area Extraction with Regularized Boundary in Remote Sensing Imagery for Mapping Application

Songlin Liu [1,2,*], Li Zhang [2], Wei Liu [2], Jun Hu [3], Hui Gong [2], Xin Zhou [4] and Danchao Gong [2]

[1] School of Remote Sensing Information Engineering, Wuhan University, Wuhan 430079, China
[2] State Key Laboratory of Geo-Information Engineering, Xi'an 710054, China
[3] School of Electronics and Communication Engineering, Sun Yat-sen University, Guangzhou 510006, China
[4] Tian-Hui Satellite Center of China, Beijing 100094, China
* Correspondence: songlinliu@whu.edu.cn; Tel.: +86-18992830817

Citation: Liu, S.; Zhang, L.; Liu, W.; Hu, J.; Gong, H.; Zhou, X.; Gong, D. RERB: A Dataset for Residential Area Extraction with Regularized Boundary in Remote Sensing Imagery for Mapping Application. *Electronics* **2022**, *11*, 2790. https://doi.org/10.3390/electronics11172790

Academic Editor: Stefano Ferilli

Received: 1 August 2022
Accepted: 2 September 2022
Published: 5 September 2022

Publisher's Note: MDPI stays neutral with regard to jurisdictional claims in published maps and institutional affiliations.

Abstract: Due to the high automaticity and efficiency of image-based residential area extraction, it has become one of the research hotspots in surveying, mapping, and computer vision, etc. For the application of mapping residential area, the extracted contour is required to be regular. However, the contour results of existing deep-learning-based residential area extraction methods are assigned accurately according to the actual range of residential areas in imagery, which are difficult to directly apply to mapping due to the extractions being messy and irregular. Most of the existing ground object extraction datasets based on optical satellite images mainly promote the research of semantic segmentation, thereby ignoring the requirements of mapping applications. In this paper, we introduce an optical satellite images dataset named RERB (Residential area Extraction with Regularized Boundary) to support and advance end-to-end learning of residential area mapping. The characteristic of RERB is that it embeds the prior knowledge of regularized contour in the dataset. In detail, the RERB dataset contains 13,892 high-quality satellite images with a spatial resolution of 2 m acquired from different cities in China, and the size of each image is approximately 256 × 256 pixels, which covers an area of more than 3640 square kilometers. The novel published RERB dataset encompasses four superiorities: (1) Large-scale and high-resolution; (2) well annotated and regular label contour; (3) rich background; and (4) class imbalance. Therefore, the RERB dataset is suitable for both semantic segmentation and mapping application tasks. Furthermore, to validate the effectiveness of the RERB, a novel end-to-end regularization extraction algorithm of residential areas based on contour cross-entropy constraints is designed and implemented, which can significantly improve the regularization degree of extraction for the mapping of residential areas. The comparative experimental results demonstrate the preponderance and practicability of our public dataset and can further facilitate future research.

Keywords: residential area extraction; mapping requirement; contour regularization; end-to-end deep model; Tian-Hui 1 satellite images

1. Introduction

Topographic map data with 1:50,000 scale are one of the most basic geographic information data, which play a significant and strategic role in national economy and national defense construction [1,2]. With the rapid development of society, users have increasingly higher requirements for the current situation of topographic maps, and the updating of topographic maps [3,4] has become the primary and urgent work. The content of topographic maps mainly includes ground objects and the undulating form of terrain. Because the changes in terrain data are generally relatively small, the updated objects of topographic map mainly consist of ground objects.

Optical remote sensing imagery is one of the main data sources for updating ground objects in topographic maps. Ground object mapping mainly refers to the acquisition of object information in remote sensing imagery according to the corresponding scale graphic specification [1]. At present, ground object mapping using remote sensing imagery is mainly completed manually, with high precision but low efficiency, making this kind of fashion tedious, expensive, and labor intensive, so it is difficult to meet the needs of rapid applications, such as land planning and automatic driving. According to official statistics [5], it takes at least 40 days and 10 thousand RMB to produce 1:50,000 topographical map data. The number of global 1:50,000 topographical maps is about 400 thousand, and each update requires an investment of about 4 billion RMB.

With the launch of Zi-Yuan 3 (ZY-3) and Tian-Hui 1 (TH-1) stereo mapping satellites [6,7], China has the capability of measuring and updating 1:50,000 topographic maps with satellite remote sensing imagery. Among the ground feature elements, residential areas are one of the most important elements in topographic map content. A survey demonstrated that in most areas, the workload of extracting residential areas accounts for more than 60% of all the work of extracting ground features [2]. Therefore, studying the automatic extraction method of residential areas for mapping applications is of considerable significance to improving the efficiency of mapping work.

With the development of this automatic extraction technology, many institutions around the world have developed digital mapping systems integrated with automatic technology for recognizing ground feature elements [2]. For examples, both the eCognition [8] of Definiens and the EasyFeature [9] of Handleray have integrated the ground feature recognition technology. Specifically, this kind of method mainly includes two steps: extraction and post-processing.

Semantic segmentation [10,11] is a typical and an efficient technology to accomplish the extraction step, which indicates, dividing the image into pixel groups with specific semantics and recognizing each region's category. In recent years, the development of deep learning techniques, such as convolution neural network (CNN), has injected new vitality into the study of semantic segmentation. However, due to the complexity of ground features and background in remote sensing imagery, the classification results of residential areas extracted by semantic segmentation method are usually not perfect, especially at residential area boundaries [12], which are irregular contours. Consequently, these classification results cannot be directly employed in mapping applications. In addition, the post-processing technology is exploited to obtain the regularized object boundary contour. The popular operations used to identify the boundary of a raster dot group include smoothing, line segment fitting, denudation under complex constraints, and conditional random field (CRF) method, etc. In addition, there are also some methods using an end-to-end network to process the boundary of objects. The abovementioned innovative works focus on improving extraction accuracy but without consideration of the matching degree between the extraction results and the mapping requirement.

Obviously, the mapping method with two steps is cumbersome, and the post-processing step also greatly reduces the overall intelligence of the mapping method. End-to-end fashion can realize intelligent mapping without manual intervention. To promote the end-to-end mapping method, we present and introduce an optical satellite images dataset named RERB (Residential area Extraction with Regularized Boundary). To the best of the authors' knowledge, there is no dataset released for the application of mapping residential area, which limits the research of end-to-end residential area regularization extraction. Compared to existing datasets, the contour of label image in RERB dataset consists of regular line segments. Given this point, it can facilitate the research for end-to-end training of residential area regularized extraction. Specifically, the public RERB dataset consists of 13,892 satellite images in 256 × 256 size, covering an area of more than 3640 square kilometers.

To summarize, our contributions are as follows:

(1)	According to the specifications for cartographic symbols of 1:50,000 topographic map, our work summarizes the requirements of regular extraction in the residential area mapping application.

(2) We construct a residential area mapping dataset called RERB with regular contour labels based on TH-1 [7] satellite images, which is the first dataset released for the residential feature mapping application. Furthermore, in order to measuring the compliance of the extraction results with the mapping requirements when using RERB dataset, we design a special evaluation index named CMI (contour matching index) based on contour matching. Extensive experiments demonstrate the superiority of RERB dataset.

(3) We sufficiently explore the contour constraint with regular contours in label images by integrating the contour cross-entropy constraint and the original loss function into an end-to-end network, which can significantly improve the regularization degree of extraction results for the mapping of residential areas.

The remainder of this paper is organized as follows: Section 2 introduces the related works. Section 3 presents the constructed RERB dataset in detail. Section 4 details the experimental results along with in-depth analysis. Section 5 finishes the paper with conclusions and our future perspective.

2. Related Works

In this section, we first describe the development of datasets for ground object extraction based on optical image and then introduce semantic segmentation methods. Finally, we introduce post-processing technology.

2.1. Datasets for Ground Object Extraction

Recently, with the advancement of deep learning technology, datasets have played an important part in ground object extraction. Any effective deep learning model is obtained by training with many original images and their corresponding labels. As shown in Table 1, the widely used open-source datasets with optical image pixel level annotation include WHU [13], LandCoverNet [14], GID [15], LoveDA [16], SSD [17], etc.

Table 1. Overall comparison of some satellite image datasets.

	Year	Resolution (m)	Image Size	Samples	Categories	Task-Semantic Segmentation	Task-Mapping
WHU	2019	0.45	512 × 512	17,388	2	√	√ (building)
	2019	0.3–2.3	512 × 512	204	2	√	√ (building)
LandCoverNet	2020	10	256 × 256	1980	7	√	
GID	2020	4	7200 × 6800	150	6	√	
LoveDA	2021	0.3	1024 × 1024	5987	7	√	
SSD	2021	2	7400 × 4950	23	5	√	
RERB (ours)	2022	2	256 × 256	13,892	2	√	√

WHU dataset is released by Wuhan University, and it includes one land-cover category, namely, buildings. WHU dataset can be used to construct a building extraction model in a topographic map with a scale of 1:10,000 or larger and cannot be directly applied to residential area mapping in 1:50,000 scale topographic maps.

The Gaofen image dataset (GID) contains 150 high-quality GF-2 images from more than 60 cities in China, with a spatial resolution of 4 m. The size of each image is approximately 7200 × 6800 pixels, and it includes six land cover categories, namely, built-up, farmland, forest, meadow, water, and others, which represents all categories other than the former five categories. Similarly, LandCoverNet, LoveDA, and SSD are also constructed for land use and land cover (LULC) classification. If they are used for topographic mapping, post-processing steps still need to be added after model inference.

To study end-to-end regularized extraction technology of residential area, we propose the RERB dataset in this paper.

2.2. Semantic Segmentation

Semantic segmentation is a long-standing research topic that assigns a label to each pixel, known as pixel-level classification. In 2015, Long et al. [18] proposed full connected networks (FCNs), whose excellent performance led researchers to change their understanding of semantic segmentation from regional clustering to pixel classification. At present, CNN-based methods have completely exceeded the segmentation accuracy of traditional methods. However, the training steps of FCNs are complex, and it is easy to lose pixel position information during up-sampling. After that, U-Net [19], SegNet [20], PSPNet [21], the DeepLab family [22–24], and FastFCN [25] were developed. U-Net can effectively fuse multilevel feature maps, and small objects and large objects are processed by using shallow and deep information, respectively. U-Net is essentially a structure based on multiscale context and multilevel feature fusion. SegNet improves the segmentation accuracy by recording the position of pooled values in the original feature map and accurately mapping the relevant values to the corresponding positions in the up-sampling step. However, SegNet still fails to recover the object boundary very well. PSPNet integrates the multiscale background information with a pyramid pooling module. To obtain a larger receptive field, PSPNet improves the backbone network by using dilated convolutions [26]. Furthermore, additional losses can provide the intermediate supervision information in PSPNet. The DeepLab series leads research on semantic segmentation. DeepLab v3+ [24], which integrates more local information in low-level features and replaces the feature extractor with a more complex Xception network [27], performs well on several public datasets. In addition, the atrous spatial pyramid pooling (ASPP) structure proposed by the DeepLab network has been widely employed in semantic segmentation research literature. FastFCN uses the joint pyramid up-sampling (JPU) module to improve the dilated convolution and obtains faster speed and higher accuracy.

Especially, semantic segmentation technology has been applied to remote sensing imagery and medical image [28,29] in recent years, which has greatly improved the research level of methods used to automatically extract ground feature elements. For example, Ying Sun et al. [30] used optical images and light detection and ranging (LiDAR) data to construct multichannel input data and designed a convolution neural network (CNN) model with multiscale encoder–decoder architecture to achieve enhanced segmentation results. Cui et al. [31] also improved the accuracy of building extraction by using the multiscale information of images. Y. Liu et al. [32] jointly used LiDAR data and introduced a higher-order CRF to increase the accuracy of ground object segmentation. In addition, several researchers designed two-stage training approaches [33], modified loss function [34], self-attention modules [35,36], edge information [37], or both self-attention and edge enhancement modules [17] to fully exploit the context information of remote sensing imagery from a larger perspective.

2.3. Post-Processing Technology

The processing object of post-processing technology is the raster dot group, which is obtained by semantic segmentation. Traditional operations used to identify the boundary of a raster dot group include smoothing, line segment fitting, and denudation under complex constraints [38,39]. Most of these methods belong to the field of traditional image processing, and the degree of automation and intelligence is low.

Moreover, CRF [40–42] methods are also widely used in post-processing of semantic segmentation results. Using CRF, the segmentation results can be corrected, especially at ground object borders. However, these CRF methods require the introduction of samples to the CRF control process, and this operation cause the CRF methods to lose their end-to-end characteristics. For mapping tasks based on automatic extraction technology, when the end-to-end characteristics are lost, the ground object mapping work must add an additional manual post-processing operation, which greatly reduces the overall intelligence of mapping tasks. Hence, new end-to-end methods must be introduced to solve this problem.

Ying Sun et al. [43] first constructed multichannel input data using optical images and LiDAR data and then achieved a better result than SegNet by designing an end-to-end encoding–decoding structure. Meanwhile, the object boundary is strengthened. There are also some methods using an end-to-end network to process the boundary of objects, such as ACE2P [44], Gated-SCNN [45], and EaNet [46]. The ACE2P model realizes end-to-end high-precision training by fully integrating the bottom characteristics, global contextual information, and edge details in the human body parsing task. Gated-SCNN is a double branch structure, in which the target shape information is embedded into the semantic segmentation network by a shape branch. Except for traditional semantic segmentation labels, image boundary labels are also needed in Gated-SCNN. To effectively separate confusing objects with sharp contours, EaNet is constructed based on a large kernel pyramid pooling (LKPP) module and a dice-based edge-aware loss function.

3. The RERB Dataset and Model Construction

This section first describes the contour requirements for mapping applications and then introduces the RERB dataset. Finally, we analyze the statistics for RERB dataset and describe the construction of a residential area regularized extraction model.

3.1. Contour Requirements for Mapping Applications

Different topographic maps are distinguished by scale and commonly used scales generally include 1:2000, 1:5000, 1:10,000, and 1:50,000. The 1:50,000 topographic map data are one of the most basic geographic information data. At present, ground object mapping using remote sensing imagery is mainly completed manually, with high precision but low efficiency, making this kind of fashion tedious, expensive, and labor intensive [5]. As a result, it is very important to analyze the mapping requirements and build a mapping dataset to improve the intelligence of mapping work.

Topographic maps of different scales are constrained by corresponding graphic specifications, which mainly stipulates the symbols, annotations, and contour decoration of various ground objects and geomorphic elements represented on topographic maps, as well as the methods and basic requirements of using these symbols. This paper mainly focuses on 1:50,000 scale, and its corresponding current national standard [1] was issued on 14 October 2017 and implemented on 1 May 2018.

Ground object mapping in surveying and mapping field mainly refers to the collection of the ground object information from remote sensing imagery according to the corresponding specification for cartographic symbols [1]. Figure 1 is an example of residential area extraction and mapping based on optical images. Figure 1b is an illustration of a residential area in the 1:50,000 topographic map corresponding to the original image in Figure 1a. Mapping is to obtain the contour of ground objects that meet the requirements of graphic specifications from remote sensing imagery.

Residential areas [2] refer to houses that are contiguous to each other in cities, towns, and villages. There are obvious outer contours and primary and secondary streets in residential areas. The graphic specification [1] stipulates that the convex and concave parts should be comprehensively represented when their length is less than 0.5–1 mm on the maps. In the 1:50,000 topographic map, 1 mm on the map represents the actual 50 m, and the length of 50 m is 25 pixels in the image with a resolution of 2 m. Therefore, the graphic specification requires that the convex and concave parts should be smoothed when their length is less than 12.5–25 pixels.

Figure 1c is a direct extraction result of residential areas based on semantic segmentation algorithms. The contour line is messy and has a high degree of border redundancy. Figure 1b shows an illustration of the residential area layer in a topographic map, and it is a standard representation corresponding to the cartographic symbols used in topographic mapping. Its outer contour is multiple straight-line segments. The comparison indicated that the results of traditional semantic segmentation algorithms are different from the requirements of the cartographic symbols, and the contour of the extracted results must be regularized as much as possible.

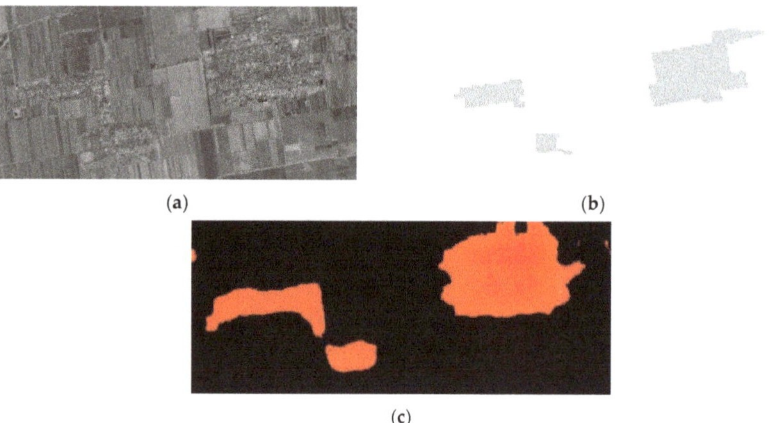

(a)   (b)

(c)

Figure 1. An example of residential area mapping based on images: (**a**) Original image; (**b**) the residential area layer in topographic map; and (**c**) direct extraction results of residential areas.

To sum up, the extracted contour is required to be regular when images are used for residential area mapping. Each segment of extracted contour is generally a straight-line segment, which is relatively regular. Therefore, when building a dataset that supports the end-to-end regularization extraction of residential areas, it is necessary to ensure that the label image contour meets the regularization requirements.

3.2. Overview and Data Properties

In order to create RERB dataset, we collected 13,892 high-resolution TH-1 images [7], and the size of each image is approximately 256 × 256 pixels. Figure 2 shows the label visualization result in this dataset. The TH-1 satellite is the first stereo mapping transmission satellite in China, and its goal is to achieve topographic mapping at a 1:50,000 scale without using ground control points. It consists of a high-resolution camera with ground pixel size of 2 m and a multispectral camera with a ground pixel size of 10 m. Images with a spatial resolution of 2 m are applied in this dataset, and these images cover a geographical area of more than 3640 square kilometers.

The proportions of residential area and other land cover categories in RERB dataset are shown in Table 2. It is obvious that the proportion of the residential area is lower than that of the other categories, which is consistent with the distribution of large-scale remote sensing imagery scenes.

Table 2. The proportion of residential area in our dataset.

	Proportion (%)	Label Number	Color
Residential area	15.89	1	(255,255,255)
Background	84.11	0	(0,0,0)

The labels used for traditional semantic segmentation usually do not have regularization characteristics, as shown in Figure 3b. This kind of label is assigned accurately according to the actual range of residential areas in the image [42]. Different from semantic segmentation labels, according to the contour regularization requirements in mapping application, we need to ensure the labels of residential areas into a regular format.

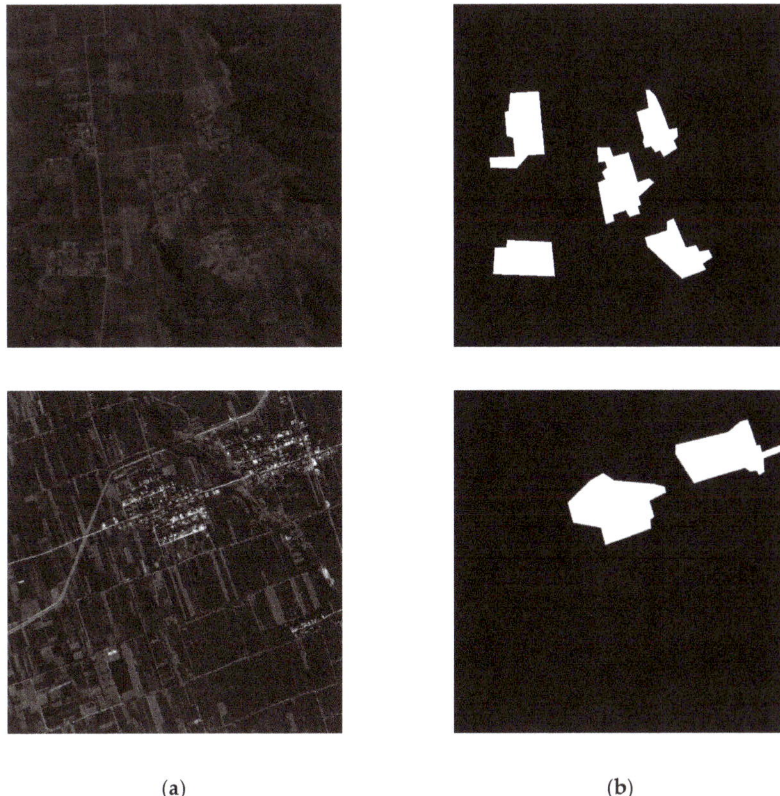

(a)   (b)

Figure 2. Label visualization for a remote sensing image: (**a**) Original image and (**b**) label image.

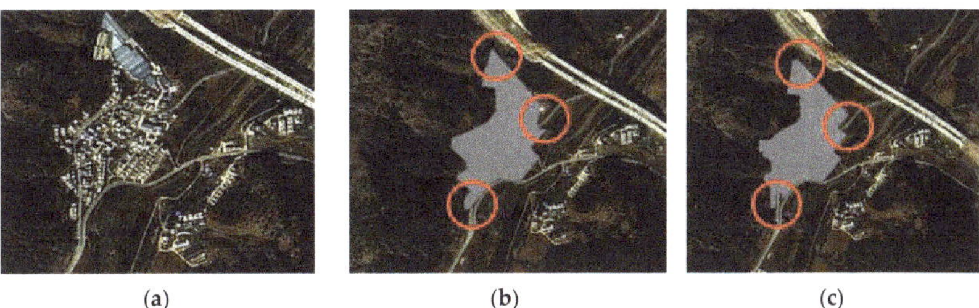

(a)   (b)   (c)

Figure 3. Comparison of traditional semantic segmentation labels and mapping application labels: (**a**) Original image; (**b**) semantic segmentation labels; and (**c**) mapping application labels.

In addition to the regularization requirements of contour line segments, special attention should also be paid to the treatment of the included angle between line segments when labeling. The main principles include small contour protrusion removal and small contour concave part filling. As shown in Figure 4, using the interior of the patch as the reference direction, the contour protrusion and the concave part of the contour are defined when the angle between contour segments is too small (<45°) and excessively large (>90°), respectively. These situations will be corrected with blunt or right angles. For example, in Figure 3b, there are small acute angles as shown in the red circles at the corner of residential areas. Therefore, as shown in Figure 3c, we edit these angles by using a right angle or an obtuse angle in mapping application labels.

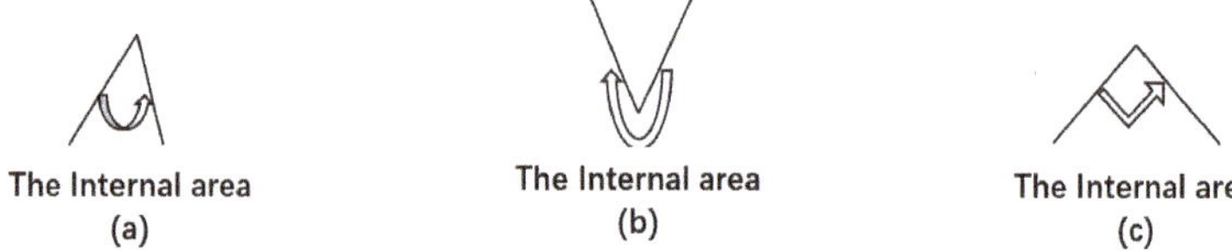

Figure 4. Diagrams of raised and sunken areas: (**a**) The angle is too small; (**b**) the angle is too large; and (**c**) the angle is 90°.

We split 85% of these images into the train set and leave the remaining 15% as the test set. As for annotation, RERB dataset provides pixel-level labels for two important categories, including background and residential area. They are labeled with black (0) and white (1).

We also analyze RERB dataset and find it has four properties: (1) Large-scale and high-resolution. As shown in Table 1, RERB contains 13,892 high-quality satellite images acquired from different cities in China. It covers an area of more than 3640 square kilometers. (2) Well annotated and regular label contour. For each satellite image, we provide accurate pixel-wise mapping application labels for two categories ('background' and 'residential' area), which are annotated by a group of experts. (3) Rich background. The remote sensing mapping task is always faced with the diverse background samples (i.e., ground objects that are not of interest). The high-resolution and different scenes bring more rich details for the background samples. (4) Class imbalance. As shown in Table 2, two categories have very different proportions, which lead to a class imbalance problem. This problem poses a special challenge for the regularization extraction of the residential areas task.

3.3. Statistics for RERB Dataset

Some statistics of the RERB dataset are analyzed in this section. The number of labeled pixels has been counted. As is shown in Table 2 and Figure 5a, the background class contains the most pixels with rich and diverse background samples, which cause special challenge for residential areas extraction.

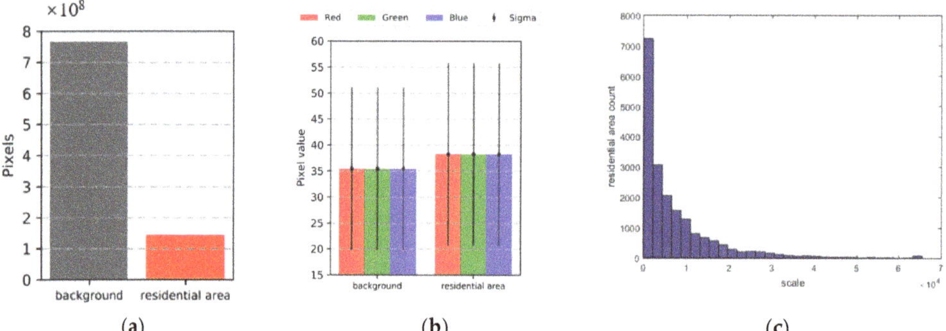

Figure 5. Statistics for the RERB dataset: (**a**) Histogram of the number of pixels for each class; (**b**) spectral statistics. The mean and standard deviation (sigma) for background and residential area are reported; and (**c**) distribution of the residential area sizes.

For the spectral statistics (Figure 5b), the background category has a lower mean value (color column) and standard deviation (vertical line). Because of the high-resolution images of TH-1 satellite are single channel, the values of red, green, and blue are same. As is shown in Figure 5c, most of the residential areas have relatively small scales. Through calculation, the average size of the minimum 30% residential areas is about 479.71 pixels, and the average size of the maximum 30% residential areas is about 18,851 pixels. The multiscale residential areas require the models to have multiscale capture capabilities.

3.4. Construction of Residential Area Regularized Extraction Model

The common semantic segmentation network is generally a symmetric network with encoding–decoding structure [19,20]. The encoding operations mainly include convolution and pooling. Convolution is used to extract high-dimensional features of the input image, and pooling is used to make the image smaller. The decoding operations mainly include deconvolution and up-sampling. Deconvolution makes the features of the image reappear after classification, and up-sampling can restore the original size of the image. Finally, the classification results of each pixel are output. In terms of loss function, cross-entropy [46] has been the most widely used loss function in semantic segmentations of images.

In order to test the effectiveness of the RERB dataset, we designed an end-to-end regularized extraction network by analyzing the regularization characteristics of label contour and the constraints of loss function.

As shown in Figure 6, compared with the traditional semantic segmentation network in Figure 7, our method extracts the contour of the label image first, and realizes regularization extraction by adding the cross-entropy constraint of the label contour image and model prediction image to the original loss function. The baseline method chosen in this article can be any semantic segmentation network, such as U-Net [19] or DeepLab v3+ [24].

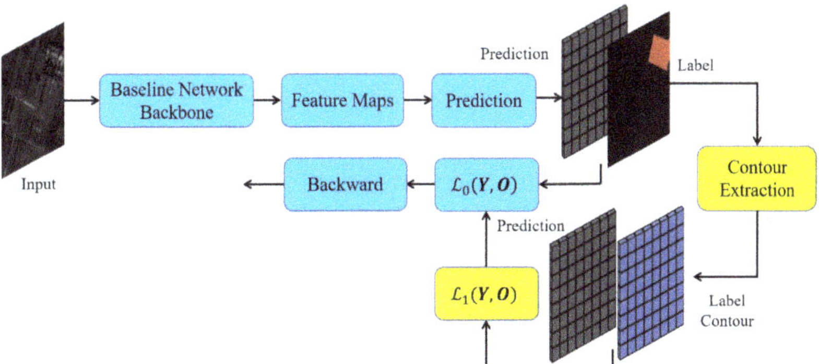

Figure 6. Regularized extraction model structure of residential area (ours).

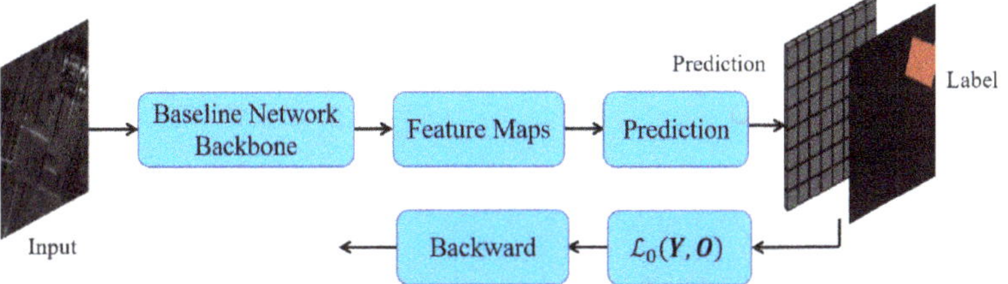

Figure 7. Traditional semantic segmentation network (previous).

The cross-entropy loss function can make the predicted image learned from the training data similar to the real label image. Considering that the label image contour of RERB dataset already has good, regularized contour characteristics, we first extract the contour of label image and then constrain the contour regularization degree of the network prediction image by calculating the cross-entropy loss between the label contour image and the network prediction image, as shown in Equation (1):

$$\mathcal{L}_1(Y, O) = F_{ce}(Y, G_{con}(O)) \tag{1}$$

where $O \in \{0, 1\}^{W \times H}$ denotes the label image, $W \times H$ represents the image size, Y is the network inference result image, which has the same size as image O. F_{ce} and $G_{con}(O)$ represent the cross-entropy loss function and the contour extraction function, respectively.

$\mathcal{L}_1(Y, O)$ represents the degree of inconsistency between the contours of the two images. Through the calculation and back propagation of $\mathcal{L}_1(Y, O)$ in the training process, the contour of the prediction image can be made more and more regular. The cross-entropy loss function is expressed as follows:

$$F_{ce}(Y, G) = -\sum_{x} g_x \log(y_x) + (1 - g_x) \log(1 - y_x) \tag{2}$$

where y_x and g_x denotes the value at position x in the image Y and G, respectively.

The contour of the label image can be extracted by the corrosion of a 3×3 structuring element. Corrosion is a commonly used morphological operation in an image processing file, and it can be expressed as follows:

$$G_{con}(O) = |O - Erosion_{3 \times 3}(O)| \tag{3}$$

In the above formula, the corrosion operation $Erosion_{3 \times 3}(O)$ can remove the area contour in the image O, and then the contour image can be obtained by subtracting the corroded image from the original image.

In the training stage, the Adam [47] optimizer is adopted, and it is a first-order optimization algorithm. The best model can be obtained by minimizing joint loss function $\mathcal{L}(Y, O)$, which is shown in the following formula:

$$\mathcal{L}(Y, O) = \alpha \mathcal{L}_0(Y, O) + \beta \mathcal{L}_1(Y, O) \tag{4}$$

where $\mathcal{L}_0(Y, O)$ is the original loss function of the baseline network, and the functions used in this paper include cross-entropy and Lovász [48]. The existence of $\mathcal{L}_0(Y, O)$ can ensure the segmentation accuracy of the original semantic segmentation network. α and β are the two weights of loss functions, which are experimentally determined.

4. Experiment and Analysis

In this section, we carried out experimental verification and tested the effectiveness of the RERB dataset by using the model constructed in Sections 3 and 4. We first introduced the evaluation metrics. Then, we performed an ablation study to determine some parameters. In the contrast experiment, the baseline networks were U-Net and DeepLab v3+. All experiments were carried out on a platform with an Intel Core (TM) i9 3.60 GHz CPU, 32 GB RAM, GeForce GTX 2080 GPU, and 11 GB video memory. These algorithms were implemented using PyTorch 1.0 and Python 3.7.

4.1. Design of Evaluation Metrics

The traditional semantic segmentation evaluation indexes, such as mean intersection over union (*mIoU*) [18], mainly evaluated the extraction accuracy in pixel units, which cannot reflect the regularization degree of contours as a whole. In detail, the calculation of *mIoU* was based on the confusion matrix, as shown in Table 3. There were n_{cl} different classes in total, including backgrounds, where n_{ij} was the number of pixels of class i predicted to belong to class j and $t_i = \sum_j n_{ij}$ was the total number of pixels of class i.

Table 3. Confusion matrix.

Confusion Matrix		Ground Truth Labels				
		class 1	$\ldots$	class i	$\ldots$	class n_{cl}
prediction	class 1	n_{11}	$\ldots$	n_{1i}	$\ldots$	$n_{1n_{cl}}$
	$\ldots$	$\ldots$	$\ldots$	$\ldots$	$\ldots$	$\ldots$
	class i	n_{i1}	$\ldots$	n_{ii}	$\ldots$	$n_{in_{cl}}$
	$\ldots$	$\ldots$	$\ldots$	$\ldots$	$\ldots$	$\ldots$
	class n_{cl}	$n_{n_{cl}1}$	$\ldots$	$n_{n_{cl}i}$	$\ldots$	$n_{n_{cl}n_{cl}}$

Therefore, *mIoU* is calculated as follows.

$$mIoU = (1/n_{cl}) mathlarger\sum_i n_{ii} / (t_i + mathlarger\sum_j n_{ji} - n_{ii}) \tag{5}$$

To quantitatively evaluate the regularization extraction results, a contour matching index (CMI) was designed to measure the performance of the algorithm in this paper. The specific steps of the CMI calculation are as follows.

(1) The contours of the model prediction image (Figure 8b) and the label image (Figure 8c) were extracted, and the results are shown in Figure 8d,e;
(2) The distance transform of the contour of label image was computed, as shown in Figure 8f;
(3) A contour matching value was obtained by matching the contour of the model prediction image (Figure 8d) with the distance transformed image (Figure 8f);
(4) The CMI value of this image was obtained by dividing the matched value by the number of pixels in the contour of the label image.

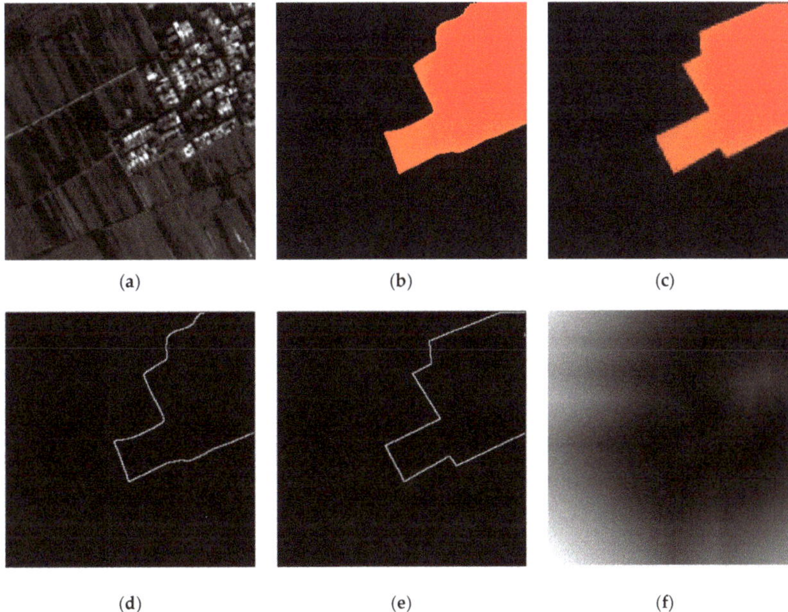

Figure 8. Schematic illustration of the procedure used to calculate the CMI: (**a**) Image; (**b**) model prediction image; (**c**) label image; (**d**) contour image of (**b**); (**e**) contour image of (**c**); and (**f**) transformed image of (**e**).

The critical factor of the distance transform [49] was the definition of distance. In this paper, a block distance transform was adopted. The pixel value of the true contour point was 0 in the image after the distance transform was computed. The farther away from the true contour point, the larger the pixel value of the transformed image. Thus, the matching value can be obtained by calculating the sum of pixel values in the transformed image corresponding to the position of the contour points in the model prediction image. Since the contour image was a binary image that contained only the residential area point (pixel value 1) and the background point (pixel value 0), the matching values between Figure 8d,f can be calculated as follows:

$$S = \sum_{(i,j)} C_{i,j} \cdot e_{i,j} / sum(G_t) \tag{6}$$

where (i, j) is the pixel coordinate, $C_{i,j}$ and $e_{i,j}$ are pixel values at (i, j) in Figure 8d,f, respectively, and $sum(G_t)$ represents the total number of contour points in label contour image G_t (Figure 8e).

The value S reflected the matching degree between the prediction result and the value image. The smaller the value was, the higher the matching degree. Furthermore, the average CMI of all images was used as the evaluation result when a whole test set was evaluated.

Considering that the background class occupied most of the image, we removed the background class in the calculation of mIoU to prevent it from affecting the evaluation of other ground features. Therefore, the evaluation indexes included the CMI and IoU of residential areas.

4.2. Parameters Settings and Ablation Study

4.2.1. Parameters Settings

In the experiment, we divided the training set and test set according to the ratio of 17:3. Finally, the training set and the test set contained approximately 13,611 image slices and 281 image slices, respectively. To verify the adaptability of the proposed method to different loss functions, Lovász was used for $\mathcal{L}_0(Y, O)$ when the baseline network was U-Net, and cross-entropy was used for $\mathcal{L}_0(Y, O)$ when the baseline network was DeepLab v3+. The number of ground feature elements c was set as 2.

The polynomial learning rate policy was employed where the initial learning rate was multiplied by $(1 - iter/total\_iter)^{1.5}$ after each iteration. The maximum number of training cycles was 100 epochs, and thus, $total_{iter} = 100$. The optimal model was determined by testing the model epoch by epoch during training. The weight decay coefficient was set to 0.0005. In terms of the optimization method, the Adam [47] optimizer was used for training.

Batch size value had a great impact on model training and quality of results. Usually, we selected the maximum value according to the network parameters and the hardware configuration (mainly the video memory of GPU). In this paper, we carried out experiments with a batch size of 8, which was determined by model size and video memory. The selection principle was to make the video memory not overflow.

4.2.2. Ablation Study

In this section, we first studied the influence of the initial learning rate on the test set of the RERB dataset. To perform this ablation study, we adopted the semantic segmentation network and the metric mIoU. We evaluated the performance pertaining to the abovementioned parameters, as described in Table 4.

Table 4. Ablation study for the initial learning rate.

	U-Net (Lovász)	DeepLab v3+ (Cross-Entropy)
The Initial Learning Rate	**mIoU**	**mIoU**
1×10^{-3}	0.76099	0.76919
1×10^{-4}	0.77895	0.79395
5×10^{-5}	0.77924	0.79749
2×10^{-5}	0.78533	0.80450
1×10^{-5}	0.77406	0.79527
1×10^{-6}	0.77384	0.77298

The experiments specified in Table 4 were conducted with training batch size = 8. As shown in Table 4, the $mIoU$ peaked when the initial learning rate was 2×10^{-5}.

Next, we studied the influence of the weight α and β on the test set of the RERB dataset. The cross-entropy loss between the label contour image and the network prediction image was inserted to the above semantic segmentation models. The metric CMI was adopted in these experiments.

The experiments specified in Table 5 were conducted with training batch size = 8 and the initial learning rate 2×10^{-5}. As shown in Table 5, the CMI index of the proposed method reached the minimum value when $\alpha = 0.3$ and $\beta = 0.7$.

Table 5. Ablation study for α and β.

	Ours (U-Net) Lovász+$\mathcal{L}_1(Y,O)$	Ours (DeepLab v3+) Cross-Entropy+$\mathcal{L}_1(Y,O)$
(α, β)	CMI	CMI
(0.1, 0.9)	50.647	69.378
(0.2, 0.8)	51.083	72.654
(0.3, 0.7)	39.687	32.074
(0.4, 0.6)	46.174	58.461
(0.5, 0.5)	54.378	36.794
(0.6, 0.4)	56.376	49.586
(0.7, 0.3)	72.545	50.277
(0.8, 0.2)	53.475	40.433
(0.9, 0.1)	43.602	44.169

4.3. Results and Analysis

We parameter tuned some parameters of U-Net, DeepLab v3+ and our proposed method, and the quantitative evaluation results on the test set of RERB dataset are shown in Table 6.

Table 6. Training parameters and quantitative evaluation results.

	U-Net	Ours (U-Net)	DeepLab v3+	Ours (DeepLab v3+)
loss	Lovász	Lovász+ $\mathcal{L}_1(Y, O)$	Cross-Entropy	Cross-Entropy $+ \mathcal{L}_1(Y, O)$
initial lr	2×10^{-5}	2×10^{-5}	2×10^{-5}	2×10^{-5}
batch size	8	8	8	8
(α, β)	--	(0.3, 0.7)	--	(0.3, 0.7)
IoU	0.7853	0.7813 (-0.51%)	0.7953	0.8003 ($+0.63\%$)
CMI	65.638	39.686 ($+39.54\%$)	43.051	32.074 ($+25.50\%$)
train epoch	18	41	41	52
test time	15.37 s	15.55 s	42.79 s	42.78 s

The contrasting experimental results are shown in Figure 9. As seen from Table 6 and Figure 9, the regularization level of residential area contours extracted by our proposed method had increased greatly, especially those areas marked by white circles. When the baseline network was U-Net, the IoU of residential areas decreased by 0.51%, but the CMI increased by 39.54%. Moreover, both the IoU of residential areas and the CMI increased by 0.63% and 25.5%, respectively, when the baseline network was DeepLab v3+.

Compared with the semantic segmentation dataset, the label image contour in RERB dataset had the regularization characteristic and provided additional information, so it could support the construction and training of end-to-end regularization extraction model of residential areas. The experimental results demonstrated the preponderance and practicability of the RERB dataset.

In terms of computational complexity, according to the model construction method in Sections 3 and 4, the increased calculation amount of this method compared with the basic network mainly included label image edge extraction and contour cross-entropy loss calculation during training. The operations of contour extraction included a corrosion operation with 3×3 structuring element and a subtraction. Contour cross-entropy loss calculation included logarithmic calculation and accumulation, which was the same as the original cross-entropy function. Therefore, during the training phase, the computational complexity of the proposed method was slightly larger than that of the traditional semantic segmentation network. Consequently, the runtime of training and the optimal epoch number of our method were higher than those before model modification. In the case of test time, our proposed method was at the same level with traditional models.

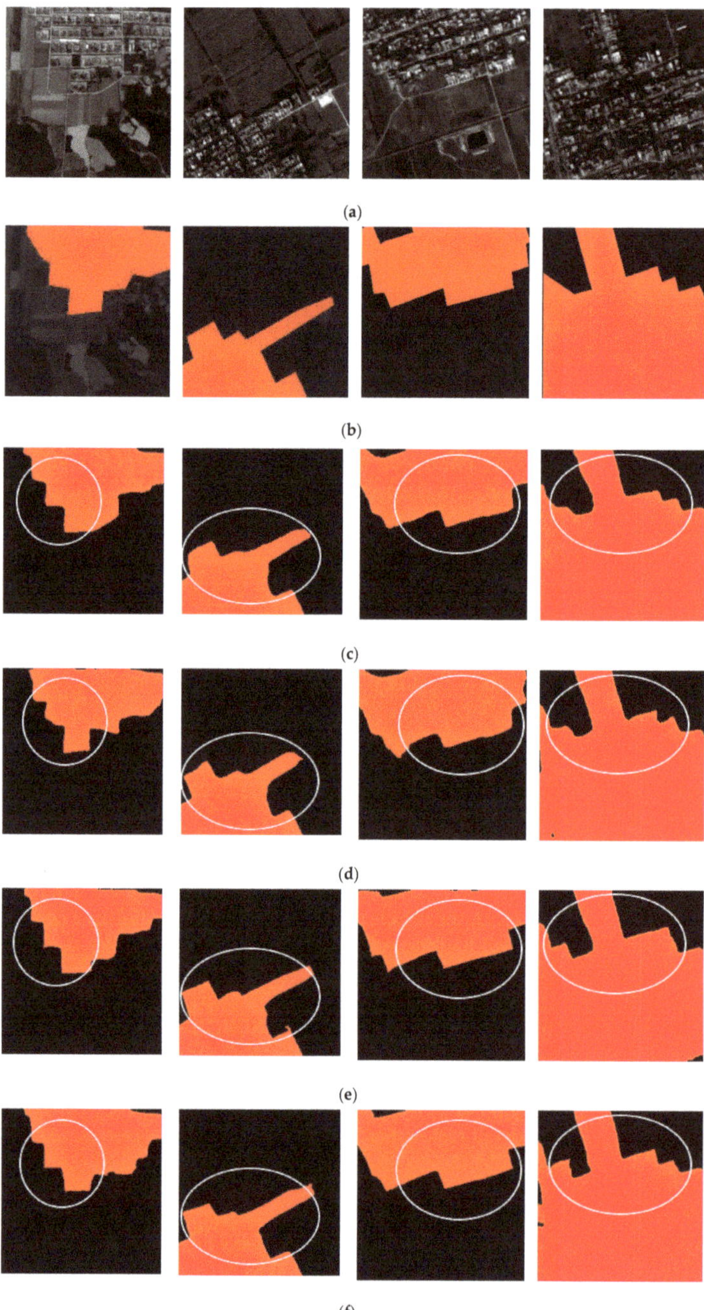

Figure 9. Extraction results of some residential areas: (**a**) Original images; (**b**) label images; (**c**) U-Net; (**d**) DeepLab v3+; (**e**) our model (the baseline network is U-Net); and (**f**) our model (the baseline network is DeepLab v3+).

5. Conclusions

For residential areas, the difference between semantic segmentation labels and mapping application labels limits the possibility of end-to-end regularization extraction training. In order to address this problem, we built a dataset named RERB (Residential area Extraction with Regularized Boundary) for the end-to-end regularization extraction of residential areas. To ensure the rationality of RERB dataset, we analyzed the contour representation requirements for residential area mapping according to the graphic specification of 1:50,000 topographic map, and then transformed it into the following annotation requirements: the contour of label image should be regular, and the included angle of contour line segments should be as right angle as possible. Based on these principles, we have completed the annotation of residential areas in 13,892 image patches based on TH-1 images. The size of each image is approximately 256 × 256 pixels. The RERB dataset encompasses four properties: (1) Large-scale and high-resolution; (2) well annotated and regular label contour; (3) rich background; and (4) class imbalance. In reality, high resolution, complex background, and category imbalance represent three challenges in residential area mapping. Finally, a residential area regularization extraction model is constructed with a contour cross-entropy constraint by using the regular contour label of a residential area. Experimental results showed that the proposed algorithm can improve the regularization degree of the extracted contour of residential areas while maintaining nearly the same extraction accuracy. This fully proves the effectiveness of RERB dataset. In the future, we will expand and improve the dataset of mapping residential area and conduct in-depth research on the end-to-end model for mapping.

Author Contributions: Conceptualization, L.Z., S.L. and D.G.; methodology, S.L., W.L. and D.G.; resources, D.G.; data curation, S.L., H.G., X.Z. and J.H.; writing—original draft preparation, S.L. and J.H.; writing—review and editing, S.L., W.L. and J.H.; supervision, L.Z. and D.G. All authors have read and agreed to the published version of the manuscript.

Funding: This research was funded by the National Natural Science Foundation of China (NSFC) under grant 62101395 and the independent research project of State Key Laboratory of Geo-Information Engineering (Grants No. SKLGIE2020-ZZ-1).

Institutional Review Board Statement: Not applicable.

Informed Consent Statement: Not applicable.

Data Availability Statement: The data presented in this study are available on request from the corresponding author. The data are not publicly available, as the research group's mapping model construction related research is still being carried on.

Conflicts of Interest: The authors declare no conflict of interest.

References

1. *GB/T 20257.3-2017*; Cartographic Symbols for National Fundamental Scale Maps-Part 3: Specifications for Cartographic Symbols 1:25000 1:50000 & 1:100000 Topographic Maps. Standardization Administration of the P.R.C.: Beijing, China, 2017; pp. 19–21. (In Chinese)
2. Feng, W. *Remote Sensing Image Interpretation*, 1st ed.; Science Press: Beijing, China, 1998; pp. 162–165. (In Chinese)
3. Peterle, J. A concept for topographic map updating using digital orthophotos. *Photogrammetria* **1985**, *40*, 87–94. [CrossRef]
4. Holland, D.A.; Boyd, D.S.; Marshall, P. Updating topographic mapping in Great Britain using imagery from high-resolution satellite sensors. *ISPRS J. Photogramm. Remote Sens.* **2006**, *60*, 212–223. [CrossRef]
5. Ministry of Finance of the PRC; State Bureau of Surveying and Mapping of the PRC. *Detailed Rules for Quota Calculation of Surveying and Mapping Production Costs*; Ministry of Finance of the PRC: Beijing, China, 2009; p. 24. (In Chinese)
6. Li, D. China's first civilian three-line-array stereo mapping satellite: ZY-3. *Acta Geod. Cartogr. Sin.* **2012**, *41*, 317–322. (In Chinese)
7. Wang, J.; Wang, R.; Hu, X.; Su, Z. The on-orbit calibration of geometric parameters of the Tian-Hui 1 (TH-1) satellite. *ISPRS J. Photogramm. Remote Sens.* **2017**, *124*, 144–151. [CrossRef]
8. eCognition. Available online: https://geospatial.trimble.com/products-and-solutions/ecognition (accessed on 26 August 2021).
9. Han, F.; Su, Y.; Zheng, J. Research on method of extracting discovery based on EasyFeature elements. *Geomat. Spat. Inf. Technol.* **2020**, *43*, 234–236. (In Chinese)

10. Alokasi, H.; Ahmad, M.B. Deep learning-based frameworks for semantic segmentation of road scenes. *Electronics* **2022**, *11*, 1884. [CrossRef]

11. Mo, Y.; Wu, Y.; Yang, X.; Liu, F.; Liao, Y. Review the state-of-the-art technologies of semantic segmentation based on deep learning. *Neurocomputing* **2022**, *493*, 626–646. [CrossRef]

12. Fu, G.; Liu, C.J.; Zhou, R.; Sun, T.; Zhang, Q.J. Classification for high resolution remote sensing imagery using a fully convolutional network. *Remote Sens.* **2017**, *9*, 498. [CrossRef]

13. Ji, S.; Wei, S. Building extraction via convolutional neural networks from an open remote sensing building dataset. *Acta Geod. Cartogr. Sin.* **2019**, *48*, 448–459. (In Chinese) [CrossRef]

14. Alemohammad, H.; Booth, K. LandCoverNet: A global benchmark land cover classification training dataset. *arXiv* **2020**, arXiv:2012.03111.

15. Tong, X.Y.; Xia, G.S.; Lu, Q.; Shen, H.; Li, S.; You, S.; Zhang, L. Land-cover classification with high-resolution remote sensing images using transferable deep models. *Remote Sens. Environ.* **2020**, *237*, 111322. [CrossRef]

16. Wang, J.; Zheng, Z.; Ma, A.; Lu, X.; Zhong, Y. LoveDA: A remote sensing land-cover dataset for domain adaptive semantic segmentation. In Proceedings of the Thirty-Fifth Conference on Neural Information Processing Systems Datasets and Benchmarks Track, Virtual, 6–14 December 2021. [CrossRef]

17. Liu, S.; Gao, K.; Qin, J.; Gong, H.; Wang, H.; Zhang, L.; Gong, D. SE^{2}Net: Semantic segmentation of remote sensing images based on self-attention and edge enhancement modules. *J. Appl. Remote Sens.* **2021**, *15*, 026512. [CrossRef]

18. Long, J.; Shelhamer, E.; Darrell, T. Fully convolutional networks for semantic segmentation. In Proceedings of the IEEE Conference on Computer Vision and Pattern Recognition, Boston, MA, USA, 7–12 June 2015; pp. 3431–3440. [CrossRef]

19. Ronneberger, O.; Fischer, P.; Brox, T. U-Net: Convolutional networks for biomedical image segmentation. In *Medical Image Computing and Computer-Assisted Intervention*; Springer: Cham, Switzerland, 2015; pp. 234–241. [CrossRef]

20. Badrinarayanan, V.; Kendall, A.; Cipolla, R. Segnet: A deep convolutional encoder-decoder architecture for scene segmentation. *IEEE Trans. Pattern Anal. Mach. Intell.* **2017**, *39*, 2481–2495. [CrossRef] [PubMed]

21. Zhao, H.; Shi, J.; Qi, X.; Wang, X.; Jia, J. Pyramid Scene Parsing Network. In Proceedings of the IEEE Conference on Computer Vision and Pattern Recognition (CVPR), Honolulu, HI, USA, 21–26 July 2017; pp. 6230–6239. [CrossRef]

22. Chen, L.C.; Papandreou, G.; Kokkinos, I.; Murphy, K.; Yuille, A.L. Semantic image segmentation with deep convolutional nets and fully connected CRFs. In Proceedings of the 3rd International Conference on Learning Representations (ICLR), San Diego, CA, USA, 7–9 May 2015.

23. Chen, L.C.; Papandreou, G.; Kokkinos, I.; Murphy, K.; Yuille, A.L. DeepLab: Semantic image segmentation with deep convolutional nets, atrous convolution, and fully connected CRFs. *IEEE Trans. Pattern Anal. Mach. Intell.* **2018**, *40*, 834–848. [CrossRef]

24. Chen, L.C.; Zhu, Y.; Papandreou, G.; Schroff, F.; Adam, H. Encoder-decoder with atrous separable convolution for semantic image segmentation. In Proceedings of the European Conf. Computer Vision (ECCV), Part VII, Munich, Germany, 8–14 September 2018; pp. 833–851. [CrossRef]

25. Wu, H.; Zhang, J.; Huang, K. FastFCN: Rethinking dilated convolution in the backbone for semantic segmentation. *arXiv* **2019**, arXiv:1903.11816.

26. Yu, F.; Koltun, V. Multi-Scale context aggregation by dilated convolutions. In Proceedings of the 4th International Conference on Learning Representations (ICLR), San Juan, Puerto Rico, 2–4 May 2016.

27. Chollet, F. Xception: Deep learning with depthwise separable convolutions. In Proceedings of the IEEE Conference Computer Vision and Pattern Recognition (CVPR), Honolulu, HI, USA, 21–26 July 2017; pp. 1251–1258. [CrossRef]

28. Hasan, A.H.; Al-Kremy, N.A.R.; Alsaffar, M.F.; Jawad, M.A.; Al-Terehi, M.N. DNA Repair Genes (APE1 and XRCC1) Polymorphisms-Cadmium Interaction in Fuel Station Workers. *J. Pharm. Negat. Results* **2022**, *13*, 32–37.

29. Alsaffar, M.F. Elevation of Some Biochemical and Immunological Parameters in Hemodialysis Patients Suffering from Hepatitis C Virus Infection in Babylon Province. *Indian J. Forensic Med. Toxicol.* **2021**, *15*, 2354–2362.

30. Sun, Y.; Zhang, X.; Xin, Q.; Huang, J. Developing a multi-filter convolutional neural network for semantic segmentation using high-resolution aerial imagery and LiDAR data. *ISPRS J. Photogramm. Remote Sens.* **2018**, *143*, 3–14. [CrossRef]

31. Cui, W.; Xiong, B.; Zhang, L. Multi-scale fully convolutional neural network for building extraction. *Acta Geod. Cartogr. Sin.* **2019**, *48*, 597–608. (In Chinese) [CrossRef]

32. Liu, Y.; Piramanayagam, S.; Monteiro, S.T.; Saber, E. Semantic segmentation of multisensory remote sensing imagery with deep ConvNets and high-order conditional random fields. *J. Appl. Remote Sens.* **2019**, *13*, 016501. [CrossRef]

33. Ding, L.; Bruzzone, L. A deep architecture based on a two-stage learning for semantic segmentation of large-size remote sensing images. In Proceedings of the IEEE International Geoscience and Remote Sensing Symposium, Yokohama, Japan, 28 July–2 August 2019; pp. 5228–5231. [CrossRef]

34. Zheng, X.; Huan, L.; Xia, G.; Gong, J. Parsing very high resolution urban scene images by learning deep ConvNets with edge-aware loss. *ISPRS J. Photogramm. Remote Sens.* **2020**, *170*, 15–28. [CrossRef]

35. Liu, J.; Xiong, X.; Li, J.; Wu, C.; Song, R. Dilated residual network based on dual expectation maximization attention for semantic segmentation of remote sensing images. In Proceedings of the IEEE International Geoscience and Remote Sensing Symposium, Waikoloa, HI, USA, 26 September–2 October 2020; pp. 1825–1828. [CrossRef]

36. Zhang, X.; Du, L.; Tan, S.; Wu, F.; Zhu, L.; Zeng, Y.; Wu, B. Land use and land cover mapping using RapidEye imagery based on a novel band attention deep learning method in the Three Gorges reservoir area. *Remote Sens.* **2021**, *13*, 1225. [CrossRef]
37. Abdollahi, A.; Pradhan, B. Integrating semantic edges and segmentation information for building extraction from aerial images using UNet. *Mach. Learn. Appl.* **2021**, *6*, 100194. [CrossRef]
38. Liu, J.; Zhang, J.; Li, Z.; Zhang, G.; Du, W.; Zhao, W.; Liu, J. Technical framework of 1:10000 cartographic element extraction based on GF-7 satellite. *Geomat. World* **2018**, *25*, 58–61. (In Chinese) [CrossRef]
39. Zhao, M.; Liu, S.; Xu, G.; Yang, M. A method of residential area contours regularization in remote sensing image based on straight line segment fitting. *Geomat. Sci. Eng.* **2019**, *39*, 29–33. (In Chinese)
40. Pan, X.; Zhao, J.; Xu, J. An end-to-end and localized post-processing method for correcting high-resolution remote sensing classification result images. *Remote Sens.* **2020**, *12*, 852. [CrossRef]
41. Maggiori, E.; Tarabalka, Y.; Charpiat, G.; Alliez, P. Fully convolutional neural networks for remote sensing image classification. In Proceedings of the 2016 IEEE International Geoscience and Remote Sensing Symposium (IGARSS), Beijing, China, 10–15 July 2016; pp. 5071–5074. [CrossRef]
42. He, C.; Fang, P.Z.; Zhang, Z.; Xiong, D.H.; Liao, M.S. An end-to-end conditional random fields and skip-connected generative adversarial segmentation network for remote sensing images. *Remote Sens.* **2019**, *11*, 1604. [CrossRef]
43. Sun, Y.; Zhang, X.; Zhao, X.; Xin, Q. Extracting building boundaries from high resolution optical images and LiDAR data by integrating the convolutional neural network and the active contour model. *Remote Sens.* **2018**, *10*, 1459. [CrossRef]
44. Ruan, T.; Liu, T.; Huang, Z.; Wei, Y.; Wei, S.; Zhao, Y. Devil in the details: Towards accurate single and multiple human parsing. In Proceedings of the AAAI Conference on Artificial Intelligence, Honolulu, HI, USA, 28–30 January 2019; pp. 4814–4821.
45. Takikawa, T.; Acuna, D.; Jampani, V.; Fidler, S. Gated-SCNN: Gated shape CNNs for semantic segmentation. In Proceedings of the IEEE/CVF Conference on Computer Vision, Seoul, Korea, 27 October–2 November 2019; pp. 5228–5237. [CrossRef]
46. YA, D.M.; Liu, Q.; Qian, Z.B. Automated image segmentation using improved PCNN model based on cross-entropy. In Proceedings of the 2004 International Symposium on Intelligent Multimedia, Video and Speech Processing, Hong Kong, China, 20–22 October 2004; pp. 743–746. [CrossRef]
47. Kingma, D.; Ba, J. Adam: A Method for Stochastic Optimization. In Proceedings of the 3th International Conference on Learning Representations (ICLR), San Diego, CA, USA, 7–9 May 2015.
48. Berman, M.; Triki, A.R.; Blaschko, M.B. The Lovász-Softmax loss: A tractable surrogate for the optimization of the intersection-over-union measure in neural networks. In Proceedings of the IEEE Conference on Computer Vision and Pattern Recognition (CVPR), Salt Lake City, UT, USA, 18–23 June 2018; pp. 4413–4421. [CrossRef]
49. Embrechts, H.; Roose, D. Parallel Algorithms for the Distance Transformation. In *Parallel Processing: CONPAR 92—VAPP V*; Springer: Berlin/Heidelberg, Germany, 1992; pp. 387–391. [CrossRef]

Article

Vision-Based Quadruped Pose Estimation and Gait Parameter Extraction Method

Zewu Gong, Yunwei Zhang *, Dongfeng Lu and Tiannan Wu

Faculty of Information Engineering and Automation, Kunming University of Science and Technology, Kunming 650500, China

* Correspondence: zhangyunwei72@gmail.com

Abstract: In the study of animal behavior, the prevention of sickness, and the gait planning of legged robots, pose estimation, and gait parameter extraction of quadrupeds are of tremendous importance. However, there are several varieties of quadrupeds, and distinct species frequently have radically diverse body types, limb configurations, and gaits. Currently, it is challenging to forecast animal pose estimation with any degree of accuracy. This research developed a quadruped animal pose estimation and gait parameter extraction method to address this problem. A computational framework including three components of target screening, animal pose estimation model, and animal gaits parameter extraction, which can totally and efficiently solve the problem of quadruped animal pose estimation and gait parameter extraction, makes up its core. On the basis of the HRNet network, an improved quadruped animal keypoint extraction network, RFB-HRNet, was proposed to enhance the extraction effect of quadruped pose estimation. The basic concept was to use a DyConv (dynamic convolution) module and an RFB (receptive field block) module to propose a special receptive field module DyC-RFB to optimize the feature extraction capability of the HRNet network at stage 1 and to enhance the feature extraction capability of the entire network model. The public dataset AP10K was then used to validate the model's performance, and it was discovered that the proposed method was superior to alternative methods. Second, a two-stage cascade network was created by adding an object detection network to the front end of the pose estimation network to filter the animal object in input images, which enhanced the pose estimation effect of small targets and multitargets. The acquired keypoints data of animals were then utilized to extract the gait parameters of the experimental objects. Experiment findings showed that the gait parameter extraction model proposed in this research could effectively extract the gait frequency, gait sequence, gait duty cycle, and gait trajectory parameters of quadruped animals, and obtain real-time and accurate gait trajectory.

Keywords: quadrupeds; pose estimation; HRNet network; gait parameters

Citation: Gong, Z.; Zhang, Y.; Lu, D.; Wu, T. Vision-Based Quadruped Pose Estimation and Gait Parameter Extraction Method. *Electronics* **2022**, *11*, 3702. https://doi.org/10.3390/electronics11223702

Academic Editor: Chiman Kwan

Received: 26 September 2022
Accepted: 10 November 2022
Published: 11 November 2022

1. Introduction

Animal pose estimation (APE) and behavior research have received increasing attention along with the ongoing advancements in computer vision. Pose estimation and behavior research go hand in hand. Pose estimation can obtain keypoints from animal pose and offer information and practical assistance for behavioral research. An essential problem that needs to be solved in behavior research is the further automatic recognition and acquisition of animal gait parameters based on pose estimation. As a result, the two taken together can help us quantitatively analyze, comprehend, and grasp the behavioral laws of quadruped animals. This has important scientific significance and practical application value for monitoring animal behavior, predicting livestock diseases, and designing and developing quadruped robots.

Currently, deep-learning-based systems and conventional artificial labeling are the two main methods used in research on the animal pose and behavior. The former requires the installation of sensor gear on the quadruped animal to gather data; this is a simple process that does not call for any learning or research-based thinking, but it does require the cooperation of the quadruped animal. This method is simple to use and efficient when the research subject is a person, but for a variety of animals, it is easy to lose sensors. It may even harm the experimental subjects due to the randomness of their movement patterns, so it cannot obtain the desired impact. There are fewer scene restrictions in the deep-learning-based scheme because data are mostly captured through the recording of video images, which does not require moving close to the research object. Methods based on deep learning have been applied quickly in recent years with the emergence of diverse neural network models, and they have been used in a few real-world situations. In order to directly forecast position coordinates for tasks involving human pose estimation, Toshev et al. [1] offered a DNN regression approach that combines multistage regression to predict keypoint coordinates directly for human pose estimation applications. The algorithm has two benefits: (1) DNN captures all the context information about key body parts, and each of those keypoints is regressed using a full human image. (2) Theoretically, it is feasible to extract keypoint information from any CNN network, irrespective of the topological relationship between keypoints; however, the error is significant due to the "concentration" of the distribution region of coordinate points. Fan et al. [2] used a dual-source deep convolution neural network (DSCNN) to expand a single RCNN to a dual-source model (DSCNN) to predict coordinate information of human keypoints. The inputs of Fan et al. are image block and entire image, while the outputs are joint detection results of sliding windows and joint placement of coordinate points. Pfisher et al. [3] proposed a keypoint detection method by fusing optical flow with a deep convolutional network. The keypoints were improved as a heatmap, and by integrating the image data from the front and back frames, the keypoint prediction of the human pose had a higher accuracy. Chen Li et al. [4] proposed an unsupervised adaptive approach to animal pose estimation, utilizing a multiscale domain adaptive module (MDAM) to design a pseudo label update strategy based on the memory effects of deep networks that enable networks to learn from clean samples early and noisy samples later to reduce the domain gap between synthetic and real data. Newell et al. [5] developed an hourglass network following a multistage design pattern, which had a detection effect of 94.1% on the MPII dataset. Mu et al. [6] trained their models using synthetic animal data generated from CAD models, which they subsequently utilized to create pseudo labels for unlabeled real animal images. The generated pseudo labels are gradually included into the model training process using three consistency check criteria. Yuhui Yuan et al. [7] proposed a method that combines vision transformer (ViT), traditional deep convolution networks, HRFormer, and improved transformer encoder. This method outperformed pure convolution networks and significantly improved performance for the COCO dataset human keypoint detection task. Cao et al. [8] presented a cross-domain adaptation scheme to learn a common shared feature space between human and animal images. An object detection algorithm, CenterNet, based on Anchor-free, was proposed by Xingyi Zhou et al. [9], and its operations of deleting ineffective and complicated anchors further improved the detection algorithm's performance. As a result, with a few simple adjustments, it may be applied to tasks involving human keypoint identification, as well as 3D object detection and 2D keypoint activities. In contrast to the usual method of returning keypoints through a heatmap or directly returning keypoints, Microsoft Asia Research Institute proposed a new algorithm called DEKR (disentangled keypoint regression) [10]. It adopted a method of decoupling multibranch regression keypoint positions and used an adaptive convolution module to perform an affine transformation on each branch to obtain the pixel offset of each keypoint. Feng Zhang et al. [11] proposed the Distribution-Aware coordinate Representation of Keypoint (DARK). It includes two parts: (1) a more principled distribution-aware decoding method; (2) a coordinate encoding process (transforming truth coordinates to heatmaps) by generating unbiased heatmaps. In extensive experiments, it was shown that DARK is the best on two common benchmarks, MPII and the COCO keypoint detection dataset. Izonin et al. [12] proposed an RBF-based input doubling method for small medical data processing based on the classical iterative RBF neural network. This method solves the problem of data processing in the medical field when the amount of data is not large enough, and they are difficult to collect. The proposed method achieves the highest accuracy compared to some existing improved RBF networks.

Daou et al. [13] employed piezoelectric sensors to obtain the gait parameters of turtles and created a motion simulation model with regard to the gait parameter extraction in animal behavior research. To extract temporal and spatial information, including the movement behavior and present condition of pigs with various body shapes, Yang et al. [14] used the optical flow frame method. An underactuated quadruped robot named BabyBot with a flexible spine and elastic joints was created by Zhang XL et al. [15] after being inspired by the systematic swing of infants' arms when they hit the ground with their knees during the crawling process. In bionic robots, Xue Bin Peng et al. [16] captured 3D motion data by placing position sensors in the limbs and joints of dogs and then applying reinforcement learning (RL) to enable the robots to mimic and learn the movements of real animals. In addition, they applied a dog gait to the Unitree Laikago robot, allowing it to mimic dog behavior such as walking, trotting, and rotating. However, this strategy still has several technical obstacles, including high artificial expenses and expertise. Kim et al. [17] proposed a spatially based joint gait and used it with lizard kinematics models after collecting data on lizard movement using infrared cameras. However, this strategy is currently only relevant to a single species, and trials for other animals must be developed to collect data. Chapinal et al. [18] demonstrated that evaluating cow walking speed with five 3D accelerometers is simple and straightforward, but does not capture more complex gait parameters such as gait frequency and gait sequence.

To sum up, the problem of pose estimation has been the estimation of numerous theoretical studies and theories, and human pose estimation is more advanced than animal pose estimation. Due to the significant differences in anatomy and behavior between humans and quadruped animals, human pose estimation and quadruped animal pose estimation are very different. Animal pose estimation has the following issues more so than human pose estimation.

1. Because there are many distinct kinds of quadrupeds and because their body kinds, limbs, and behavior modes frequently change substantially, the results of applying a model originally applied using data from one animal to others are very subpar.
2. Animal fur comes in various hues, and if the color is close to the backdrop color, it is easy for the background to distort the image, making it difficult to extract an accurate animal pose information from complicated surroundings.
3. Because quadruped animals often have their limbs below the body, it is easy for the limbs to become mutually occluded during exercise, which makes it more challenging to extract the keypoints.

Because of these issues, it is challenging to apply human pose estimation methods to animal pose estimation directly. Additionally, the present method for extracting quadruped animal gait parameters mainly involves placing sensors in strategic locations on the limbs to track pertinent motion parameters. The problem of how to obtain precise motion gait parameters through noncontact computer vision perception and automatically assess and identify the motion features of quadrupeds based on pose estimation remains unresolved.

This research proposes a quadruped animal pose estimation model and gait parameter extraction method to address these problems, using target screening, an animal pose estimation model, and animal gait parameter extraction, which can completely and effectively solve the problem of animal pose estimation and obtain animal gait parameters. Among them, an improved animal pose estimation network based on the HRNet network is proposed to improve the performance of animal pose estimation, aiming for the extraction efficiency of animal keypoint information. The basic idea behind it is to use a special receptive field block called DyC-RFB to improve the first branch's feature extraction capability in stage 1 of the HRNet network and to improve the network model's overall feature extraction capability. This will result in a better keypoint extraction effect for the animal pose keypoint. Despite being a simple enhancement to the HRNet network model, it has improved the performance of keypoint extraction of animal pose estimation without appreciably increasing network parameters and computation. Additionally, an object detection network may also considerably improve the animal pose estimation effect of some small targets, multitargets, and increase the stability and reliability of pose estimation by being added to the front end of the pose estimation task for target screening. The calculation models of keypoint gait parameters, including gait frequency, gait sequence, and duty cycle, were established with the goal of realizing the automatic extraction of gait parameters and obtaining real-time and accurate foot gait trajectory. This method was based on pose estimation and by analyzing the relationship between the temporal and spatial changes in keypoints of animal pose and gait parameters.

The chapters of this study are organized as follows. In Section 2, we mainly introduce the animal pose estimation model and animal gait parameter extraction methods. Section 3 describes the experiments and results in this paper. Finally, we provide a conclusion and discussion of this paper in Section 4.

2. Models and Methods

2.1. Computational Framework

To comprehensively and effectively handle the problem of animal pose estimation and gait parameter extraction, this research proposes a computational framework based on computer vision that consists of three sections, as depicted in Figure 1.

(1) Target screening: The individual animal in the input image was targeted after the raw resolution video image data were passed through an object detection network, and the upper left and lower right corner coordinates of the selected box, the confidence of the box, and the animal species were output. In the experiment, if the confidence score >0.9, confidence was correctly selected, but the gait parameters collected were not continuous, possibly. Even though the parameters that were extracted were correct, some valid information was lost at the same time. If the confidence score <0.7 was correctly selected for the animal, more mutant noise was added to the final gait parameters, and the processing was difficult. We set this confidence threshold at 0.8 to balance accuracy and data integrity. The animal was correctly identified when the frame had a confidence score greater than 0.8. Following a cropping process based on the coordinates of the frame, the resolution was adjusted to 288 × 384 without changing the aspect ratio of the raw picture. If the detection confidence was less than 0.80, we skipped the current image, detected the next image, and entered the results into the subsequent RFB-HRNet network.

(2) Animal pose estimation model: The image sliced by the pre-object detection network was resized to a fixed size, 288 × 384 × 3 resolution, and input into the keypoint extraction network RFB-HRNet to acquire the heatmap of the animal's keypoints in the image, and the heatmap measured 72 × 96 × 17. The coordinate information of the keypoints in the feature map was calculated based on the predicted heatmap, and the resulting coordinate information was restored to the ground truth coordinates in the original image space and connected sequentially.

(3) Animal gaits parameter extraction: After obtaining the coordinates of the keypoints of the animal under the original resolution, a set of time series of the coordinates of the animal keypoint was obtained to process the obtained data, and the gait information of the quadruped animal was acquired. This information included gait frequency, gait sequence, gait duty cycle, and gait trajectory. Section 2.3 contains information regarding the particular technique of implementation.

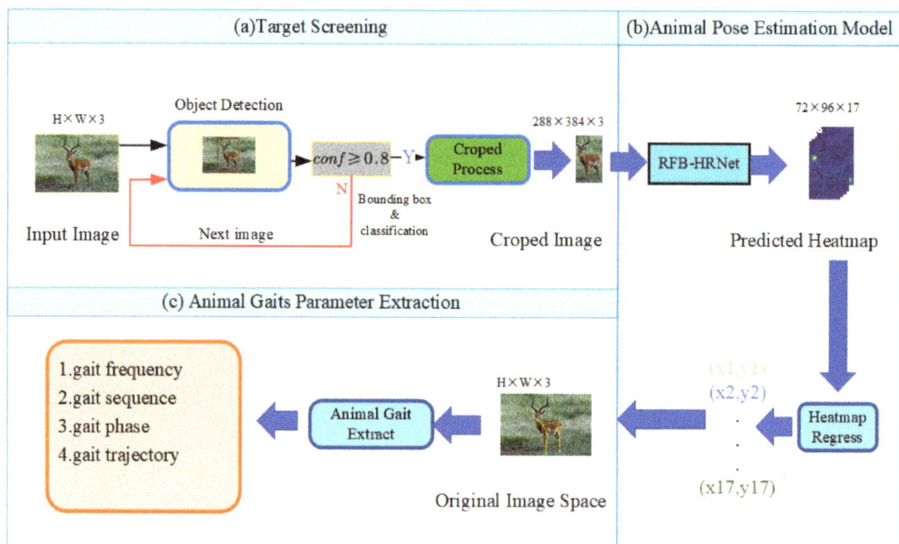

Figure 1. Quadruped animal pose estimation and animal gait parameter extraction computational framework.

2.2. Animal Pose Estimation Model

2.2.1. Improved Animal Pose Estimation Network (RFB-HRNet)

The animal pose estimation task is position-sensitive. Common low-resolution representation model methods, such ResNet [19], VGG [20], and the MobileNet [21] series, perform badly for this kind of task. These multilayer models compress feature map resolution to accomplish semantic aggregation, but do so at the expense of details, making them better suited for image classification tasks. Regarding tasks involving pose estimation, the current main research hypothesis is to fuse the low-resolution feature map with the high-resolution retention network in order to increase the independence of the semantic information between the resolutions while retaining the large-scale feature map data extracted from the front layer using the high-resolution retention network. High-resolution maintenance networks have become the backbone networks for the majority of common pose estimation models in recent years. The performance of keypoints extraction significantly improved after the HRNet high-resolution maintenance network [22] was introduced. In comparison to the human pose estimation task, the animal pose estimation task is more easily obscured by the background, every keypoint's motion relationship is intricate, and mutual occlusion occurs frequently, all of which place higher demands on the ability to extract keypoints. In order to further enhance the extraction performance of important details of quadruped animal poses, this paper uses the HRNet network as the backbone network and enhances the network model on this basis.

In order to preserve high-resolution features and integrate multiscale low-resolution features, HRNet adopted the idea of a multibranch structure, with each branch corresponding to a feature map of different scale. As the network depth expanded, the network branches would also rise, but the branches corresponding to the larger-scale feature map would remain and move in reverse. The body portion of the HRNet network consisted primarily of four stages in order to actualize this idea. Each stage had several network branches, with the number of network branches increasing by one with each stage; the number of channels in the lower network branch was twice that of the upper network branch, and the feature map resolution was one-half that of the upper network branch. Thus, the number of network branches would gradually rise with each successive stage. Due to the unique multibranch structure of the HRNet, it can be seen that the ability to extract features from the first branch of stage 1 has a direct effect on the ability to extract features from each branch in the subsequent three stages, which in turn affects the extraction performance of the entire network model for keypoint information. If the capacity to extract features from the first branch of stage 1 of the HRNet network can be increased, the ability to extract features from the entire network model will be enhanced as a result of the transfer impact of each branch in subsequent stages.

Inspired by this, and based on HRNet, this research proposes an enhanced keypoint extraction network, RFB-HRNet, as shown in Figure 2, to further improve the performance of keypoint extraction of quadruped animals. The basic idea is to use a special receptive field block module to improve the feature extraction ability of the first branch of stage 1 of the HRNet network, and then to improve the overall feature extraction ability of the network model with the aid of the transmission effect of each branch in the later stage, thereby achieving improved quadruped animal keypoint extraction effects. To provide a receptive field mechanism that resembles human vision, a DyC-RFB (dynamic convolution receptive field block) with a dynamic convolutional layer was introduced after the first branch of stage 1 of the HRNet-W48 backbone network. The ability of the first branch to extract features was improved, and this effect was propagated to the subsequent stage, ultimately enhancing the extraction performance of the entire network for keypoint information. In addition, the computational complexity of the DyC-RFB module was minor after it was placed in the first branch of stage 1, which accounted for the trade-off between model complexity and computational speed and prevented a significant influence on network processing speed.

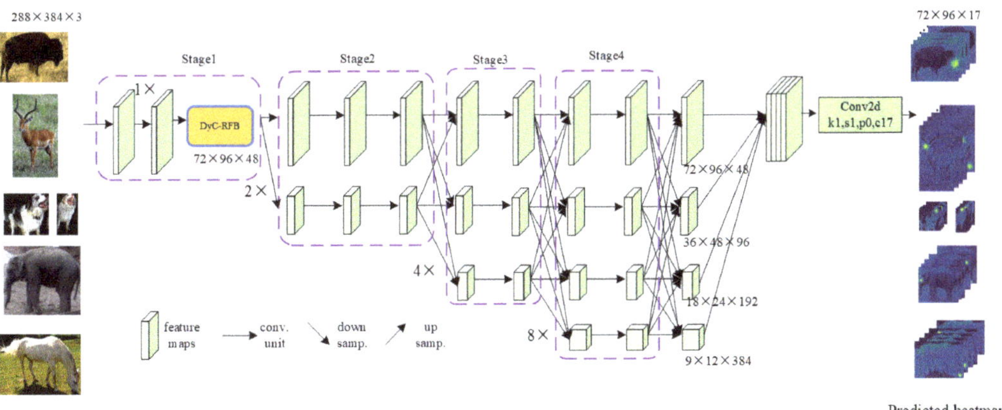

Figure 2. RFB-HRNet network structure.

The model's computation procedure is as follows: Firstly, the image resolution obtained by the object detection network is kept at its original aspect ratio of 288 × 384 × 3. The process is not a simple scaling of the original image, but rather is the filling of one dimension with 0 to compensate for another dimension's deficiency while guaranteeing that the resolution of 288 × 384 is met. A feature diagram with measurements of 72 × 96 × 48 was presented after the updated DyC-RFB module. After successively completing stage 2, stage 3, and stage 4, four branches' feature maps were obtained. Later, following multiscale feature fusion, the resolution was adjusted to 72 × 96 and the number of feature map channels was extended to 384. After a convolution with kernel size = 1, stride = 1, and padding = 0, the number of channels was resized to 17, and the heatmap was regressed 17 times, thus obtaining the two-dimensional coordinates of the quadruped's keypoint, which were then scaled in succession. Finally, it was returned to the coordinate position of the keypoint according to the original resolution space.

To reduce the deviation of decoding from the heatmap to the keypoints, the formula $k = Decoding(Encoding(k))$ should be satisfied as much as possible, where k represents the keypoint coordinates marked by the dataset. This paper adopts the DARK method proposed in [11], and the process is shown in Formula (1):

$$p = m - \left(\mathcal{D}''(m)^{-1} \mathcal{D}'(m) \right), \tag{1}$$

where m represents the position coordinates of the first maximum in the heatmap, and $\mathcal{D}'$ and $\mathcal{D}''$ are the first and second derivatives of the log-likelihood function of the predicted heatmap coding formula, respectively. The original formula is shown as follows:

$$\mathcal{D}(x; p, \Sigma) = -\ln(2\pi) - \frac{1}{2}\ln(|\Sigma|) - \frac{1}{2}(x - p)^T \Sigma^{-1}(x - p), \tag{2}$$

where Σ is the covariance matrix $diag(\sigma^2)$, x is the two-dimensional position coordinate of the heatmap, and p is the keypoint position of the predicted heatmap.

The final coordinate position at the original resolution space can be expressed:

$$\hat{p} = \lambda p, \tag{3}$$

where λ represents the scaling factor, and $\hat{p}$ is the coordinate position of the keypoint under the original resolution space.

2.2.2. DyC-RFB Module

The RFB module is primarily utilized by the SSD [23] algorithm. By mimicking the receptive field of human vision, the RFB module can improve the network's capacity for feature extraction. As depicted in Figure 3a, its structure is primarily based on the concept of an inception algorithm [24]. It adds a dilated convolution layer [25] to inception, which increases the model's ability to extract features while taking into account the amount of computation required. Compared to static convolution, dynamic convolution offers a greater capacity for feature expression and can improve the model's expression capability without expanding the network's depth or width. Its calculation volume is just 4% greater than that of static convolution, but its performance in feature extraction surpasses that of static convolution. In order to further improve the model's ability to express features, this research employs the dynamic convolution (DyConv) module [26] to upgrade the RFB module. It then offers an improved DyC-RFB module, as depicted in Figure 3b. Specifically, it replaces the 3×3 convolution of the final layer of the RFB module with dynamic convolution. The size of the dynamic convolution kernel remains as 3×3, and the respective dilation rates are 1, 3, and 5.

Figure 4 depicts the dynamic convolution (DyConv) module structure. Its structure has k convolution kernels and the attention module calculates the convolution kernel weight. Each convolution kernel's size and output dimension are identical. The average pooling layer is carried out in the attention layer before being projected to the k dimension via two fully connected dense layers. Softmax normalization is then applied to the k convolution kernel weights and convolution kernels to aggregate them. After the aggregate convolution, the batch normalization is applied and the ReLU activation function is employed to generate the dynamic convolution module. Its calculation formula is as follows:

$$\left\{ \begin{array}{l} y = g\left(\widetilde{W}^T x + \widetilde{b} \right) \\ \widetilde{W} = \sum\limits_{k=1}^{K} \pi_k(x) \widetilde{W}_k \\ \widetilde{b} = \sum\limits_{k=1}^{K} \pi_k(x) \left(\widetilde{b} \right)_k \\ s.t. 0 \leq \pi_k(x) \leq 1, \sum\limits_{k=1}^{K} \pi_k(x) = 1 \end{array} \right. \tag{4}$$

where $\widetilde{W}_k$ and $\widetilde{b}_k$ are the weight vector and bias vector of the k th convolution kernel, respectively, and $\pi_k(x)$ is the attention weight learned by the attention module.

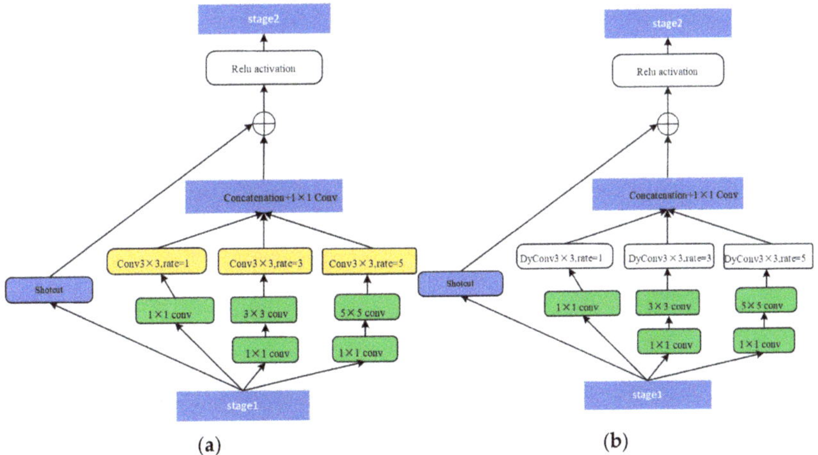

Figure 3. (**a**) RFB module; (**b**) DyC-RFB module.

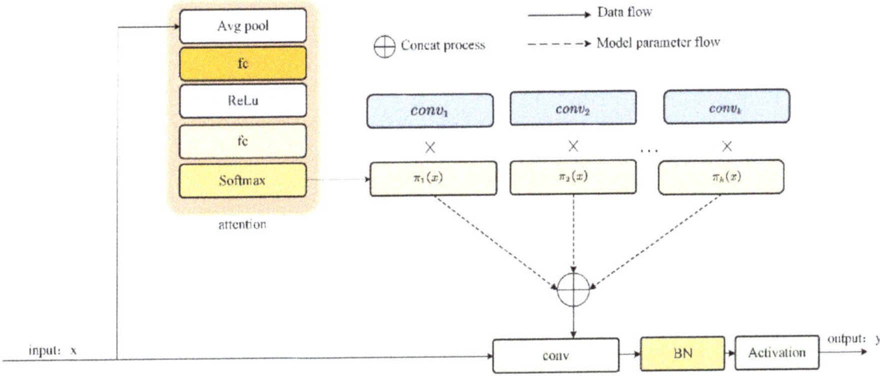

Figure 4. Dynamic convolution (DyConv) module.

2.2.3. Evaluation Metric

The standard evaluation metric is based on object keypoint similarity (OKS), and average precision (AP), mean average precision (mAP), and average recall (AR) are the main evaluation metrics of the pose estimation task.

(1)   OKS

In this paper, after the pose estimation model obtains the prediction heatmap of 17 keypoints, the position coordinates corresponding to each keypoint are obtained by regression processing. Then the prediction effect is evaluated by comparing the position coordinates to the ground truth keypoint. The OKS calculating formula can be denoted as:

$$OKS = \frac{\sum\limits_{i} \exp\left(\frac{-d_i^2}{2s^2\sigma_i^2}\right)\delta(v_i > 0)}{\sum\limits_{i}^{n}\delta(v_i > 0)}, \tag{5}$$

where σ_i^2 represents the normalization factor of the ith keypoint, which is the standard deviation between the manually labeled keypoints and the actual keypoints in all image data. The larger the standard deviation, the larger the labeling error for this type of keypoint and the more difficult it is to label. Labeling this type of keypoint becomes easier as its labeling error decreases. The calculation formula is: $\sigma_i^2 = E[d_i^2/s_i^2]$. The constants of each keypoint are different; the specific values were set according to the AP10k dataset processing code in the mmpose codebase. The normalization factors for 17 keypoints, including the nose, eye, neck, tail, shoulder, knee, and ankle, are [0.025, 0.025, 0.026, 0.035, 0.035, 0.079, 0.072, 0.079, 0.072, 0.062, 0.107, 0.087, 0.089, 0.107, 0.087, 0.087, 0.089]. d_i represents the Euclidean distance between the detected keypoint and the corresponding ground truth. s is the object scale. $\delta_i(v_i)$ represents the visibility mark of the keypoint; when $v_i > 0$, this keypoint is visible, and $\delta(v_i) = 1$. When $v_i < 0$, the keypoint is invisible, and $\delta(v_i) = 0$. Each keypoint in the formula has a similarity between 0 and 1. Ideally, the predicted keypoint's OKS = 1 when its coordinates match its actual keypoint.

(2)　AP

The test dataset Formula (6) depicts the AP of each keypoint. AP^{50} and AP^{75} represent the average precision when IoU equals 0.50 and 0.75, respectively. AP^M indicates the average precision of animals on medium objects, whereas AP^L represents the average precision of animals on large objects.

$$A_p = \frac{\sum\limits_{p} \delta(O_{oks} > s)}{\sum p_1}, \tag{6}$$

where s is the OKS threshold. When $O_{oks} > s$, $\delta(O_{oks}) = 1$.

(3)　mAP

The mAP is the mean value of AP calculated for different types of keypoints in the entire dataset, reflecting the performance of the detection algorithm. The formula is $P_{mAP} = mean\{A_P@s(0.50 : 0.05 : 0.95)\}$.

(4)　AR

AR reflects the proportion of the keypoints in the image that are successfully found, so the AR is calculated as follows:

$$AR = \frac{TP}{TP + FN}, \tag{7}$$

where TP is the number of keypoints correctly identified in the image, and FN is the number of keypoints not correctly identified in the image.

2.3. Animal Gait Parameter Extraction Model

2.3.1. Animal Gait Frequency Extraction Model

The motion state of quadruped animals can be separated into two phases: the swing phase and the support phase [27]. The swing phase is the motion state of the animal's feet from leaving the ground to the next landing during the gait cycle, whereas the support phase is the movement state of the feet during a gait cycle when the feet are on the ground. The swing phase and support phase of a quadruped animal are shown in Figure 5.

Periodically, the limbs of the quadrupeds swing when they move. During this brief period of time, the relatively flat segment of the curve shows that the limbs are in the support phase, while the steeper segment (Figure 6a) indicates that the limbs are in the swing phase. Through the first-order differential processing of the corner points of the limbs, the differential frame (former frame t_{frame}, later frame $t_{frame+1}$) between the two adjacent slopes is taken to obtain the minimum value, and the time t is mapped to frame

F(fps). The video used in this paper is 30 frames, that is, the minimum time scale of a single frame is $\Delta t = 1/F$. The formula for calculating the gait frequency of a quadruped animal is:

$$f = \frac{1}{\left(t_{frame+1} - t_{frame}\right)\Delta t},\tag{8}$$

where t_{frame} is the previous gait cycle slope minimum frame; $t_{frame+1}$ is the next gait cycle slope minimum frame; and Δt means a minimum frame time unit, which is a constant of $1/30$ in this paper.

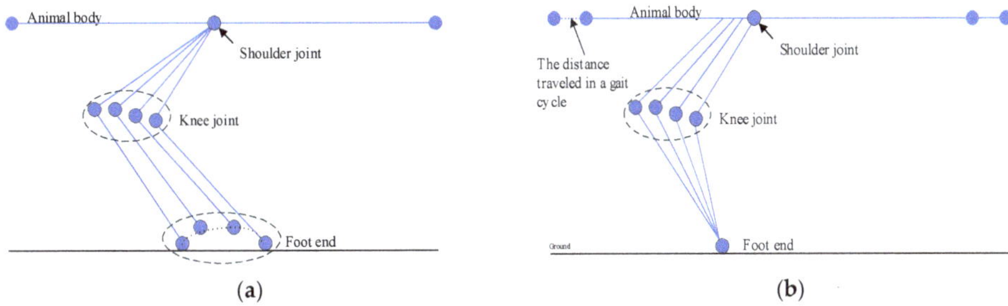

Figure 5. (**a**) The swing phase of a quadruped animal; (**b**) the support phase of a quadruped animal.

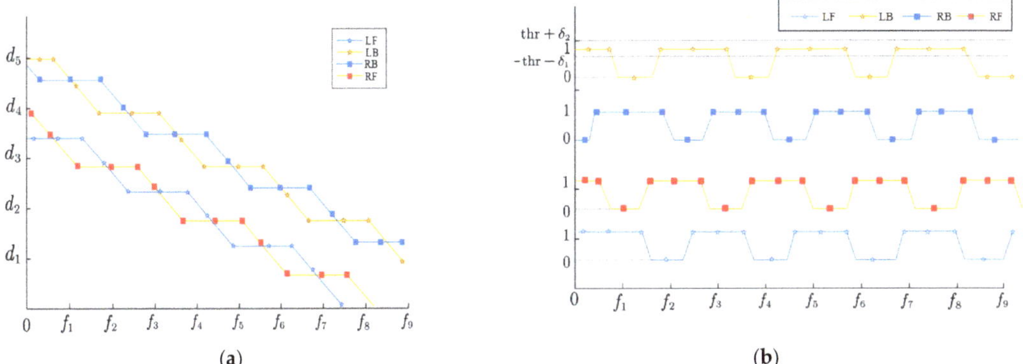

Figure 6. (**a**) The feet-end displacement curve of a quadruped animal; (**b**) the first-order differential curve of feet-end displacement.

In an ideal state, the first-order differential curve of feet-end displacement should produce a signal consisting of a square wave; however, this is not possible in practice. Therefore, the first-order differential curve depicted in Figure 6b is not a complete first-order differential processing of Figure 6a, but is a reasonable assumption.

2.3.2. Gait Sequence Recognition Model

During a gait cycle, the gait sequence of a quadruped animal is also an important feature, but it is easily affected by unfavorable circumstances such as environmental occlusion and shot angle when attempting to extract it. Based on the extracted relative position parameters of the foot end, this paper analyzes the gait sequence features of the quadruped animal over time, then binarizes it to obtain the gait phase diagram displayed in Figure 7. By the concept presented in Section 2.3.1, the first-order differential curve of the corner displacement of the quadruped animal is obtained and binarized to produce the gait phase diagram.

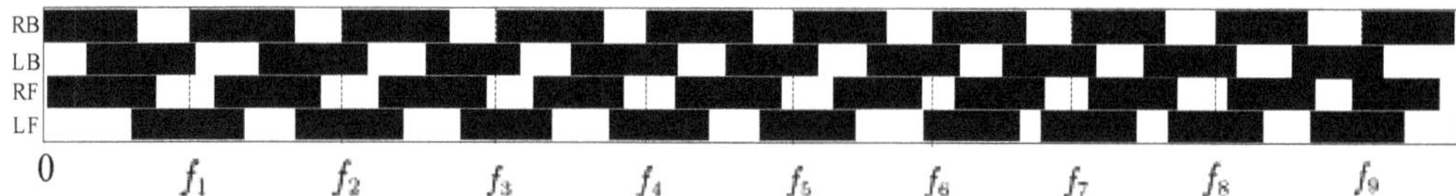

Figure 7. Gait phase diagram of each limb's gait (left front leg, right front leg, left hind leg, and right hind leg).

The actual first-order differential curve frequently has a transition process during the switching process between the support phase and the swing phase, i.e., the foot end has lifted a part but has not yet completely left the ground; therefore, a threshold value must be designed to fault-tolerantly control the process. If the binarization threshold is set as *thr*, then whether the quadruped animal is in the swing phase or the support phase can be judged by the following formula:

$$a = \begin{cases} 0, \dot{f}_i \in else \\ 1, \dot{f}_i \in (-thr - \delta_1, thr + \delta_2) \end{cases}, \tag{9}$$

where a is the label of the support phase and the swing phase, 0 represents the swing phase, and 1 represents the support phase. *thr* is the threshold of the swing phase and the support phase, δ_1, δ_2 are the threshold margins. The specific size can be set according to the actual first-order differential curve. The degree of oscillation is set on demand so that the threshold can be dynamically adjusted according to the actual situation.

Figure 7 depicts the quadruped animal gait cycle for each limb. According to the color-coded blocks in the figure, the limb is currently in the support phase. When examining and calculating the order in which each limb's support phase occurs throughout a gait cycle T, the gait sequence characteristics of each quadruped animal can be obtained appropriately.

2.3.3. Duty Cycle Parameter Extraction Model

The quadruped gait occupancy parameter describes the percentage of time that the pendulum support phase occupies throughout the gait cycle when quadruped animals move. We selected a gait phase diagram of one of the limbs to serve as a method for the extraction of occupancy parameters in a quadruped. From the Figure 8, we can calculate a gait cycle of T for each member of the quadruped: t_1 for the moment the foot initially contacts the ground, t_2 for the moment it leaves the ground, and t_3 for the moment it contacts the ground again.

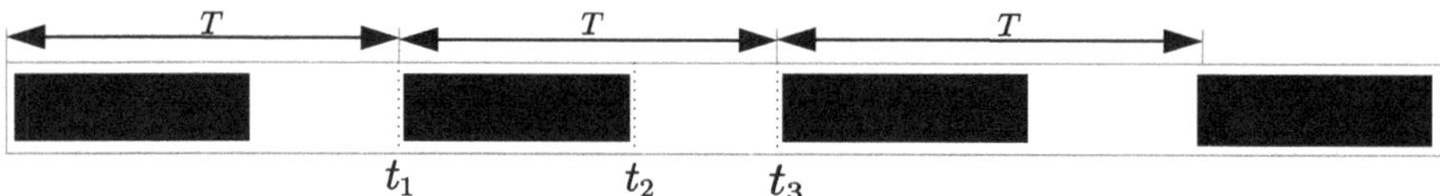

Figure 8. A gait phase diagram of one of the limbs of a quadruped animal.

Gait duty cycle parameter extraction is shown in Formula (10):

$$\alpha = \frac{t_g}{T}, \tag{10}$$

where t_g is the duration of the limbs in the support phase during a gait cycle. The specific calculation formula is $t_g = t_2 - t_1$, and T is the gait cycle.

The gait cycle of each limb varies slightly during quadruped animal movement, even at constant speeds. Hence, it is necessary to average all the occupancy parameters for each leg.

2.3.4. Gait Trajectory Extraction Model

Due to their distinct body composition and gait, quadruped animals can walk freely in various challenging ground environments, including plains, hills, and mountains. This inspires the design of legged bionic robots. Legged robot gait planning has traditionally been the subject of research in this area.

The recorded video data of the experimental animals are broken down into images frame by frame using the model method in this research, and the keypoints are then identified. The foot-end trajectory characteristics of the experimental animals can then be obtained by connecting the keypoint locations of each limb's foot end in a systematic manner in accordance with the time sequence.

2.3.5. Evaluation Index of Gait Parameter Extraction

We analyze the gait parameter method proposed in this study by comparing it with manual computation to verify the practical effect of the gait parameter extraction method. The three different types of experimental animals include buffalo, horse, and dog. It consists mainly of the following two indicators.

(1) Relative error

The gait parameters of the target animals were extracted by the method presented in this paper and compared with the actual value of the manually labeled, as follows:

$$\begin{cases} E_{rel} = \frac{|M_c - O_A|}{O_A} \times 100\% \\ M_c = \frac{1}{4} \sum M_i \\ O_A = \frac{1}{4} \sum O_i \end{cases}, \tag{11}$$

where M_c represents the average measurement of an individual animal; O_A is he actual value manually marked by individual animals; M_i represents the measured value of the ith leg; O_i represents the manual labeled value of the ith leg.

(2) Gait sequence consistency judgment index

The indication C for the gait sequence consistency judgment index is used to calculate a quadruped animal sequence, as shown in Formula (12):

$$C = \frac{\sum\limits_{i} \delta(S_{ic} = S_{ia})}{N}, \tag{12}$$

where N is the total number of steps the subject took throughout the experiment; S_{ic} represents the leg the experimental object used in step I, and S_{ia} represents the leg the experimental object actually used in step i. The four limbs of the experimental subjects can be represented in this work by the letters LF, RF, LB, and RB, which stand for the respective left front leg, right front leg, left hind leg, and right hind leg.

3. Experiments and Results

3.1. Animal Pose Estimation Experiment

3.1.1. Dataset Source and Preprocessing

In this study, we adopted the AP10K dataset [28], a public animal keypoint detection dataset. The dataset combines previously published datasets and arranges all animals into a collection of 54 species belonging to 23 families in accordance with the biological concepts of family and species. All 54 different animals are eventually labeled, yielding 10,105 labeled images and 13,028 pose annotations. The dataset animals' entire bodies were split up into 17 keypoints, and Table 1 shows animal keypoint definitions. K-fold cross-validation was used to validate the dataset to fairly assess the model's performance and fully utilize the data to prevent overfitting. Firstly, 10 slices of each species were cut, and the slices were then separated into three groups: a training set, a validation set, and a test set, in that order: 7:1:2.

The dataset was rerandomized three times using the same methodology, and the evaluation score of the model was taken as the average of the three results.

Table 1. Definitions of animal keypoints.

Keypoint	Definition	Keypoint	Definition
1	Left Eye	10	Right Elbow
2	Right Eye	11	Right Front Paw
3	Nose	12	Left Hip
4	Neck	13	Left Knee
5	Root of Tail	14	Left Back Paw
6	Left Shoulder	15	Right Hip
7	Left Elbow	16	Right Knee
8	Left Front Paw	17	Right Back Paw
9	Right Shoulder		

The keypoint annotations format is (x, y, z), in which x and y are the pixel coordinates of the keypoint relative to the upper left corner of the image, and v represents the visibility of the keypoint. If $v = 0$ means that the keypoint position does not appear in the image, the keypoint is not marked, and the corresponding (x, y) is also set as $(0, 0)$; $v = 1$ indicates that the keypoints appear in the image, but there is occlusion; $v = 2$ indicates that the keypoints are visible in the image and are not occluded. The connection order of the quadruped animal's skeleton is as follows: [1 2], [1 3], [2 3], [3 4], [4 5], [4 6], [6 7], [7 8], [4 9], [9 10], [10 11], [5 12], [12 13], [13 14], [5 15], [15 16], [16 17]. The keypoints are connected in order, and the labeled information is shown in Figure 9.

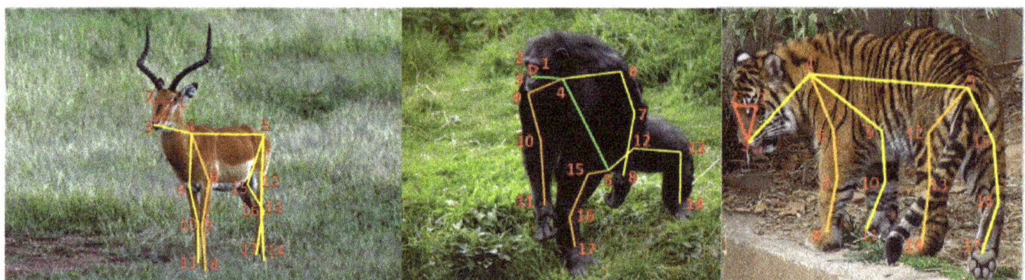

Figure 9. The keypoints of the quadruped animal.

3.1.2. Experimental Design

The experimental setup employed for this paper's experiment is equipped with an Intel i5 10600kf core CPU, 32 GB of memory, and an NVIDIA RTX3060 graphics card. The deep learning framework chooses Pytorch1.8.2, Adam [29] optimizer, 16 for the batch size, 210 epochs, 0.001 for the learning rate, and 0.0001 for the learning rate after scaling the original image resolution to 288×384 resolution. This paper employs the pretraining model pose_hrnet_w48_288 $\times$ 384 to transfer the model, which speeds up the training process and lowers the time complexity of training the network from scratch.

This paper's three sets of comparative ablation study were set up to contrast with the original HRNet network. The first group consisted of the original HRNet network by repeating the training network three times. The results were then averaged, and the acquired dataset was analyzed and confirmed. The second group added an RFB module, and the subsequent processing was the same as for the first group of studies. The DyC-RFB module and the same evaluation approach were introduced in the third group of studies.

Some blocked image data were introduced at random while the network model's keypoint predictions were being made in the process to assess the network's capacity to handle complicated networks. After discussing the similarities between human pose estimation and those of four-legged animals, a number of images from the COCO2017 dataset were chosen for experimental validation.

3.1.3. Experimental Results on the AP10K Dataset

Using the AP10K dataset, this experiment initially examined the performance of RFB-HRNet before testing the performance of alternative model approaches.

To further verify the prediction performance of the network model of this paper on 17 keypoints, we randomly selected 1280 images from the test dataset (the test data contained all 54 animal species in the dataset used in this paper) and randomly flipped, cut, added noise, random rotation ($[-45°, 45°]$), and random scale ($[0.65, 1.35]$). We obtained the accuracy of 17 types of quadruped animal keypoints, as shown in Table 2.

Table 2. Average precision of keypoints.

Keypoint	Average Precision	Keypoint	Average Precision
Left Eye	0.810	Right Elbow	0.701
Right Eye	0.799	Right Front Paw	0.693
Nose	0.777	Left Hip	0.691
Neck	0.652	Left Knee	0.718
Root of Tail	0.694	Left Back Paw	0.685
Left Shoulder	0.732	Right Hip	0.766
Left Elbow	0.728	Right Knee	0.671
Left Front Paw	0.704	Right Back Paw	0.722
Right Shoulder	0.728		

In Table 2, the average precision is better for feature keypoint with apparent characteristics such as the nose, eyes, and shoulder. The accuracy of recognition is only somewhat high for other feature points, such as wrist joints and tails, because these keypoints are influenced by the environment, the animal's pose, and the characteristics of its hair, such as color and texture.

We report the results of our method and other methods in Table 3. The outcomes demonstrate that our method produced the best outcomes. The AP^{50}, AP^{75}, AP^M, AP^L, AR, and mAP increased by 2.2%, 1.6%, 4.4%, 1.7%, 1.9%, and 2.1%, respectively, compared to the original HRNet-w48 network. As a result, our method was much improved when compared to other models.

Table 3. Comparison on the AP10K validation set.

Methods	Input Size	GPLOPs	#Params	mAP	AP^{50}	AP^{75}	AP^M	AP^L	AR
ResNet50 [19]	288 × 384	5.396	23.508M	0.681	0.926	0.738	0.552	0.687	0.718
ResNet101 [19]	288 × 384	10.272	42.500M	0.683	0.921	0.751	0.545	0.690	0.719
HRNet-W48 [21]	288 × 384	21.059	63.595M	0.729	0.936	0.802	0.577	0.736	0.762
RFB-HRNet	288 × 384	22.612	63.972M	0.750	0.958	0.818	0.621	0.753	0.781

Our network RFB-HRNet, trained from scratch with input size 288 × 384, achieved a 0.75 mAP score more than other models with the same input size. (i) Our method improved mAP by 2.1 points compared to HRNet-W48, while the numbers of parameters are similar to HRNet-w48, and ours is slightly larger. (ii) Compared to ResNet101 and ResNet50, our method improved mAP by 6.7 points, compared to ResNet50, which improved by 6.9 points, respectively.

3.1.4. Ablation Study

Analysis and comparison were compared on the parameters and mAP score of the three different HRNet networks: the original HRNet-w48 network, the HRNet network with the RFB module added, and the HRNet network with DyC-RFB module. According to the findings, which are compared in Table 4, the RFB module's network parameters increased by 0.32 M in comparison to the original HRNet network, and the mAP score increased by almost 1.3 points. The number of parameters increased by 0.057 M after adding the improved DyC-RFB module, and the performance of the mAP score improved by 0.8 points as a result.

Table 4. Comparison of experimental results under different methods.

HRNet	DyConv	RFB	#Params	mAP
√	×	×	63.595 M	0.729
√	×	√	63.915 M	0.742
√	√	√	63.972 M	0.750

We selected 10 animal species randomly to compare the recognition impact of our method with the original HRNet network and to compare our method's prediction effect with the original HRNet-w48 network. Among the animals selected, some like to dwell in groups, while others prefer to move alone. The sampled images contain single-target images, multitarget images, small-area occlusions, and large-area occlusions. The detection results of the original HRNet-w48 network are shown in the even-numbered rows of Figure 10, and the detection results of the proposed front object detection network + RFB-HRNet network in this paper are shown in the odd-numbered rows.

Figure 10. Some qualitative results of our method (the odd rows) vs. original HRNet-w48 (the even rows) on AP10K.

As observed in Figure 10, the two networks can produce identical detection results for a single animal when the scale of the animal is large. However, our method is more accurate in some feet-end positions while the original HRNet network exhibits some bias. The detection performance of our method is noticeably superior to that of the original HRNet-w48 network for small target animals. For instance, our method can identify the facial features of the last group of cats in the image which are running. In contrast, the original HRNet-w48 network has a significant offset when identifying their faces and has a lower identification effect on the limbs than our method. Our method can completely detect numerous targets and effectively separate each individual animal, in contrast to the original HRNet-w48 network, which produces the scenario of missed detection and mutual adhesion and is unable to separate multiple targets effectively. This is controlled by the input–output structure of the pose estimation network; thus, if numerous animals are close together, it is simple to connect the coordinates of keypoints amongst different individuals, as demonstrated by the effect of the pose estimation network on the sheep in row 4. In contrast, due to the inclusion of a prior object detection network in the method proposed in this paper, as many animals as feasible in a picture including numerous animals can be detected, so that the task is broken into recognizing many single targets, making it significantly more effective than the original HRNet network.

In a sense, humans are also a special "four-legged animal". Therefore, to verify the generalization ability of the proposed model, several human images were randomly selected from the COCO2017 dataset and their keypoints were predicted. The results are shown in Figure 11.

Figure 11. Some qualitative results of our method on the COCO2017 human keypoint task.

According to the above figure, quadruped animals have a body structure similar to humans. This is especially true in terms of facial feature recognition, which has a high level of accuracy. However, effective recognition is not possible when dealing with complex human body poses and varied scenes.

3.2. Gait Parameter Extraction Experiment
3.2.1. Video Data Collection

The quadruped animal gait parameter extraction experiment, which made up the second portion of the study, sought to confirm the reliability and validity of the suggested quadruped animal gait parameter extraction model. The gait parameter extraction experiment requires continuous motion video footage of different quadruped animals with multi-gait cycles. In this study, the experimental dataset for state parameter extraction was a motion video of quadruped animals that was shot in real time. Figure 12 depicts the quadruped animal video capture system, which was shot using a tripod-fixed shooting gear. The camera's position is roughly 1 m above the ground, and its coordinates are in the O–XYZ space. The included angle between the motion direction of the quadruped animal and the X–Y plane is defined as the motion direction angle β. The motion direction is positive when it is far from the X–Y plane and negative when it is close to the X–Y plane. The captured video had 720×1280 pixels and a frame rate of 30 FPS, and the range of β was $-33°\sim33°$. Quadruped animals include buffalo, horse, and dog, representing typical quadruped animal of different sizes. Among them, buffalo and horse have a walking gait, and dog has a trot gait. The buffalo moved from right to left, while the horse and the dog moved from left to right. During the shooting, the walking process of the quadruped animal was recorded as completely as possible to ensure that the video contained multiple gait cycles.

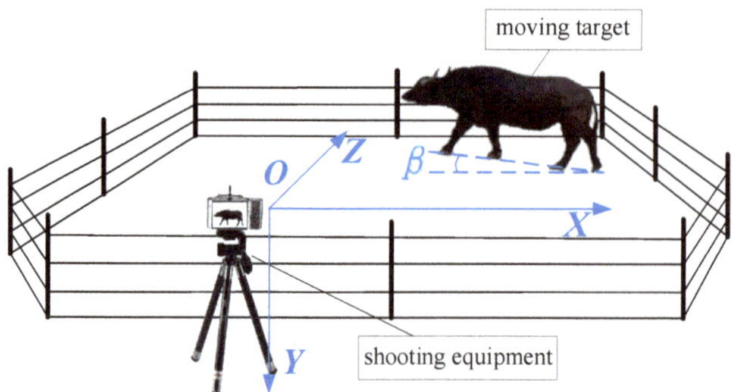

Figure 12. Quadruped animal video capture system in our study.

3.2.2. Gait Frequency Extraction Experiment

Figures 13–15 depict the foot distance curves of a buffalo, a horse, and a dog in the experimental animal's x direction (where x indicates the animal's heading direction and y represents the direction perpendicular to the ground).

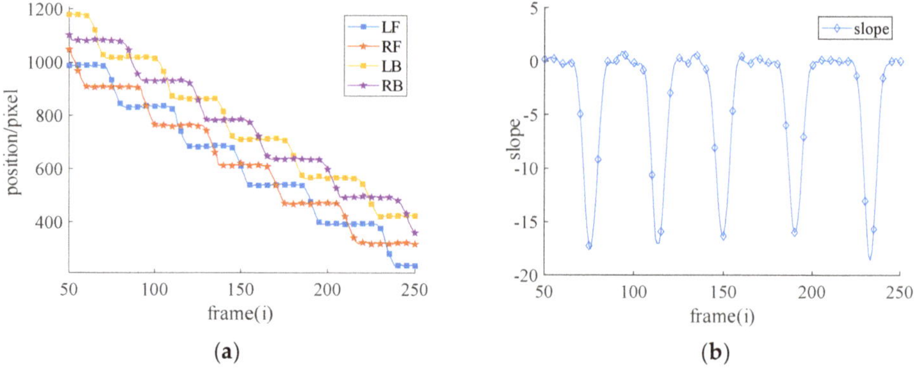

(a)   (b)

Figure 13. (**a**) Buffalo's feet displacement curve; (**b**) first−order differential curve of buffalo's feet displacement.

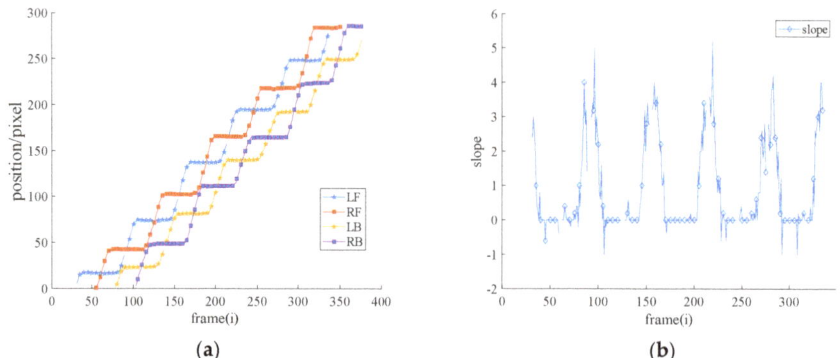

(a)   (b)

Figure 14. (**a**) Horse's feet displacement curve; (**b**) first−order differential curve of horse's feet displacement.

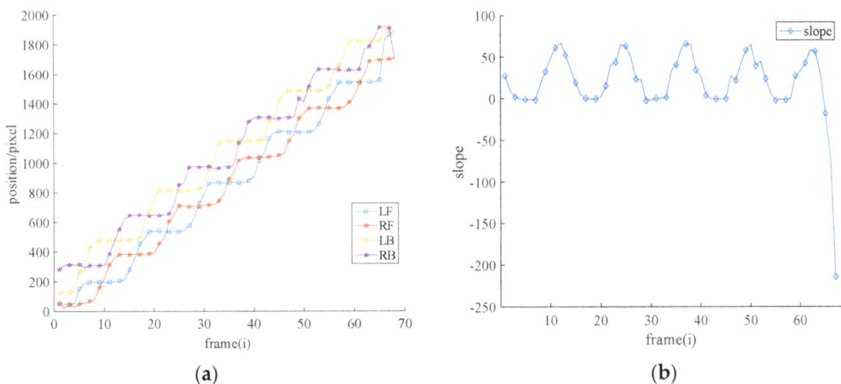

Figure 15. (**a**) Dog's feet displacement curve; (**b**) first−order differential curve of dog's feet displacement.

The quadruped animal's left front leg, right front leg, left hind leg, and right hind leg, as well as the first-order differential slope, are represented in the above image by the letters LF, RF, LB, and RB, respectively. Since the camera is fixed during the recording of the video material, it is clear from the figure that when quadrupeds move forward, their limbs alternately swing. Therefore, the experimental animal's foot displacement direction should be monotonic, moving from right to left when its features are monotonically reducing, and from left to right when they are monotonically growing.

Because the quadruped animal's forward movement process involves the four legs swinging in essentially the same movement, the difference between the starting frame and the end frame must take this into account. However, the order of the legs' movements does alter slightly. As a result, one leg's pixel change curve can be chosen as the reference, and the first derivative can then be determined. The gap frame period of a gait cycle is the difference between the minimum values of the first-order differential curve. Figure 13b, Figure 14b, and Figure 15b display the first-order differential curves. The gait frequency of the buffalo, horse, and dog may be estimated using Formula (8), as shown in Table 5.

Table 5. Gait frequency of buffalo, horse, and dog.

Species	Mean Interval Frame (Frames)	This Article (Hz)	Manual Method	Relative Error/%
Buffalo	38.25	0.792	0.812	2.46
Horse	59.8	0.507	0.498	1.81
Dog	12.75	2.377	2.331	1.97

Table 5 above shows that when compared to the manual method, our method has a maximum frequency relative error of 2.46%. In comparison to horses and dogs, buffalo showed frequency errors that were noticeably larger. This resulted from the buffalo we photographed having a darker background, which had a detrimental impact on the experimental results, but overall, the error was within the acceptable range. The gait frequency of the buffalo and horse is substantially lower than that of the dog. In general, there is a correlation between body size and the gait of quadruped animals. The frequency of gait decreases as body size increases.

3.2.3. Experiment of Gait Sequence Recognition

The leg stride order of quadruped animals is a crucial component of a gait cycle, but it can be easily influenced by a number of negative conditions during the extraction process, such as shooting angle and ambient occlusion. This paper examined the quadruped animals' gait sequence features over time and binarized them to obtain the gait phase diagram, as shown in Figure 16. This analysis was based on the relative position parameters of the foot end that were extracted above.

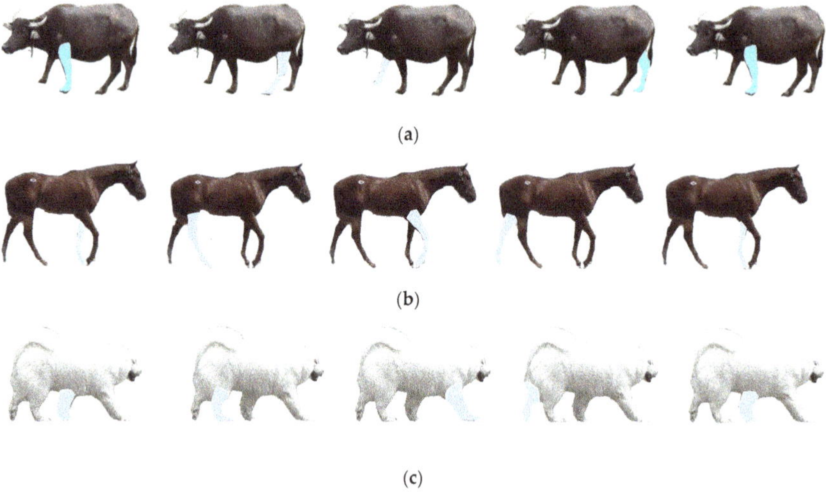

Figure 16. (**a**) The gait phase diagrams of a buffalo; (**b**) the gait phase diagrams of a horse; (**c**) the gait phase diagrams of a dog.

Figure 17 depicts the gait patterns of a buffalo, horse, and dog. It is clear that the gait patterns of the two animals are identical. The left front leg comes first, followed by the left front leg- > right hind leg- > right front leg- > left hind leg- > left front leg throughout a full gait cycle. There will be more than two legs in the support phase because there is overlap in each leg's movement. The initial motion state may differ from the shot scene, but it will eventually be in a stable state. With the exception of the swing phase, which accounts for a variable duty cycle throughout the gait cycle, the dog's gait sequence is essentially in the same order as the first two. In most cases, only two legs are in the support phase or swing phase, and both legs belong to the same side.

Figure 17. (**a**) Buffalo gait sequence; (**b**) horse gait sequence; (**c**) dog gait sequence.

The gait sequence of buffalo, horse, and dog is shown in Table 6.

Table 6. Gait sequence of buffalo, horse, and dog.

Species	S1	S2	S3	S4	S5	Gait Sequence Consistency Judgment Index (Formula (12)
Buffalo	LF	RB	RF	LB	LF	100%
Horse	LF	RB	RF	LB	LF	100%
Dog	RF	LB	LF	RB	RF	100%

Table 6 depicts that the gait sequence of buffalo, horse, and dog can be accurately extracted by the method proposed in this paper.

3.2.4. Gait Duty Cycle Extraction Experiment

Using the gait phase diagrams of buffalo, horse, and dog obtained in Section 3.2.3, we can plot the corresponding gait cycle curves, respectively, by using Formula (9) in Section 2.3.2. As shown in Figures 18–20, where 0 represents the support phase and 1 represents the swing phase, the gait duty cycle data of each animal can be calculated by calculating the time occupied by the swing phase in a gait cycle.

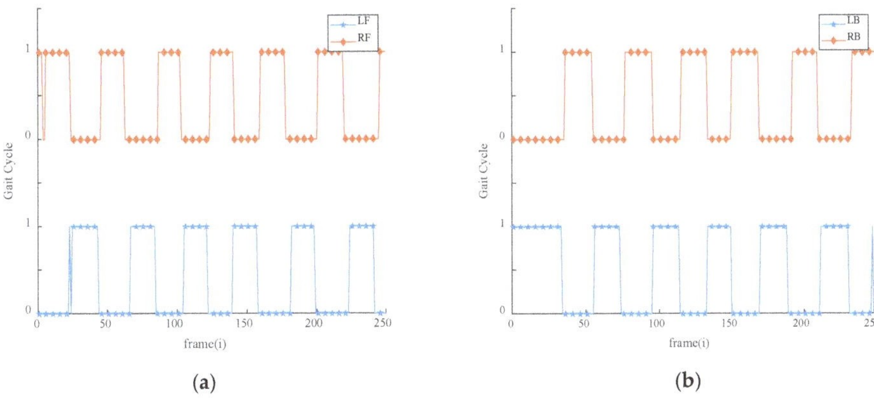

(a)   (b)

Figure 18. (**a**) Buffalo's gait cycle curves of LF, RF leg; (**b**) buffalo's gait cycle curves of LB, RB leg.

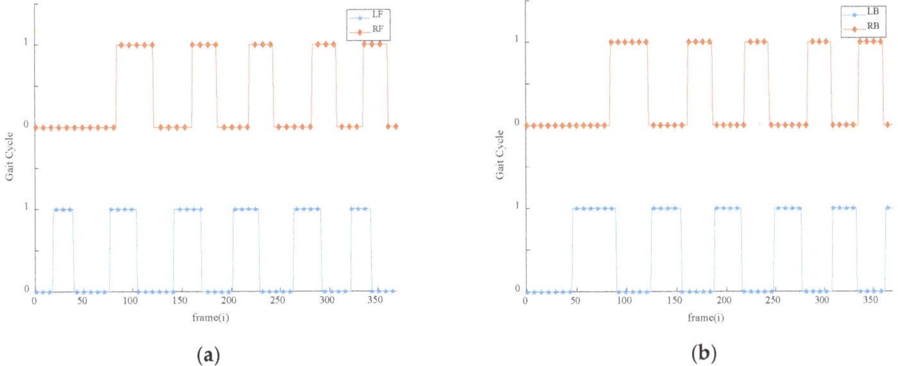

(a)   (b)

Figure 19. (**a**) Horse's gait cycle of LF, RF leg; (**b**) horse's gait cycle of LB, RB leg.

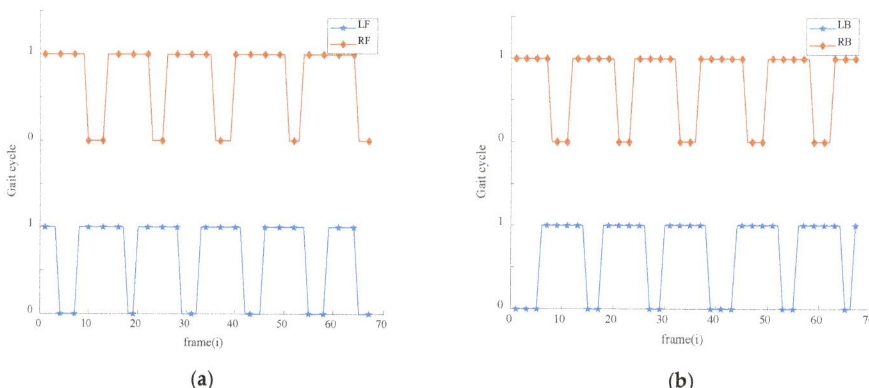

Figure 20. (**a**) Dog's gait cycle of LF, RF leg; (**b**) dog's gait cycle of LB, RB leg.

We can calculate the gait duty cycle parameters for the buffalo, horse, and dog by using Formula (10) and Figures 18–20, and the results are shown in Table 7.

Table 7. Gait duty cycle of buffalo, horse, and dog.

Limb	Buffalo		Horse		Dog	
	Our Method	Manual Calculation	Our Method	Manual Calculation	Our Method	Manual Calculation
LF	0.543	0.512	0.615	0.608	0.372	0.356
RF	0.580	0.563	0.630	0.622	0.310	0.330
LB	0.546	0.555	0.597	0.556	0.360	0.321
RB	0.540	0.525	0.631	0.642	0.308	0.288
Average value	0.552	0.539	0.618	0.607	0.338	0.324
Relative error/%	2.41		1.81		4.3	

According to the above table, the greatest inaccuracy of the suggested technique is 4.3% when compared to the manual calculation method, demonstrating that it is capable of reliably calculating the quadrupeds' duty cycle parameters. It is clear that dogs have a relatively modest gait duty cycle in comparison to huge quadrupeds such as the buffalo and horse.

3.2.5. Gait Trajectory Extraction Experiment

In a complete gait cycle, the foot trajectory of the fore and hind limbs of the image sequence of buffalo, horse, and dog is calculated, and the experimental findings are depicted in Figures 21–23. Comparing the foot trajectory of the forelimbs and hindlimbs of the buffalo and the horse, respectively, it can be seen that the trajectory of quadrupeds is not a complete polynomial curve during the movement process, but a downward movement occurs after reaching the highest point of its trajectory, so its trajectory has a "concave" segment, and there is a certain difference in the movement of the forelimbs and hindlimbs, that is, for the forelimbs, the phase difference between the support phase and the swing phase is 180 degrees in most cycles. In other words, when one limb is in the support phase, the other is in the swing phase and moves repeatedly in opposition. For the limbs on the same side, there is also a 180-degree difference between them. Therefore, it can be deduced that, for a complete gait cycle, the movement of the two limbs on the neighboring diagonals can be considered substantially synchronized, but not entirely synchronous. At some point, three legs are in the support phase, while the last leg is in the swing phase, allowing the animal to retain movement stability at moderate speeds. Both the buffalo and the horse move in essentially the same ways, promoting the body's forward motion during the support phase and having a fairly identical swinging order. In conclusion, multiple limb joints in the movement process of quadrupeds coordinate and create smooth movement by varying their movement phase.

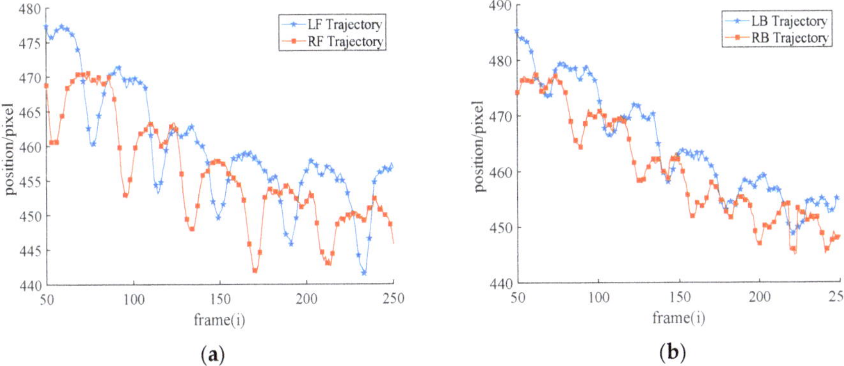

Figure 21. (**a**) Foot trajectory of buffalo's forelimbs; (**b**) foot trajectory of buffalo's hindlimbs.

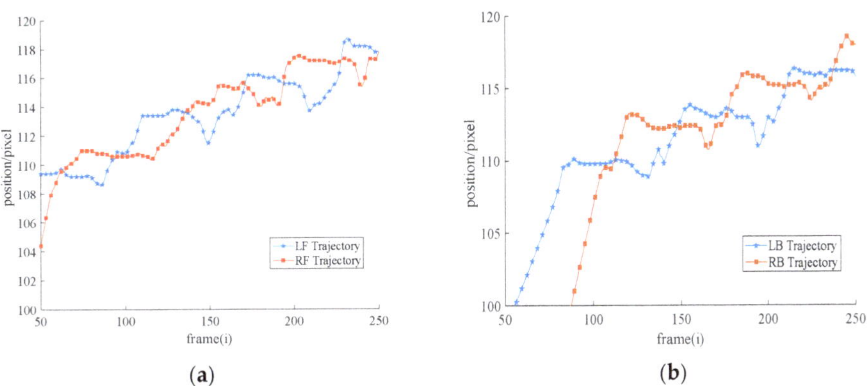

Figure 22. (**a**) Foot trajectory of horse's forelimbs; (**b**) foot trajectory of horse's hindlimbs.

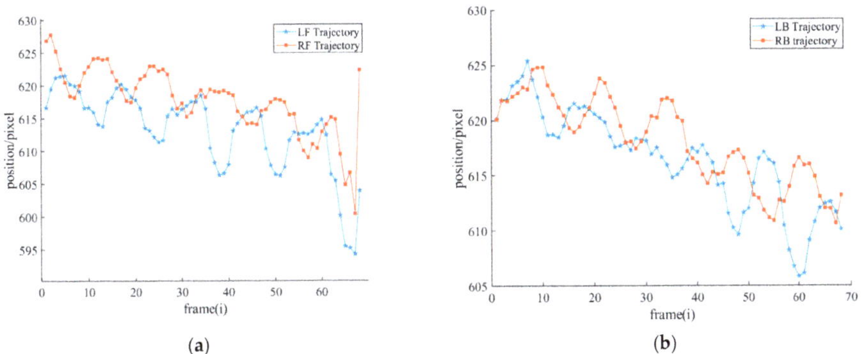

Figure 23. (**a**) Foot trajectory of dog's forelimbs; (**b**) foot trajectory of dog's hindlimbs.

4. Conclusions and Discussion

A method for quadruped animal pose estimation and gait parameter extraction was presented in this paper. Target screening, animal pose estimation model, and animal gait parameter extraction are its three main components, which together form a vision-based computational framework that can fully and successfully address the issues of quadruped animal pose estimation and gait parameter extraction. The fundamental idea can be broken down into the following steps.

(1) We converted the original video data to images and transmitted them to the network for object detection to obtain the location anchor box of the animal in the image.

(2) The position anchor frame obtained in the first stage was used to crop the image and feed it into the RFB-HRNet network to obtain quadruped animal keypoints in original resolution space.

(3) Various quadruped animal gait characteristics were obtained through computational research.

The test results using the public dataset AP10K showed that, in comparison to the original HRNet-w48 network and other methods, our method yielded the best results for the keypoint extraction of quadruped animal poses. The mAP was 2.10% more than the original HRNet-w48, while the AR increased by 1.9%. As for gait parameter extraction, three typical quadruped animals representing different body sizes, buffalo, horse, and dog, were tested experimentally, and the results demonstrated that the gait parameters, including gait frequency, gait sequence, and duty cycle, and foot trajectory could be automatically extracted, and the real-time and accurate gait trajectory could be obtained. The greatest error of gait frequency was 2.46%, the maximum error of duty cycle was 4.33%, and the detection of the gait sequence was accurate.

Based on the research work of this paper, the following conclusions were drawn.

(1) The feature extraction capability of the network model as a whole could be significantly improved by using a special receptive field module DyC-RFB to improve the feature extraction capability of the first branch of stage 1 of the HRNet network and by using the transfer effect of each branch in the later stage. Even if there was only a small improvement, it was still possible to improve the performance of quadrupeds in terms of keypoint extraction without significantly increasing the network parameters or the computation.

(2) A two-stage cascade network was created by adding an object detection network to the front end of the animal pose estimation model for target screening. This network could significantly improve the animal pose estimation effect of some small targets and multitargets, as well as the stability and reliability of pose estimation.

However, some deficiencies still need to be improved in the subsequent work. The three-dimensional gait data of quadrupeds cannot currently be calculated since the detection of animal joints still relies on two-dimensional plane estimation and the absence of depth data. Therefore, its potential will be investigated and improved upon in the future.

Author Contributions: Data curation, Z.G. and D.L.; investigation, Z.G., D.L., and Y.Z.; methodology, Z.G. and T.W.; software, Z.G.; visualization, Z.G. and Y.Z.; validation, Z.G.; writing—original draft preparation, Z.G. and T.W.; writing—review, Y.Z. and Z.G. All authors have read and agreed to the published version of the manuscript.

Funding: This research was funded by the National Natural Science Foundation of China (No. 51365019).

Institutional Review Board Statement: Ethical review and approval were waived for this study. For the following reasons. The research performed a non-invasive, quadruped animals gait extraction study based on computer vision. All of the data we use is from publicly available dataset and video taken in zoos, without contact with any animals. The study will not cause any harm to animals. According to the type of procedure used, no formal ethical approval was required.

Informed Consent Statement: Not applicable.

Data Availability Statement: The dataset (AP10K) used in this paper is publicly available and can be downloaded from AlexTheBad/AP-10K: NeurIPS 2021 Datasets and Benchmarks Track (github.com, accessed on 11 November 2022).

Acknowledgments: The authors greatly acknowledge the financial support by the National Natural Science Foundation of China. We thank Y.Z. for inspiring this paper in the methodology of the paper, and T.W. and D.L. for their contributions in software and data processing. Finally, thanks to the editors for their hard work.

Conflicts of Interest: The authors declare no conflict of interest.

References

1. Toshev, A.; Szegedy, C. Deeppose: Human pose estimation via deep neural networks. In Proceedings of the IEEE Conference on Computer Vision and Pattern Recognition, Columbus, OH, USA, 23–28 June 2014; pp. 1653–1660.
2. Fan, X.; Zheng, K.; Lin, Y.; Song, W. Combining local appearance and holistic view: Dual-source deep neural networks for human pose estimation. In Proceedings of the IEEE Conference on Computer Vision and Pattern Recognition, Boston, MA, USA, 7–12 June 2015; pp. 1347–1355.
3. Pfister, T.; Charles, J.; Zisserman, A. Flowing convnets for human pose estimation in videos. In Proceedings of the IEEE International Conference on Computer Vision, Santiago, Chile, 7–13 December 2015; pp. 1913–1921.
4. Li, C.; Lee, G.H. From synthetic to real: Unsupervised domain adaptation for animal pose estimation. In Proceedings of the IEEE/CVF Conference on Computer Vision and Pattern Recognition, Nashville, TN, USA, 20–25 June 2021; pp. 1482–1491.
5. Newell, A.; Yang, K.; Deng, J. Stacked hourglass networks for human pose estimation. In Proceedings of the European Conference on Computer Vision, Amsterdam, The Netherlands, 11–14 October 2016; Springer: Berlin, Germany, 2016; pp. 483–499.
6. Mu, J.; Qiu, W.; Hager, G.D.; Yuille, A.L. Learning from synthetic animals. In Proceedings of the IEEE/CVF Conference on Computer Vision and Pattern Recognition, Seattle, WA, USA, 13–19 June 2020; pp. 12386–12395.
7. Yuan, Y.; Fu, R.; Huang, L.; Lin, W.; Zhang, C.; Chen, X.; Wang, J. Hrformer: High-resolution vision transformer for dense predict. *Adv. Neural Inf. Process. Syst.* **2021**, *34*, 7281–7293.
8. Cao, J.; Tang, H.; Fang, H.S.; Shen, X.; Lu, C.; Tai, Y.W. Cross-domain adaptation for animal pose estimation. In Proceedings of the IEEE/CVF International Conference on Computer Vision, Seoul, Korea, 27 October–2 November 2019; pp. 9498–9507.
9. Zhou, X.; Wang, D.; Krähenbühl, P. Objects as points. *arXiv* **2019**, arXiv:1904.07850.
10. Geng, Z.; Sun, K.; Xiao, B.; Zhang, Z.; Wang, J. Bottom-up human pose estimation via disentangled keypoint regression. In Proceedings of the IEEE/CVF Conference on Computer Vision and Pattern Recognition, Nashville, TN, USA, 20–25 June 2021; pp. 14676–14686.
11. Zhang, F.; Zhu, X.; Dai, H.; Ye, M.; Zhu, C. Distribution-aware coordinate representation for human pose estimation. In Proceedings of the IEEE/CVF Conference on Computer Vision and Pattern Recognition, Seattle, WA, USA, 13–19 June 2020; pp. 7093–7102.
12. Izonin, I.; Tkachenko, R.; Fedushko, S.; Koziy, D.; Zub, K.; Vovk, O. RBF-Based Input Doubling Method for Small Medical Data Processing. In Proceedings of the International Conference on Artificial Intelligence and Logistics Engineering, Kyiv, Ukraine, 22–24 January 2021; Springer: Berlin, Germany, 2021; Volume 82, pp. 23–31.
13. Daou, H.E.; Libourel, P.A.; Renous, S.; Bels, V.; Guinot, J.C. Methods and experimental protocols to design a simulated bio-mimetic quadruped robot. *Int. J. Adv. Robot. Syst.* **2013**, *10*, 256. [CrossRef]
14. Yang, A.; Huang, H.; Zhu, X.; Yang, X.; Chen, P.; Li, S.; Xue, Y. Automatic recognition of sow nursing behaviour using deep learning-based segmentation and spatial and temporal features. *Biosyst. Eng.* **2018**, *175*, 133–145. [CrossRef]
15. Dissanayake, G. Infrastructure robotics: Opportunities and challenges. Assistive robotics. In Proceedings of the 18th International Conference on CLAWAR 2015, Hangzhou, China, 6–9 September 2015; p. 3.
16. Peng, X.B.; Coumans, E.; Zhang, T.; Lee, T.W.; Tan, J.; Levine, S. Learning agile robotic locomotion skills by imitating animals. *arXiv* **2020**, arXiv:2004.00784.
17. Kim, C.H.; Shin, H.C.; Lee, H.H. Trotting gait analysis of a lizard using motion capture. In Proceedings of the 2013 13th International Conference on Control, Automation and Systems (ICCAS 2013), Gwangju, Korea, 20–23 October 2013; pp. 1247–1251.
18. Chapinal, N.; de Passille, A.M.; Pastell, M.; Hänninen, L.; Munksgaard, L.; Rushen, J. Measurement of acceleration while walking as an automated method for gait assessment in dairy cattle. *J. Dairy Sci.* **2011**, *94*, 2895–2901. [CrossRef] [PubMed]
19. He, K.; Zhang, X.; Ren, S.; Sun, J. Deep residual learning for image recognition. In Proceedings of the IEEE Conference on Computer Vision and Pattern Recognition, Las Vegas, NV, USA, 27–30 June 2016; pp. 770–778.
20. Simonyan, K.; Zisserman, A. Very deep convolutional networks for large-scale image recognition. *arXiv* **2014**, arXiv:1409.1556.
21. Howard, A.G.; Zhu, M.; Chen, B.; Kalenichenko, D.; Wang, W.; Weyand, T.; Andreetto, M.; Adam, H. Mobilenets: Efficient convolutional neural networks for mobile vision applications. *arXiv* **2017**, arXiv:1704.04861.
22. Sun, K.; Xiao, B.; Liu, D.; Wang, J. Deep high-resolution representation learning for human pose estimation. In Proceedings of the IEEE/CVF Conference on Computer Vision and Pattern Recognition, Long Beach, CA, USA, 15–20 June 2019; pp. 5693–5703.

23. Liu, W.; Anguelov, D.; Erhan, D.; Szegedy, C.; Reed, S.; Fu, C.Y.; Berg, A.C. Ssd: Single shot multibox detector. In Proceedings of the European Conference on Computer Vision, Amsterdam, The Netherlands, 11–14 October 2016; Springer: Berlin, Germany, 2016; pp. 21–37.
24. Szegedy, C.; Vanhoucke, V.; Ioffe, S.; Shlens, J.; Wojna, Z. Rethinking the inception architecture for computer vision. In Proceedings of the IEEE Conference on Computer Vision and Pattern Recognition, Las Vegas, NV, USA, 27–30 June 2016; pp. 2818–2826.
25. Yu, F.; Koltun, V. Multi-scale context aggregation by dilated convolutions. *arXiv* **2015**, arXiv:1511.07122.
26. Chen, Y.; Dai, X.; Liu, M.; Chen, D.; Yuan, L.; Liu, Z. Dynamic convolution: Attention over convolution kernels. In Proceedings of the IEEE/CVF Conference on Computer Vision and Pattern Recognition, Seattle, WA, USA, 13–19 June 2020; pp. 11030–11039.
27. Golubitsky, M.; Stewart, I.; Buono, P.L.; Collins, J.J. Symmetry in locomotor central pattern generators and animal gaits. *Nature* **1999**, *401*, 693–695. [CrossRef] [PubMed]
28. Yu, H.; Xu, Y.; Zhang, J.; Zhao, W.; Guan, Z.; Tao, D. AP-10K: A Benchmark for Animal Pose Estimation in the Wild. *arXiv* **2021**, arXiv:2108.12617.
29. Kingma, D.P.; Ba, J. Adam: A method for stochastic optimization. *arXiv* **2014**, arXiv:1412.6980.

Article

Cyclic Federated Learning Method Based on Distribution Information Sharing and Knowledge Distillation for Medical Data

Liang Yu [1,2] and Jianjun Huang [1,2,*]

1   College of Electronic and Information Engineering, Shenzhen University, Shenzhen 518060, China; liangyu20210816@163.com
2   Guangdong Key Laboratory of Intelligent Information Processing, Shenzhen 518061, China
*   Correspondence: huangjin@szu.edu.cn

Citation: Yu, L.; Huang, J. Cyclic Federated Learning Method Based on Distribution Information Sharing and Knowledge Distillation for Medical Data. *Electronics* **2022**, *11*, 4039. https://doi.org/10.3390/electronics11234039

Academic Editor: Alberto Fernandez Hilario

Received: 11 November 2022
Accepted: 2 December 2022
Published: 5 December 2022

Publisher's Note: MDPI stays neutral with regard to jurisdictional claims in published maps and institutional affiliations.

Abstract: Federated learning has been attracting increasing amounts of attention for its potential applications in disease diagnosis within the medical field due to privacy preservation and its ability to solve data silo problems. However, the inconsistent distributions of client-side data significantly degrade the performance of traditional federated learning. To eliminate the adverse effects of non-IID problems on federated learning performance on multiple medical institution datasets, this paper proposes a cyclic federated learning method based on distribution information sharing and knowledge distillation for medical data (CFL_DS_KD). The method is divided into two main phases. The first stage is an offline preparation process in which all clients train a generator model on local datasets and pass the generator to neighbouring clients to generate virtual shared data. The second stage is an online process that can also be mainly divided into two steps. The first step is a knowledge distillation learning process in which all clients first initialise the task model on the local datasets and share it with neighbouring clients. The clients then use the shared task model to guide the updating of their local task models on the virtual shared data. The second step simply re-updates the task model on the local datasets again and shares it with neighbouring clients. Our experiments on non-IID datasets demonstrated the superior performance of our proposed method compared to existing federated learning algorithms.

Keywords: cyclic federated learning; non-IID; distribution information sharing; knowledge distillation

1. Introduction

Deep learning is widely used in clinical scenarios, such as disease screening, health management, diagnosis and treatment. Obtaining models that can perform various medical tasks well often requires a large amount of training data; however, due to privacy limitations in the medical field, it is not possible to pool data from various medical sites to form larger datasets, which isolates each medical site and means that models can only be trained with a small amount of local data, resulting in the poor performance of trained models. Federated learning [1] has been proposed as an effective solution to this problem. Firstly, as a kind of distributed machine learning, federated learning can jointly train global models for multiple medical institutions by combining data and annotations from each institution to expand the sample data volume and the number of annotations [2], thereby making it possible to solve unbalanced data distributions. Secondly, federated learning does not require data exchanges among healthcare institutions, which satisfies requirements such as patient privacy protection, data security and government regulations. Additionally, the results of federated learning can be shared among medical institutions, which can alleviate the problem of the uneven distribution of medical resources to a certain extent.

The training process of federated learning involves medical institutions training model parameters based on local datasets, then sharing model parameters among medical institutions and finally fusing all model parameters in an aggregated manner to form better-performing models. When the data distributions of medical institutions are inconsistent, i.e., the assumption of independent and identical distribution (IID) is not satisfied among medical institutions, the complexity of the problem modelling, theoretical analysis and empirical evaluation of solutions increases, resulting in the degradation of model performance [3]. A feasible idea to solve this problem is to share data distributions based on the model sharing in federated learning, i.e., share the data distribution information of different medical institutions with other medical institutions. This is similar to the sharing and exchange of treatment experiences among doctors at multiple medical institutions, which can improve treatment levels by learning from each other. In addition, there are certain requirements for data security while keeping shared data.

The initial federated learning framework was the centralised federated learning framework, which faced the problem that it is difficult to find trusted third parties to perform parameter aggregation [4]. To solve this problem, decentralised federated learning frameworks have been developed, such as peer-to-peer network structures; however, they have certain requirements for the computing power of each client. Due to the frequent information exchanges between multiple clients, the communication costs are also relatively high. The decentralised federated learning architectures remove the central server to perform task model aggregation locally and only exchange information between adjacent clients on the communication graph, which reduces the probability of network congestion and communication overheads while improving data privacy protection capabilities. Therefore, these architectures are very suitable for the model exchange framework of federated learning and the exchange of shared data.Thus, based on this, our approach is proposed to improve the task model performance of federated learning for non-IID data.

To sum up, the main contributions of this work can be summarised as follows:

1. A novel unidirectional synchronous cyclic decentralised federated learning framework and an effective evaluation of the convergence of the model;
2. A new distribution information sharing and knowledge distillation model aggregation algorithm for the federated task model, which solves the problem of data distribution inconsistency both at the algorithm level and the data level;
3. The first attempt to use federated learning to diagnose Alzheimer's disease based on medical datasets;
4. A way to measure the inconsistent distributions of data features using the maximum mean difference (MMD).

The rest of our paper is organised as follows. Section 2 introduces related work. Section 3 details our proposed approach. Section 4 describes the experimental environment and our experimental results. Section 5 concludes the paper and proposes future work.

2. Related Work

Since federated learning was first proposed, four main types of challenges have arisen: communication challenges, system challenges, statistical challenges and privacy challenges [4]. We can refer to these two articles [5,6] for the communication challenges and system challenges of a cyclic federated learning framework, which have been analyzed and solved by predecessors. For privacy challenges, we can refer to the solutions in these two articles [7,8]. The privacy security protection strategies proposed in both papers consist of a privacy protection module and an attack detection module, while the major difference between the two is that the first scheme uses a two-level privacy data protection module. This scheme uses perturbation-based privacy converts categorical values into numeric and normalizes feature values into a range of [0, 1] before transforming the data using DL-based encoder techniques, which strengthens privacy and increases the utility of DL models .The statistical challenges, e.g., the non-independent and identical distribution of data (non-IID) problem, are some of the most non-negligible challenges in the application of federated learning in the medical field. Therefore, in this paper, we mainly focus on the non-IID problems.In response to non-IID problems, existing research has mainly solved the problems at the algorithm and data levels.

The algorithm-level solutions mainly include objective function modification and solution mode optimisation. Objective function modification involves adding regularisation terms on the client side. A trade-off has been achieved between optimising local models and reducing the differences between local models and global models to solve the non-independent homogeneous distribution of data at each node [9–12]. The measure of the differences between local models and global models by the regularisation terms can be either the distance between them or the differences in model behaviour. The distance measures between local and global models are Euclidean distances [9] and weighted distances [10]. For example, the federated proximal optimisation (FedProx) algorithm that has been proposed in the literature [9] corrects the client-side drift that occurs in FedAvg by restricting the Euclidean distances between local models and global models as proximal terms. This means that the local updates do not excessively deviate from the global models, which alleviates any inconsistencies in the client-side data and improves the stability of global model convergence. The federated curvature (FedCurv) algorithm that has been proposed in the literature [10] uses Fisher information from global models obtained during the previous rounds of training to weight the distances, which can reduce excessive errors in the model parameters. The differences in model behaviour between local and global models can be measured by the degree of inconsistency in the model output distributions on local datasets or by the gradient of the global models on local datasets. For example, in the literature [11], the maximum mean discrepancy (MMD) has been used as a metric to measure the inconsistency in model output distributions on local datasets. The stochastic controlled averaging (SCAFFOLD) algorithm that has been proposed in the literature [12] improves the FedProx algorithm by adding a control variable on the client side. This control variable can take either the gradient norm of global models on local datasets or the Euclidean distances between local and global models, thus preventing local models from deviating from the globally correct training direction. These methods can improve the performance of federated learning for model learning on non-IID datasets to some extent, but the degree of improvement is limited by the consistency of the client-side data sampling [3].

In solution optimisation, the good performance of federated learning models is mainly achieved by improving the server-side aggregation method. The ideal application conditions for federated learning are IID-based datasets (such as the initially proposed FedAvg algorithm) and weights for clients that are proportional to the number of samples.The accuracy of global models is greatly degraded in the case of the inconsistent, unbalanced and non-independent distribution of client data [13]. For this reason, most scholars have aimed to improve the shortcomings of aggregation methods for federated averaging algorithms. Accuracy-based averaging (ABAvg) has been in the literature [14], in which the server-side tests the accuracy of temporary models on validation datasets to obtain the accuracy of the models on the client side and then normalises them before aggregating all parameters. The federated learning with matched averaging (FedMA) algorithm that has been proposed in the literature [15] uses Bayesian non-parametric methods to match and average weights in a hierarchical manner. The federated averaging with momentum (FedAvgM) algorithm that has been proposed in the literature [16] applies momentum when updating global models on a server. The federated normalised averaging (FedNova) algorithm that has been proposed in the literature [17] normalises local updates before averaging. However, these methods have limited success in improving the performance of global models [12], so some scholars have proposed approaches that evade this problem, such as personalised federated learning, multitask federated learning and federated meta-learning, which can also improve the performance of federated learning on non-IID data to some extent.

The source of global model performance degradation is the non-IID problem; thus, data-level approaches to sharing client-side data have become new options for solving the non-IID problem. Client-side data sharing can be divided into two types: direct data sharing and indirect data sharing. In terms of direct data sharing for federated learning, one approach is to use a global sharing strategy [18–20], in which the server-side shares small amounts of public data with the client for training to reduce the variance between trained local models, thus increasing the robustness and stability of the training process. This sharing approach relies on task-specific public datasets, and, in practice, there is a risk of privacy violation during both the acquisition and sharing of public data. Another approach is to use a local sharing strategy [21,22], in which small amounts of data are shared directly through trusted communication links between clients; however, this approach also violates the privacy preservation conventions of federated learning.

Indirectly shared federated learning does not share data directly, but rather makes the distributions of client datasets consistent by sharing data distribution information on the client side and then augmenting local training datasets with the shared distribution information [23,24]. The data distribution information can be learned using generator networks, which can be divided into global and local generators, depending on how the generators are trained. For example, a global generator shared approach has been proposed in the literature [23] that trains conditional generative adversarial network(CGAN) [25] generators on central servers and then shares the generators with clients to share distribution information. However, the data required for training CGANs using central servers are extracted from all clients, and there is a risk of privacy violation during the transmission of extracted data from the clients to the server side. A local generator shared approach has also been proposed in the literature [24] that trains bulldozer distance-based generative adversarial networks (i.e., Wasserstein generative adversarial networks, WGANs) [23] on local datasets on the client side and shares them with other clients. An image translation network is then trained using local generators and other generators to solve the federated learning problem for client-side heterogeneous data. Implicit data sharing through generators does not cause any privacy problems and is more practical than direct data sharing because it meets the need for patient privacy protection in healthcare organisations.

The data-immobile and model-immobile nature of federated learning has led to its increasingly widespread application in fields with high requirements for sensitive data protection, such as medicine. To address the problem of the degradation of federal learning performance due to inconsistent data distributions among federated learning participants, federated learning for client-side data sharing has become an effective solution strategy. Among the different options, the approach of sharing data distributions rather than the data themselves is more appropriate for application because it does not create the risk of privacy violation. Therefore, we addressed this issue by integrating solutions at both the data and algorithm levels. See Figure 1 for details of classification guidelines.

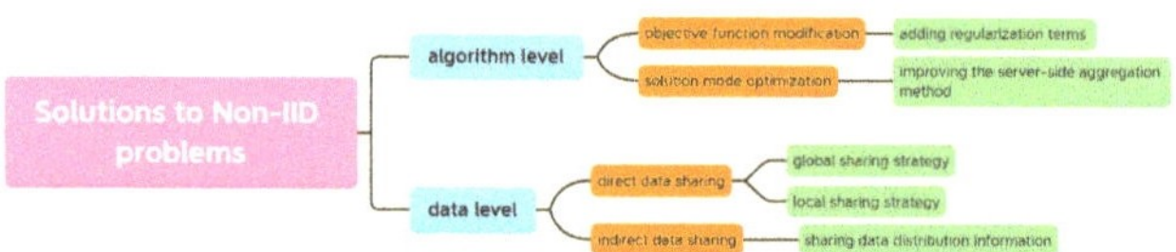

Figure 1. Solutions to non-IID problems.

3. The Distribution Information Sharing- and Knowledge Distillation-Based Cyclic Federated Learning Method

The ultimate goal of federated learning is to jointly train optimal models for multiple clients; in this paper, we refer to these as task models, which are made by multiple medical institutions to obtain target models. Task models can be for the diagnosis of diseases, lesion segmentation, etc. In federated learning, local task models tend to be consistent with global task models; however, in the case of non-IID local client data, local task models deviate from global task models. In the existing state-of-the-art circular decentralised federated learning schemes, the model parameters of nodes are updated after multiple steps of weighted summation and then averaged, which is a complex and costly communication strategy. In addition, the weighted average approach to model parameter aggregation often yields poor task model performance on non-IID datasets because the client data distributions of neighbouring nodes may differ significantly and thus, the trained task models are biased. To address this, a natural idea is to degrade this bias by sharing data distributions to generate augmented datasets while preserving data privacy and then using the augmented data to learn the data distributions of other clients to achieve the implicit aggregation of model parameters. For this purpose, we used generators to learn the data distribution information of clients and share the local task models of clients, together with the local data generators, with neighbouring clients. Since both the generators and the task models carrying the data distribution information of the neighbouring clients were trained on the same datasets, this facilitated the use of the migration learning idea to aggregate the task models of two neighbouring clients. Based on this, we proposed a teacher–student model-based migratory learning approach for task model aggregation. Figure 2 shows a general block diagram of our proposed approach.

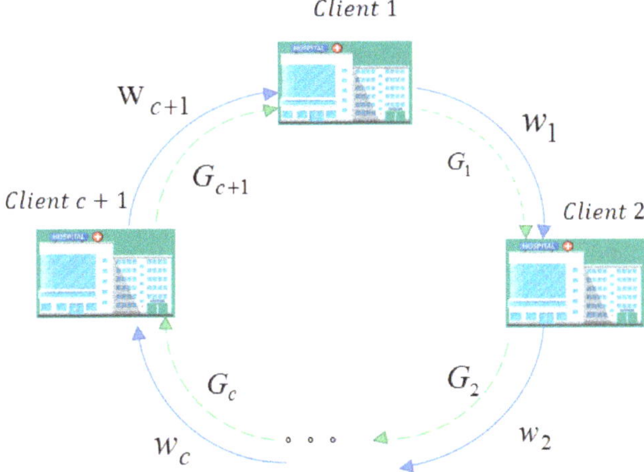

Figure 2. A schematic diagram of our cyclic federated learning method based on distribution information sharing and knowledge distillation.

Supposing that there are C clients involved in the federated learning task (where G is the shared generator model parameters that are locally trained offline, and w is the task model parameters that are dynamic shared weights), the overall process can be divided into two stages as follows:

Stage 1: The offline process. All clients participating in the federated learning task train the generator network offline on local datasets to obtain the generator network G that responds to local distribution information. Then, all clients pass the trained generator G to the next client in turn. The next client $c+1$ generates the corresponding virtually shared local data after receiving the generator from client c before.

Stage 2: The online process, which can be mainly divided into two steps. The first step is the knowledge distillation learning process, in which all clients first initialise the task model on local datasets and share it with the next client, and the next client then uses the shared task model to teach its task model on the data that were virtually shared via knowledge distillation. The second step simply re-updates the trained task model on local datasets again and shares it with the next client.

3.1. Distribution Information Acquisition Based on Deep Learning

To eliminate the adverse effects of the non-IID problem on the performance of medical institution federated learning, an effective approach is to augment the local datasets of medical institutions by sharing their data distributions. To obtain information about the data distributions of healthcare institutions, the current state-of-the-art approach is to use a generator model with deep learning. Generators are the most effective tools for data augmentation because they not only learn the distribution information of data effectively but also generate data that match the real distributions. Generative adversarial networks (GANs), as one of the current types of mainstream deep neural network generators, are powerful in terms of image enhancement and image-to-image conversion [22]. Therefore, we adopted a GAN as a data generator on the main server to obtain the data distribution information of local clients [26–28] and added conditional information to generate the type of data that we needed, i.e., the final generator model was a CGAN. Specifically, let the total number of clients (federated learning participants) participating in the federated learning task be C, let the local datasets of the c ($c = 1, 2, \cdots, C$) client be $D_c = \{x_i \mid i = 1, 2, \cdots, N_c\}$ and let $N_c = |D_c|$ be the number of clients in the training sample. The client c trains a generator and reflects the distribution information G_c of local datasets D_c. Thus, C clients are trained to obtain C generator models. The distribution of information obtained in this way is relatively safe from privacy breaches.

3.2. Distribution Information Sharing

The purpose of sharing distribution information is to enable later clients in the cyclic communication graph to have virtually shared data about the previous client's data distribution information, thus enabling two adjacent clients to achieve a consistent distribution of data to improve the performance of task models. To this end, we combined the features of a cyclic federated learning architecture and model parameters to accomplish this process. Let $c = 1, 2, \cdots, C$ and let the client c transmit the generator G_c to the client $c+1$. When $c = C$, let $c + 1 = 1$, thus forming a ring-shaped communication link. Under the condition of this cyclic communication link, let the client $c + 1$ receive the generator G_c from the client c, where $N_c = |D_c|$ is the number of local data points from the client c. Accordingly, G_c can generate N'_{c+1} virtually shared data points, i.e., $D'_{c+1} = \{x_l \mid x_l = G(z_l), l = 1, 2, \cdots, N'_{c+1}\}$. Therefore, only the client $c+1$ has the distribution information of the client c, which indirectly realises distribution information sharing while protecting patient privacy. The distribution information sharing process is schematically illustrated in Figure 3.

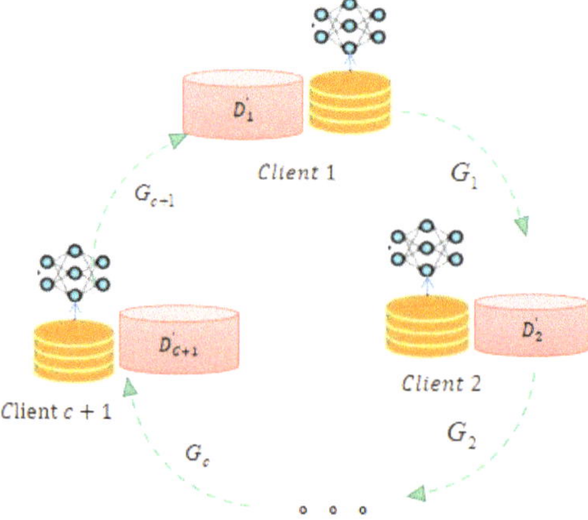

Figure 3. A schematic diagram of the distribution information sharing process.

3.3. Task Model Parameter Aggregation

The task model parameter aggregation process focuses on how to use shared distribution information for model parameter aggregation to eliminate the adverse effects of the non-IID problem on federated learning performance. In our cyclic federated learning framework, the client *c+1* not only receives the task model parameters from the client *c* through a trusted channel but also the generator model G_c. The virtually shared data D'_{c+1} can be generated locally via G_c. Since D'_{c+1} have consistent distributions across the local datasets D_c of the client *c*, the task model $f(x, w_c)$ obtained by the client *c* after training using D_c has a good performance. However, the distributions of the local datasets D_{c+1} of the client *c+1* are usually not consistent with those of D_c, such that $f(x, w_c)$ performs worse on the local datasets D_{c+1} of the client *c+1* than on D'_{c+1}. As a result, existing model aggregation algorithms, such as federated averaging and its various improvements, performed poorly in our cyclic federated learning framework. To this end, we proposed a new method for model aggregation for federation learning tasks based on knowledge distillation.

Since the locally trained task model of client *c* has a similar optimal performance on datasets D'_{c+1} and D_c, the locally trained task model $f(x, w_{c+1})$ of client *c+1* can be trained using the local task model $f(x, w_c)$ of client *c* on the datasets D'_{c+1} to improve performance. This idea could be implemented using the teacher–student model for migration learning, as shown in Figures 4 and 5.

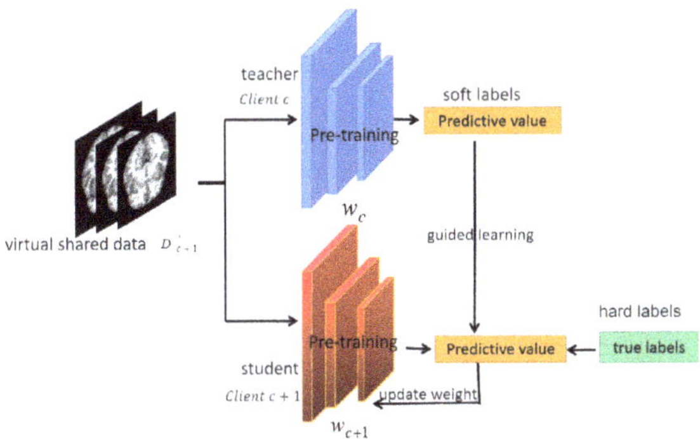

Figure 4. A schematic diagram of the teacher–student guided learning approach.

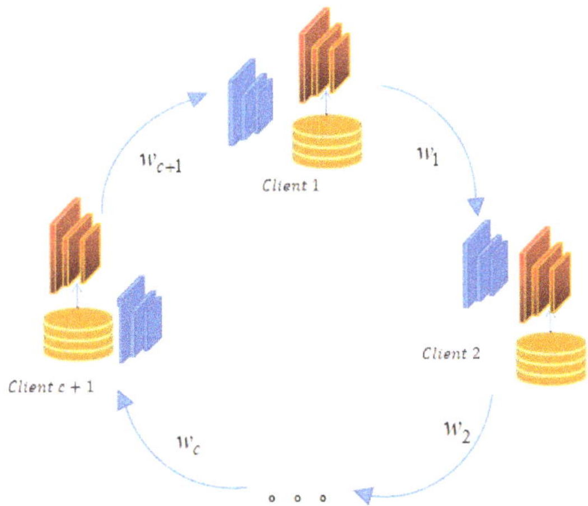

Figure 5. A schematic diagram of the teacher–student guided learning approach.

The training goal of our cyclic federated learning method based on the distribution of information sharing and knowledge distillation was the minimisation of the total loss function:

$$\ell(w_1, w_2, \cdots, w_C) = \sum_{c=1}^{C} L_{c+1}(w_{c+1}) + \lambda \sum_{c=1}^{C} R_{c+1}(w_{c+1}, w_c) \tag{1}$$

where γ is a hyperparameter that controls the propensity of the local task model, w_{c+1} is the parameter of the local task model of the client $c + 1$ (the task model to be trained can be the same or different for each client), and the loss function corresponding to $L_{c+1}(w_{c+1})$ has the following definition:

$$L_{c+1}(w_{c+1}) = \sum_{x \in D_{c+1}} l_{c+1}(x; w_{c+1}) \tag{2}$$

where $l_{c+1}(x; w_{c+1})$ is the loss of the task model $f(x, w_{c+1})$ on the data sample x and $R_{c+1}(w_{c+1}, w_c) \geq 0$ is the difference between the models of the adjacent clients c and $c+1$ in the cyclic communication graph, which is defined as follows:

$$R_{c+1}(w_{c+1}, w_c) = \alpha L_{soft}(w_{c+1}, w_c) + \beta L_{havd}(w_{c+1}) \tag{3}$$

where

$$L_{soft}(w_{c+1}, w_c) = \sum_{x \in D'_{c+1}} l_{soft}(x; w_{c+1}, w_c) \tag{4}$$

$$L_{hard}(w_{c+1}) = \sum_{x \in D'_{c+1}} l_{hard}(x; w_{c+1}) \tag{5}$$

are the knowledge distillation loss and student loss on the datasets D'_{c+1}, respectively (which are defined in the same way as in the standard teacher–student model), and α and β are two hyperparameters with values of 0 when the adjacent client models are the same and values of greater than 0 when they are different; the smaller the difference, the smaller the value (and vice versa). According to the incremental convex optimisation theory, the minimisation equation (Equation (1)) can be solved using the following iteration. At the k-th iteration, the gradient descent update is first performed on the intermediate variable u_{c+1}:

$$u_{c+1}^{(k)} = w_{c+1}^{(k-1)} - \alpha_k \nabla R_{c+1}\left(w_{c+1}^{(k-1)}, w_c^{(k-1)}\right) \tag{6}$$

where α_k is the gradient descent size, and the superscripts k and $k+1$ denote the values of the k-th and k-th$+1$ iterations, respectively. Then, the model parameters are updated as follows:

$$w_{c+1}^{(k)} = \arg\min_{w} L_{c+1}(w) + \frac{\lambda}{2\alpha_k} \left\| w - u_{c+1}^{(k)} \right\|^2 \tag{7}$$

Using Equation (6), the iteration of $u_{c+1}^{(k)}$ learns the behaviour of $f\left(x, w_c^{k-1}\right)$ on the datasets D'_{c+1}, thus optimising the performance of the local model $f\left(x, w_{c+1}^k\right)$ that was updated using Equation (7) on the datasets $D_{c+1} \cup D'_{c+1}$. After multiple further iterations of training, as shown in Figure 4, all clients can learn the features of the data distributions of other clients via this cyclic framework, i.e., the training effect of a global model is reached. Ultimately, the adverse effects of the non-IID problem on medical institution-federated learning performance can be eliminated.

The above solution process can be described in pseudo-code as shown in Algorithm 1.

Algorithm 1 Federated learning algorithm based on distribution information sharing and knowledge distillation.

Input: C clients,each with its own training datasets D_c,generator G_c and its own task model
    $f(x, w_c)$
Output: Trained model parameter set $\{w_1, w_2, \cdots, w_C\}$
  1: **for** $c = 1, 2, \cdots, C$ **do**
  2:     Client c sends G_c to Client c+1
  3:     Client c+1 generates virtual shared data D'_{c+1} with G_c
  4: **end for**
  5: **for** $k = 1, 2, \cdots, K$ **do**
  6:     **for** $c = 1, 2, \cdots, C$ **do**
  7:         Client c sends $w_c^{(k-1)}$ to Client c+1
  8:         Client c+1 updates $u_{c+1}^{(k)}$ according to (3) and (6)
  9:         Client c+1 updates $w_{c+1}^{(k)}$ according to (1) and (7)
10:    **end for**
11: **end for**

---

4. Experimental Results and Discussion

4.1. Development Environment and Datasets

Our machine learning model was built by the well-known deep learning framework PyTorch, version 1.6.0, and Python, version 3.7.1. A self-built cyclic federated learning framework was used, in which a Kafka cluster was used as the information medium for model parameter exchange. The generator network used six convolutional layers with a convolutional kernel size of 4x4.We used two datasets to validate the effectiveness of the proposed method, one of which was the Alzheimer's dataset that was used in a Kaggle competition .For this dataset, we used the pre-trained model VGG16 provided by torchvision as our classifier network. The second dataset was the MNIST dataset, which was also used to validate the generality of the proposed method, i.e., the generalisation ability of the method. For this dataset, we used the two-layer convolutional layer network used in MOON. The Alzheimer's dataset, which has a total of 5120 training data points and 1279 testing data points, has a 1:1 ratio of diseased to non-diseased data in both the testing and training sets. The MNIST dataset has a total of 60,000 images in the training set and 10,000 images in the testing set.

4.2. Experimental Parameters

There are various scenarios of non-IID data. In this study, we focused on two of them: attribute skew and label skew. To study these two different types of data distributions, we conducted experiments on the two selected datasets. For the Alzheimer's dataset, due to its high data latitude and the few types of labels, we used the maximum mean difference (MMD) to measure the attribute skew of the client data [29]. The maximum mean difference is mainly used to measure the distance between two data distributions. Given two data distributions, the square of their MMD can be expressed as:

$$MMD^2(x, y) = \|E[\varphi(x)] - E[\varphi(y)]\|^2 \tag{8}$$

where $\phi(\bullet)$ denotes the mapping to the regenerated Hilbert space (RKHS). The inconsistency of client data distributions was measured by calculating the MMD, and the entire dataset was divided according to the MMD value to measure the federated learning performance under different MMD values. As for the MNIST dataset, we used the Dirichlet distribution [30] to divide the non-IID samples because of the many types of labels. Figure 6 shows the Dirichlet distribution when $\alpha = 0.5$ and the number of clients was 10.

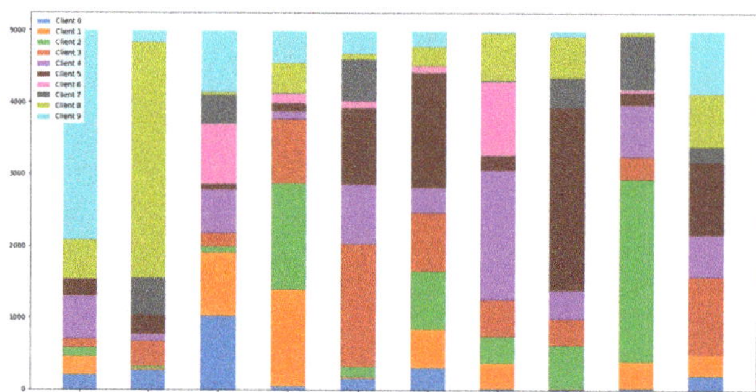

Figure 6. Dirichlet distribution.

4.3. Algorithm Evaluation

For the two distribution types (attribute skew and label skew), we effectively evaluated our algorithm on the selected datasets.

4.3.1. Attribute Skew

To study attribute skew, we conducted a correlation experiment on the Alzheimer's dataset, the results of which are shown in Table 1 and Figures 7–10. By comparing the accuracy rates on the testing set that are shown in Table 1, it can be seen that, in different MMD scenarios, the cyclic federated average-based method had a larger model loss, its performance was different from that of the centralised learning method, and the training was unstable. Our proposed method outperformed the cyclic federated average method, and the performance was close to or attained the centralised learning performance. The box line plot in Figure 10 shows that the MMD increased from the top left to the bottom right. By dynamically increasing the MMD, we could see that as the MMD increased (i.e., as the data distribution became more inconsistent), the model performance of the cyclic federated average method degraded faster and deviated greatly from the centralised learning performance, while the performance of our proposed method was better than that of the federated average, and the deviation from the centralised learning performance was slower.

Table 1. The top-1 accuracy of different MMD values on the Alzheimer's dataset.

Data Division ID	MMD	CFL_DS_KD	CFL_FedAvg	CL
1	0.514	79.95%	78.97%	
2	1.029	79.56%	78.11%	
3	1.283	78.73%	77.24%	79.22%
4	1.546	78.60%	72.20%	
5	1.803	78.77%	75.37%	
6	2.059	78.05%	70.00%	

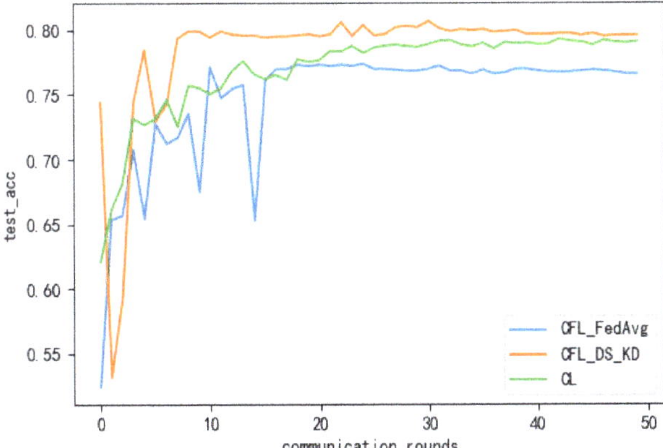

Figure 7. The accuracy of the different methods on the testing set after different amounts of communication rounds.

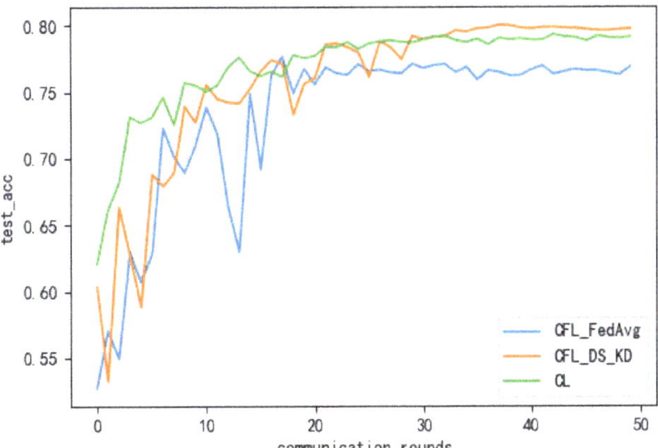

Figure 8. The accuracy of the different methods on the testing set after different amounts of communication rounds.

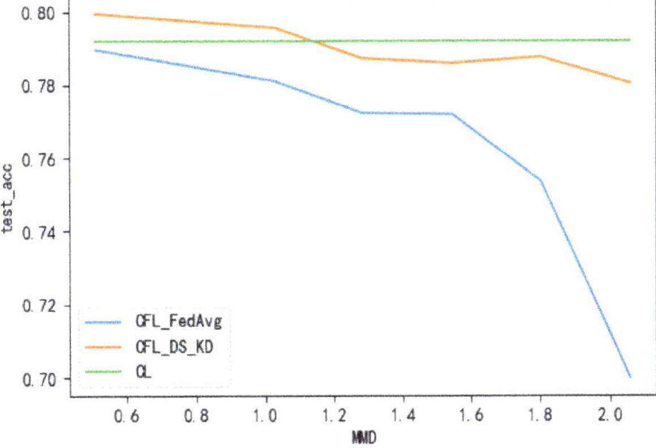

Figure 9. A performance comparison of the different methods under different MMD values.

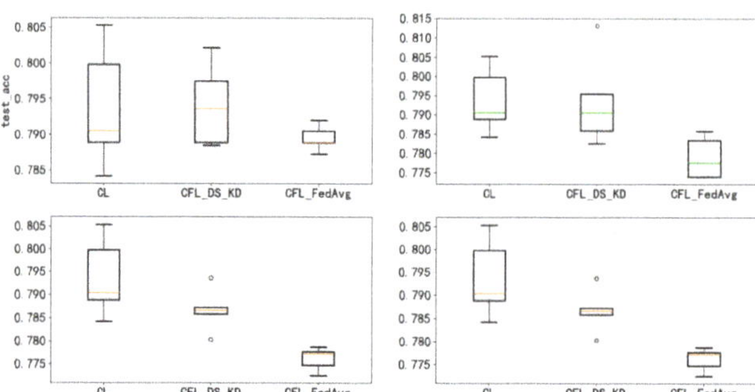

Figure 10. A performance comparison of the different methods under different MMD values.

Additionally, to further verify the effectiveness of our method, the output differences between the proposed method and the centralised learning method were analysed. When the difference was smaller, it meant that the proposed method was closer to the centralised learning performance. From Figure 11, it can be seen that the output performance difference between our method and the centralised learning method was almost 0, which effectively illustrated the beneficial effects of our method.

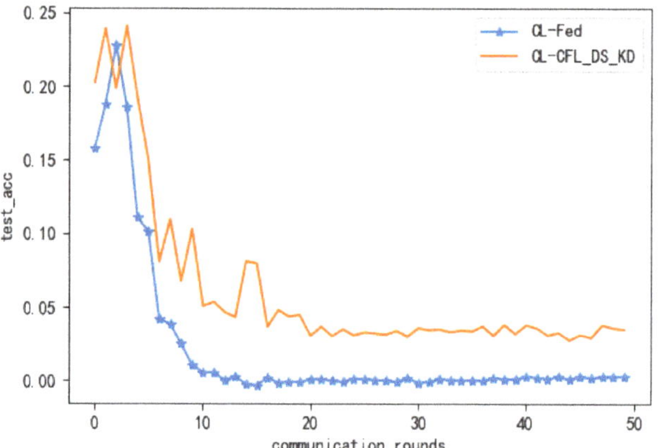

Figure 11. The difference between the CFL_DS_KD and CL method outputs.

4.3.2. Label Skew

The superior performance of our method was effectively demonstrated after several experiments on the Alzheimer's dataset. To further demonstrate the performance of our proposed method in the case of label skew, we also conducted corresponding comparative experiments on the public MNIST dataset. Comparisons were made between our proposed method and the state-of-the-art federated learning algorithm MOON and the mutual learning method Def_KT within a centreless federated learning framework. As shown in Figure 12 and Table 2, α is the Dirichlet distribution coefficient, and the smaller its value, the more inconsistent the data distribution. From the experimental results, it can be seen that the classification accuracy of our proposed method on the testing set was almost comparable to the centralised learning method and higher than those of MOON [31], Def_KT [32] and FedAvg. Thus, the superiority of our proposed method was effectively demonstrated, both on a medical dataset and a publicly available natural dataset.

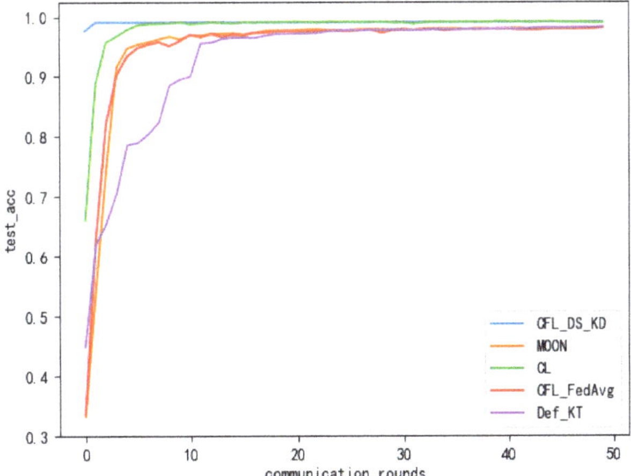

Figure 12. A performance comparison of the different methods when $\alpha = 0.5$.

Table 2. The top-1 test accuracy with $\alpha = 0.5$ and $\alpha = 0.1$ on MNIST datasets.

Method	$\alpha = 0.1$	$\alpha = 0.5$
MOON	95.7%	98.1%
CFL_FedAvg	97.1%	98.1%
Def_KT	95.2%	99.0%
CFL_DS_KT	98.9%	99.1%

4.3.3. Convergence Evaluation

The convergence of centralised federated learning has been effectively proven, whereas that of centreless cyclic federated learning frameworks has not yet been proven. Therefore, in addition to the performance of the selected methods described above, we also experimentally evaluated the convergence of our cyclic federated learning architecture. We used two parametric numbers to find the differences in weights between clients. The weight differences could be expressed as follows:

$$D_l = \frac{1}{C} \sum_{c=1}^{C} \left\| w_{c+1}^l - w_c^l \right\|_2 \tag{9}$$

$$D_l^i = \frac{1}{C} \sum_{c=1}^{C} \left\| w_{c+1}^l i - w_c^l i \right\|_2 \tag{10}$$

Equation (9) represents single-layer weight differences, and Equation (10) represents single-weight differences in each layer, where w_c^l denotes the c-th client's i-th layer weight and w_c^{li} denotes the i-th weight difference in the i-th layer of the c-th client. We conducted experiments on the Alzheimer's dataset. Figure 13 shows the single-layer weight differences, in which it can be seen that as the amounts of communication rounds increased, the weight differences degraded sequentially and eventually stabilised. Figures 14 and 15 show the single-layer single-weight differences, in which it can be seen that the largest weight difference was in the thirteenth convolutional layer, and the maximum difference was only 0.0035 (i.e., close to 0) and reached convergence.

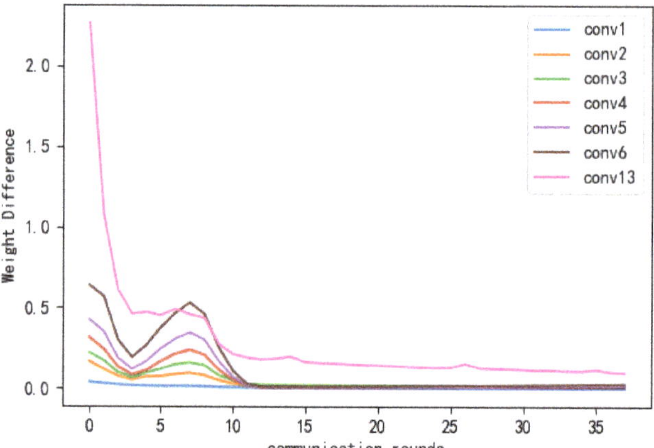

Figure 13. The single-layer weight differences.

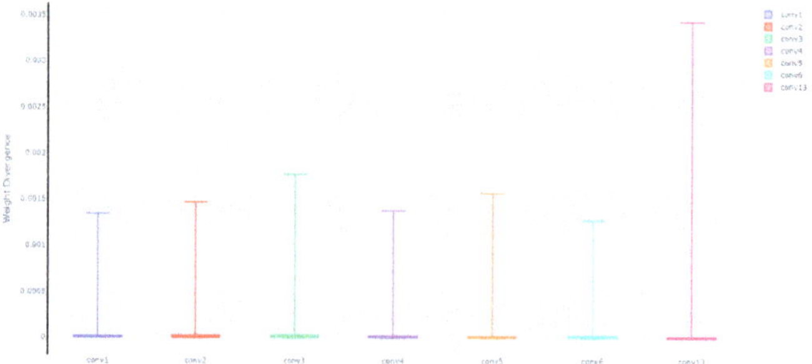

Figure 14. The single-weight differences in each layer.

Figure 15. The single-weight differences in each layer.

The output variance could be expressed as follows:

$$DM = \frac{1}{C} \sum_{c=1}^{C} \| f(w_{c+1}; x) - f(w_c; x) \|_2 \tag{11}$$

where $f(w_c; x)$ denotes the output of the model of the c-th client under the input sample x. The output difference results are shown in Figure 16, in which the diagnosis difference represents the differences between the diagnosis results and the input data. The disease output difference referred to the output differences between the results that were diagnosed as diseased, and the model output difference referred to the overall differences between the model outputs. It can be seen from the results that the output differences were small on the whole, i.e., very close to 0. Therefore, from the above experiments, we could conclude that the convergence of our proposed method was effectively evaluated. At present, the model is convergent.

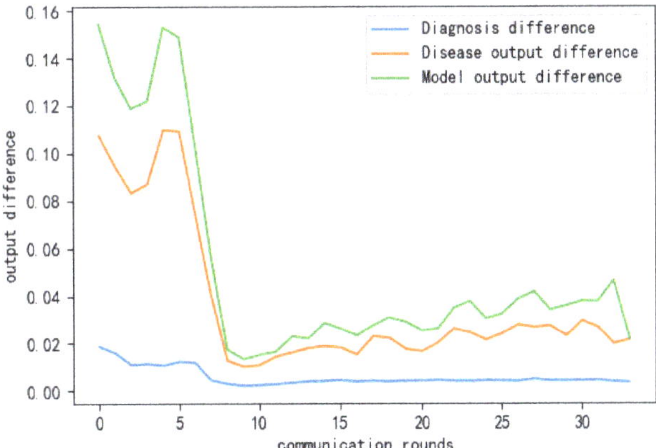

Figure 16. The output differences.

5. Summary

To address the non-IID problem in medical institution federated learning that cannot be effectively solved using existing federated learning techniques, this paper proposed a cyclic federated learning method (CFL_DS_KT) based on distribution information sharing and knowledge distillation. This is a novel and effective federated learning approach and, to the best of our knowledge, the first time we have used this unidirectional synchronous cyclic decentralised federated learning framework and effectively evaluated the convergence of a model with this structure. The experimental results also show that the task model achieves convergence under our proposed approach. Furthermore, in contrast to existing scholarly research solutions, we solve the non-IID problem by optimising the solution through the solution approach of distribution sharing and knowledge distillation. By considering both data-level and algorithm-level optimisation approaches, we achieve better performance of the federation learning model under non-IID while safeguarding client data privacy. In our extensive experiments on medical and public datasets, CFL_DS_KT shows a good improvement over various state-of-the-art methods, and its accuracy is closer to that of centralised learning. Further improvements in privacy preservation were achieved due to using a cyclic federated learning method. It also provided the idea of training federated learning models on heterogeneous data, which could eliminate data heterogeneity by transforming the data distribution information from one client to another.

However, our proposed approach has some shortcomings. When the client data is extremely heterogeneous, it is difficult to train a good generator to generate high-quality images due to the small amount of training data. Additionally, it is not suitable to train federated learning models with large numbers of clients as this could increase breakpoint failures and model training cycle times. Therefore, this method would mainly be suitable for federated learning across medical institutions.

Author Contributions: Conceptualization, L.Y. and J.H.; methodology, L.Y.; software, L.Y.; validation, L.Y. and J.H.; formal analysis, L.Y. and J.H.; investigation, L.Y.; resources, L.Y.; data curation, L.Y.; writing—original draft preparation, L.Y.; writing—review and editing, J.H.; visualization, L.Y.;

supervision, J.H.; project administration, J.H.; funding acquisition, J.H. All authors have read and agreed to the published version of the manuscript.

Funding: This research was funded by Science and Technology Planning Project of Shenzhen Municipality (No. 20200821152629001).

Institutional Review Board Statement: Not applicable.

Informed Consent Statement: Not applicable.

Data Availability Statement: Not applicable.

Conflicts of Interest: The authors declare no conflict of interest.

References

1. McMahan, B.; Moore, E.; Ramage, D.; Hampson, S.; Arcas, B.A.Y. Communication-efficient learning of deep networks from decentralized data. *arXiv* **2017**, arXiv:1602.05629.
2. Yuan, D.; Zhu, X.; Wei, M.; Ma, J. Collaborative Deep Learning for Medical Image Analysis with Differential Privacy. In Proceedings of the 2019 IEEE Global Communications Conference (GLOBECOM), Waikoloa, HI, USA, 9–13 December 2019; pp. 1–6. [CrossRef]
3. Li, Q.; Diao, Y.; Chen, Q.; He, B. Federated learning on non-iid data silos: An experimental study. *arXiv* **2021**, arXiv:2102.02079.
4. Rahman, S.A.; Tout, H.; Ould-Slimane, H.; Mourad, A.; Talhi, C.; Guizani, M. A Survey on federated learning: The journey from centralized to distributed on-site learning and beyond. *IEEE Internet Things J.* **2021**, *8*, 5476–5497.
5. Wang, Z.; Hu, Y.; Yan, S.; Wang, Z.; Hou, R.; Wu, C. Efficient Ring-Topology Decentralized Federated Learning with Deep Generative Models for Medical Data in eHealthcare Systems. *Electronics* **2022**, *11*, 1548. [CrossRef]
6. Liu, W.; Chen, L.; Zhang, W. Decentralized federated learning: Balancing communication and computing costs. *IEEE Trans. Signal Inf. Process. Netw.* **2022**, *8*, 131–143. [CrossRef]
7. Kumar, P.; Gupta, G.P.; Tripathi, R. PEFL: Deep Privacy-Encoding-Based Federated Learning Framework for Smart Agriculture. *IEEE Micro* **2021**, *42*, 33–40. [CrossRef]
8. Kumar, P.; Gupta, G.P.; Tripathi, R.; Garg, S.; Hassan, M.M. DLTIF: Deep learning-driven cyber threat intelligence modeling and identification framework in IoT-enabled maritime transportation systems. *IEEE Trans. Intell. Transp. Syst.* **2021**, 1–10.. [CrossRef]
9. Li, T.; Sahu, A.K.; Zaheer, M.; Sanjabi, M.; Talwalkar, A.; Smith, V. Federated optimization in heterogeneous networks. *Proc. Mach. Learn. Syst.* **2020**, *2*, 429–450.
10. Shoham, N.; Avidor, T.; Keren, A.; Israel, N.; Benditkis, D.; Mor-Yosef, L.; Zeitak, I. Overcoming forgetting in federated learning on non-iid data. *arXiv* **2019**, arXiv:1910.07796.
11. Yao, X.; Huang, C.; Sun, L. Two-stream federated learning: Reduce the communication costs. In Proceedings of the 2018 IEEE Conference on Visual Communications and Image Processing (VCIP), Taichung, Taiwan, 9–12 December 2018; pp. 1–4.
12. Karimireddy, S.P.; Kale, S.; Mohri, M.; Reddi, S.; Stich, S.; Suresh, A.T. Scaffold: Stochastic controlled averaging for federated learning. In Proceedings of the International Conference on Machine Learning, Virtual, 13–18 July 2020; pp. 5132–5143.
13. Li, X.; Huang, K.; Yang, W.; Wang, S.; Zhang, Z. On the convergence of FedAvg on Non-IID Data. *arXiv* **2019**, arXiv:1907.02189.
14. Xiao, J.; Du, C.; Duan, Z.; Guo, W. A Novel Server-side Aggregation Strategy for Federated Learning in Non-IID situations. In Proceedings of the 2021 20th International Symposium on Parallel and Distributed Computing (ISPDC), Napoca, Romania, 28–30 July 2021; pp. 17–24.
15. Hsu, T.M.H.; Qi, H.; Brown, M. Measuring the effects of non-identical data distribution for federated visual classification. *arXiv* **2019**, arXiv:1909.06335.
16. Wang, H.; Yurochkin, M.; Sun, Y.; Papailiopoulos, D.; Khazaeni, Y. Federated learning with matched averaging. *arXiv* **2020**, arXiv:2002.06440.
17. Wang, J.; Liu, Q.; Liang, H.; Joshi, G.; Poor, H.V. Tackling the objective inconsistency problem in heterogeneous federated optimization. *Adv. Neural Inf. Process. Syst.* **2020**, *33*, 7611–7623.
18. Zhao, Y.; Li, M.; Lai, L.; Suda, N.; Civin, D. Federated learning with non-iid data. *arXiv* **2018**, arXiv:1806.00582.
19. Bassily, R.; Thakkar, O.; Guha, T.A. Model-agnostic private learning. In Proceedings of the 32nd Conference on Neural Information Processing Systems (NeurIPS 2018), Montreal, QC, Canada, 2–8 December 2018; pp. 1–11.
20. Yao, X.; Huang, T.; Zhang, R.-X.; Li, R.; Sun, L. Federated learning with unbiased gradient aggregation and controllable meta updating. *arXiv* **2019**, arXiv:1910.08234.
21. Sun, Y.; Zhou, S.; Gündüz, D. Energy-aware analog aggregation for federated learning with redundant data. In Proceedings of the ICC 2020–2020 IEEE International Conference on Communications (ICC), Dublin, Ireland, 7–11 June 2020; pp. 1–7.
22. Wu, X.Z.; Liu, S.; Zhou, Z.H. Heterogeneous model reuse via optimizing multiparty multiclass margin. In Proceedings of the International Conference on Machine Learning, New York, NY, USA, 22–24 February 2019; pp. 6840–6849.
23. Jeong, E.; Oh, S.; Kim, H.; Park, J.; Bennis, M.; Kim, S.L. Communication-Efficient On-Device Machine Learning: Federated Distillation and Augmentation under Non-IID Private Data. *arXiv* **2018**, arXiv:1811.11479.

24. Yan, Z.; Wicaksana, J.; Wang, Z.; Yang, X.; Cheng, K.-T. Variation-aware federated learning with multi-source decentralized medical image data. *IEEE J. Biomed. Health Inform.* **2021**, *25*, 2615–2628. [CrossRef]
25. Antipov, G.; Baccouche, M.; Dugelay, J.L. Face aging with conditional generative adversarial networks. In Proceedings of the 2017 IEEE International Conference on Image Processing (ICIP), Beijing, China, 17–20 September 2017; pp. 2089–2093.
26. Chen, Y.; Wang, Y.; Kao, M.; Chuang, Y. Deep photo enhancer: Unpaired learning for image enhancement from photographs with gans. In Proceedings of the IEEE Conference on Computer Vision and Pattern Recognition, Salt Lake City, UT, USA, 18–22 June 2018; pp. 6306–6314.
27. Goodfellow, I.; Pouget-Abadie, J.; Mirza, M.; Xu, B.; Warde-Farley, D.; Ozair, S.; Courville, A.; Bengio, Y. Generative adversarial nets. In Proceedings of the Advances in Neural Information Processing Systems, Montreal, QC, Canada, 8–13 December 2014; pp. 2672–2680.
28. Creswell, A.; White, T.; Dumoulin, V.; Arulkumaran, K.; Sengupta, B.; Bharath, A.A. Generative adversarial networks: An overview. *IEEE Signal Process. Mag.* **2018**, *35*, 53–65. [CrossRef]
29. Wang, Z.; She, Q.; Ward, T.E. Generative adversarial networks in computer vision: A survey and taxonomy. *ACM Comput. Surv. (CSUR)* **2021**, *54*, 1–38. [CrossRef]
30. Chen, Y.; Lu, W.; Qin, X.; Wang, J.; Xie, X. MetaFed: Federated Learning among Federations with Cyclic Knowledge Distillation for Personalized Healthcare. *arXiv* **2022**, arXiv:2206.08516.
31. Li, Q.; He, B.; Song, D. Model-contrastive federated learning. In Proceedings of the IEEE/CVF Conference on Computer Vision and Pattern Recognition, Nashville, TN, USA, 20–25 June 2021; pp. 10713–10722.
32. Li, C.; Li, G.; Varshney, P.K. Decentralized federated learning via mutual knowledge transfer. *IEEE Internet Things J.* **2021**, *9*, 1136–1147. [CrossRef]

Article

Siamese Neural Pointnet: 3D Face Verification under Pose Interference and Partial Occlusion

Qi Wang [1,*], Wei-Zhong Qian [1], Hang Lei [1] and Lu Chen [2]

[1] School of Information and Software Engineering, University of Electronic Science and Technology of China, Chengdu 610054, China
[2] School of Aeronautics and Astronautics, University of Electronic Science and Technology of China, Chengdu 611731, China
* Correspondence: wangqi1210@std.uestc.edu.cn; Tel.: +86-155-2837-0162

Abstract: Face verification based on ordinary 2D RGB images has been widely used in daily life. However, the quality of ordinary 2D RGB images is limited by illumination, and they lack stereoscopic features, which makes it difficult to apply them in poor lighting conditions and means they are susceptible to interference from head pose and partial occlusions. Considering point clouds are not affected by illumination and can easily represent geometric information, this paper constructs a novel Siamese network for 3D face verification based on Pointnet. In order to reduce the influence of the self-generated point clouds, the chamfer distance is adopted to constrain the original point clouds and explore a new energy function to distinguish features. The experimental results with the Pandora and Curtin Faces datasets show that the accuracy of the proposed method is improved by 0.6% compared with the latest methods; in large pose interference and partial occlusion, the accuracy is improved by 4% and 5%. The results verify that our method outperforms the latest methods and can be applied to a variety of complex scenarios while maintaining real-time performance.

Keywords: chamfer distance; convolutional neural network; face verification; Siamese network

Citation: Wang, Q.; Qian, W.-Z.; Lei, H.; Chen, L. Siamese Neural Pointnet: 3D Face Verification under Pose Interference and Partial Occlusion. *Electronics* **2023**, *12*, 620. https://doi.org/10.3390/electronics12030620

Academic Editor: Giovanni Ramponi

Received: 31 December 2022
Revised: 19 January 2023
Accepted: 23 January 2023
Published: 26 January 2023

1. Introduction

Face recognition algorithms are traditionally split into two specific tasks by the computer vision community: verification and identification [1]. Different from face identification, face verification is a one-to-one comparison task; given a pair of images as input, a face verification system should predict if the input items contain faces of the same person or not [2]. The computer vision community has broadly addressed the problem in both the 2D RGB and 3D domains [3]. However, ordinary RGB cameras cannot obtain effective images in the case of a large variation of illumination. In addition, 2D RGB images lack stereo information and are more susceptible to interference from head pose and partial occlusion.

Recently, the computation of geometric descriptors of 3D shapes has played an important role in many 3D computer vision applications [4]. In general, 3D objects are mainly represented by the following four methods: mesh, voxel grid, octree, and point cloud. However, the expression of mesh is complex, the voxel grid makes the space redundant, and the octree is complicated to use. In contrast, the point cloud can be directly used to represent 3D information, and the mathematical expression is very concise. With the improvement of depth map devices, obtaining effective point clouds has become easier. Depth maps have two main advantages. Firstly, the devices are stable with illumination changes. Secondly, depth maps can be easily exploited to manage the scale of the target object in detection tasks [5]. However, compared with point clouds, depth maps have two disadvantages. First, depth maps are expressed in the form of single-channel 2D images, which cannot directly reflect the geometric characteristics of objects in a 3D space. Second, the contours of depth maps overlap with the surrounding pixels, which makes the contours unclear, and some important information will be lost. Relying on a simple coordinate transformation, depth maps can be converted into point clouds; therefore, point clouds inherit the above two advantages of depth maps and also keep clearer geometric characteristics. Furthermore, since the pioneering work of Charles et al. [6], who constructed Pointnet, which solves the sparsity and disorder of point clouds, many deep learning models have been proposed, and point clouds now have more abundant applications.

In this paper, in order to reduce interference from head pose and partial occlusion, we rely on point clouds to construct a novel Siamese network for 3D face verification. In our method, we first obtain face information from depth maps and convert it to point clouds. Secondly, we construct a Siamese network to extract features. In this step, we adopt the *farthest point sampling algorithm* to sample points and employ two set abstractions to extract local-to-global face features hierarchically. Thirdly, in order to reduce the influence of the self-generated point clouds, we employ the chamfer distance to constrain the original point clouds and design a new energy function to measure the difference between two features.

In order to verify the performance of our method, we conduct experiments on two public datasets—the Pandora dataset and the Curtin Faces dataset. We also split the Pandora dataset into groups for cross-training and testing to verify the effectiveness of our method under pose interference and partial occlusion.

The main contributions of this paper are summarized as follows:

1. We propose an end-to-end 3D face verification network, which, to the best of our knowledge, is the first attempt to construct a Siamese network with point clouds for face verification.
2. We employ the charm distance to constrain the original point clouds, which can effectively improve the accuracy, and enables our network to better cope with the interference from head pose and partial occlusion.
3. The experimental results on public datasets show that our network has good real-time performance, and the verification accuracy outperforms the latest methods, especially under pose interference and partial occlusion.

2. Related Works

In recent years, the most widely used face verification methods have mainly been based on intensity images [7]. Before neural networks became widely used for image tasks, most of the methods were based on hand-crafted features [8]. With the improvement of hardware such as GPUs, more deep learning methods in neural networks have been applied to computer vision. Benefiting from the perceptual power of deep learning, most methods outperform humans on the LFW dataset [9]. Among them, Schroff et al. [10] constructed a network, FaceNet, which takes pairs of images as inputs and introduces a triplet loss to calculate the difference between images. In [11], Phillips et al. designed a VGG-face algorithm to recognize faces as variables. Richardson et al. [12] combined CoarseNet and FineNet and introduced an end-to-end CNN framework that derives the shape in a coarse-to-fine fashion. In order to avoid noise and degradation, Deng et al. [13] explored a robust binary face descriptor, compressive binary patterns (CBP). Wu et al. [14] proposed a center invariant loss and added a penalty to the differences between each center of classes to generate a robust and discriminative face representation method. Wang et al. [15] introduced a more interpretable additive angular margin for the softmax loss in face verification and discussed the importance of feature normalization. To combat the data imbalance, Ding et al. [16] combined generative adversarial networks and a classifier network to construct a one-shot face recognition network. Likewise, in order to deal with the imbalance problem, based on margin-aware reinforcement learning, Liu et al. [17] introduced a fair loss, in which deep Q-learning is used to learn an appropriate adaptive margin for each class. Targeting racial and gender differences in face recognition, Zhu et al. [18] combined NAS technology and the reinforcement learning strategy into a face recognition task and proposed a novel deep neural architecture search network. In order to deal with low-resolution face verification, Jiao et al. [19] constructed an end-to-end low-resolution face translation and verification framework which improves the accuracy of face verification while improving the quality of face images. Recently, Lin et al. [20] proposed a novel similarity metric, called explainable cosine, which can be plugged into most of the verification models to provide meaningful explanations. Aimed at facial comparison in a forensic context, Verma et al. [21] employed an automatic approach to detect facial landmarks, and selected independent facial indices extracted from a subset of these landmarks. Cao et al. [22] introduced two descriptors and one composite operator to construct a framework named GMLM-CNN for face verification between short-wave infrared and visible light.

Compared to RGB images, depth maps lack texture detail, but they cope well with dramatic light changes. Based on the depth maps, Guido et al. [23] generated other types of pictures using a GAN network for head pose estimation. In [7,24], Ballota et al. utilized convolutional neural networks for head detection, marking the first time CNN was leveraged for head detection based on depth images. In recent years, many face verification methods based on depth maps have been proposed. Borghi et al. [3] constructed JanusNet, which is a hybrid Siamese network composed of depth and RGB images. Subsequently, Borghi et al. [2] used two fully convolutional networks to build a Siamese network, which only relies on deep images for training and testing and achieved very good results. Afterwards, Wang et al. [25] adopted a one-shot Siamese network for depth face verification which significantly improved the accuracy. In order to reduce the interference from the head pose, Zou et al. [26] projected the face features onto a 2D plane and introduced the attention mechanism to reduce interference from facial expressions. Rajagopal et al. [27] introduced a CDS feature vector and proposed three levels of networks for face expression categorization. Wang et al. [28] used $L2$ to constrain facial features and constructed an L2–Siamese network for depth face verification.

Most of the proposed related 3D methods have achieved excellent performance. In order to solve the photometric stereo for non-Lambertian surfaces and a disordered and arbitrary number of input features, Chen et al. [29] proposed a deep fully convolutional network PS-FCN to predict a normal map of the object in a fast feed-forward pass. Aiming at 3D geometry reconstruction and avoiding blurred reconstruction, Ju et al. [30] proposed a self-learning conditional network with multi-scale features for photometric stereo. Similar to depth maps, surface normal maps can also provide 3D information for relevant tasks. The pioneers Woodham et al. [31] proposed photometric stereo, which varies the direction of incident illumination between successive images while holding the viewing direction constant to recover the surface normal of each of the image points. Recently, Ju et al. [32] presented a normalized attention-weighted photometric stereo network NormAttention-PSN, which significantly improved surface orientation prediction for complicated structures.

In the field of 3D point cloud vision, building on the innovation of Pointnet, Qi et al. [6] solved the disorder and application in deep learning of point clouds; although many point cloud methods are proposed, this work only considered the global features and missed local features. Subsequently, Charles et al. [33] improved Pointnet by extracting local features from a group of Pointnets. In order to solve the application of point clouds in convolutional neural networks, Li et al. [34] proposed Pointcnn to learn X-transformation, which is the generalization of typical CNNs into learning features. Guerrero et al. [35] changed the first transformation of Pointnet and proposed PCPNet, which avoids the quality defects of the point clouds and reduces the interference of invalid points. The above works [33–35] optimize the feature extraction of point clouds and maintain good real-time performance, but they did not consider the spatial geometric characteristics of original points. In PPFNet, Deng et al. [36] applied a four-dimensional feature descriptor to describe the geometric characteristics of original point pairs. Zhou et al. [4] constructed a Siamese point network for feature extraction and measured the difference between the original point clouds. Both [4] and [36] considered the spatial geometric characteristics of the original point clouds, but they adopted a matrix for registration, which is computationally intensive and time-consuming.

As mentioned above, many point-cloud-based networks have been proposed, and they have their own advantages. Due to these advantages, many face analysis methods are proposed. Recently, Xiao et al. [37] constructed a classification network to guide the regression process of Pointnet++ for head pose estimation. Ma et al. [38] combined a deep regression forest and Pointnet for predicting head pose. Cao et al. [39] proposed a local descriptor to describe the projection of point clouds for 3D face recognition.

Face verification is a one-to-one comparison task taking into account both the effectiveness of feature extraction and real-time performance. Based on Pointnet, we construct a novel Siamese network and adopt the chamfer distance to constrain the geometric characteristics of the original point clouds.

3. Methods

For face verification with 3D point clouds, we convert the depth maps into point clouds, construct a Siamese network to extract the features of a pair of faces, and employ the chamfer distance to design the energy function to predict the similarity between the two faces.

3.1. Point Cloud Extraction

As described above, we transform depth maps to point clouds. This means converting depth data from an image coordinate system to the world coordinate system. Each pixel of a depth map represents the distance from the target to the sensor (in mm). In this step, we assume that the whole head information and head center (x', y') with its depth value D_p has been obtained (head detection and center localization are not the focus of our work). Firstly, removing the background, we set the pixel value, which is greater than $D_p + L$ to 0, where L is the general amount of space for a real head [24] (300 mm in our method). Secondly, according to Equation (1), we convert depth data to point clouds.

$$\begin{bmatrix} x \\ y \\ z \end{bmatrix} = D_p \begin{bmatrix} \frac{1}{f_x} & 0 & 0 \\ 0 & \frac{1}{f_y} & 0 \\ 0 & 0 & 1 \end{bmatrix} \begin{bmatrix} x^i \\ y^i \\ 1 \end{bmatrix} \tag{1}$$

where (x, y, z) is the point location in the world coordinate system, and (x^i, y^i) is the pixel position in the image. f_x and f_y are camera internal parameters which represent the horizontal and the vertical focal length, respectively. As shown in Equation (1), a point cloud is a list of points (represent in position (x, y, z)) in a 3D space.

3.2. Siamese Neural Network

Siamese neural networks were first proposed and applied to the signature and verification certificate tasks by Bromley et al. [40]. A Siamese network consists of two shared weight networks which accept distinct inputs and are joined by an energy function at the end. This energy function computes a metric between two high-level features. The parameters between the twin networks are tied, which can guarantee network consistency, and ensures that a pair of very similar features are not mapped to very different locations in feature space by the respective networks [41].

The structure of the Siamese neural network is shown in Figure 1. The input layer sends an object to the hidden layer which extracts object features. The ends of two networks are connected by an energy function in the distance layer which computes certain metrics between features based on task requirements. Output layers predict the result of the Siamese network.

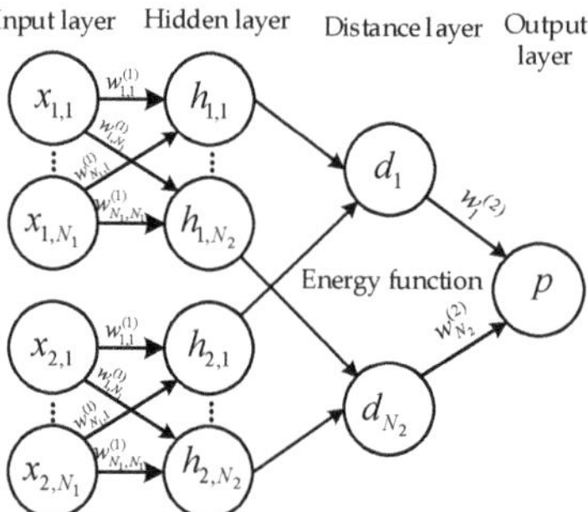

Figure 1. The framework of Siamese neural network. A pair of parallel networks extract features separately, and an energy function connects them to measure a certain relationship between two features.

3.3. Feature Extraction

As mentioned above, the essence of a point cloud is a list of points ($n \times 3$ matrix, where n is the number of points, and 3 represents (x, y, z) in the world coordinates). Geometrically, the order of points does not affect its representation of the overall shape in 3D space. As shown in Figure 2, the same point clouds can be represented by completely different matrices. In order to deal with the disorder of point clouds and their application in deep learning, Chen et al. [6], based on the idea of symmetric function, constructed a deep learning model called Pointnet. The idea is to approximate a general function by applying a symmetric function:

$$f(x_1, x_2, x_3, \cdots, x_n) \approx \gamma \circ g(h(x_1), h(x_2), h(x_3), \cdots, h(x_n)) \tag{2}$$

where f is a general function, which maps all independent variables ($x_1, x_2, x_3, \cdots, x_n$) to a new feature space $\mathbb{R}^m$. h is another general function used to map each independent variable x_i to feature space $\mathbb{R}^l$, and g is a symmetric function (the input order does not affect the result). r is also a general function which maps the result of function g to the specific feature space $\mathbb{R}^m$. According to Equation (2), the left part of the equation can be approximated by the right part. As described above, we adopt Pointnet to approximate the right part.

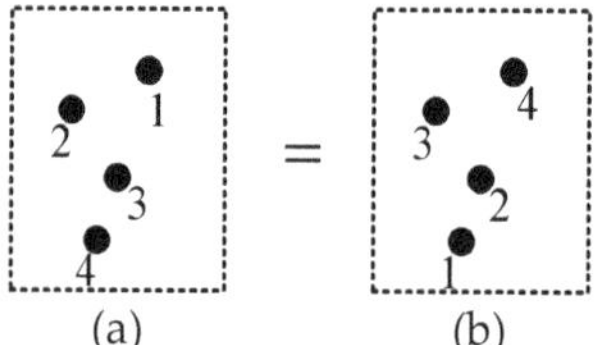

Point Cloud Representation

$$\begin{vmatrix} x_{a1} & y_{a1} & z_{a1} \\ x_{a2} & y_{a2} & z_{a2} \\ x_{a3} & y_{a3} & z_{a3} \\ x_{a4} & y_{a4} & z_{a4} \end{vmatrix} \neq \begin{vmatrix} x_{b1} & y_{b1} & z_{b1} \\ x_{b2} & y_{b2} & z_{b2} \\ x_{b3} & y_{b3} & z_{b3} \\ x_{b4} & y_{b4} & z_{b4} \end{vmatrix}$$

Figure 2. Disorder of point cloud. The point clouds in (**a**,**b**) have the same geometry, but the order of the points is different, and the expressions are also different.

The structure of the network is shown in Figure 3.

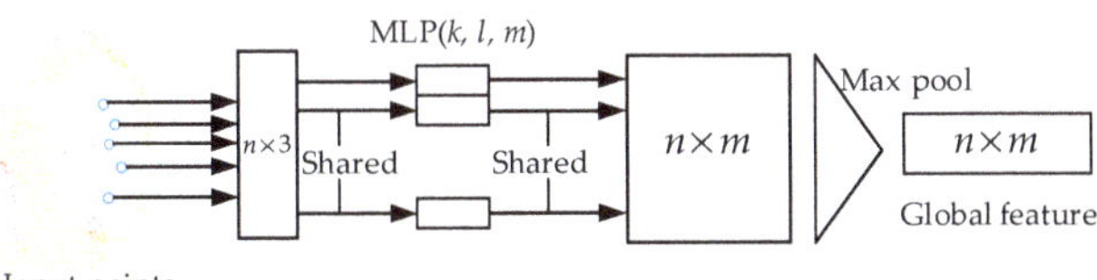

Figure 3. The structure of Pointnet. Each point is mapped into the feature space by three convolutional layers, and a max pooling layer acts as the symmetric function to extract the global feature of the point cloud.

As shown in Figure 3, where n is the total number of points. We adopt three convolutional layers as the function h in Equation (2) (convolution kernel is 1×1, the filter is k, l, m, respectively), which is used to map the feature of each point to the feature space $\mathbb{R}^3 \rightarrow \mathbb{R}^k \rightarrow \mathbb{R}^l \rightarrow \mathbb{R}^m$. Finally, according to [6,29,30], a max pooling layer is adopted as the symmetric function, g, which can solve the disorder of the features and extract the global feature in $\mathbb{R}^m$.

As described above, only the global feature of the object can be obtained by Pointnet. There is no step to extract local features. Because point clouds lack a detailed texture, only global features lead to a limited generalization ability of the network, especially in complex scenarios. In order to improve the cognitive ability of the network, according to [33], we adopt the set abstraction to extract the local-to-global features. The structure of a set abstraction is shown in Figure 4. A set abstraction consists of the following three parts: sampling, grouping, and local feature extraction. For a point cloud $\{p_1, p_2, p_3, \ldots, p_N\}$ (the feature dimension of these points is C), in order to sample uniformly, we first use the *farthest point sampling* method to sample the points. In this step, we arbitrarily select a point p_i as the starting point and find the farthest point p_{i1} from the point cloud, and put p_{i1} into a new point set. Next, we regard p_{i1} as a new starting point and find the farthest point in the rest of the points. We iterate the above steps until we obtain a new point set $\{p_{11}, p_{12}, p_{13}, \ldots, p_{1N_1}\}$ with a fixed number N_1. Compared with random sampling, *farthest point sampling* can cover the whole point set [42].

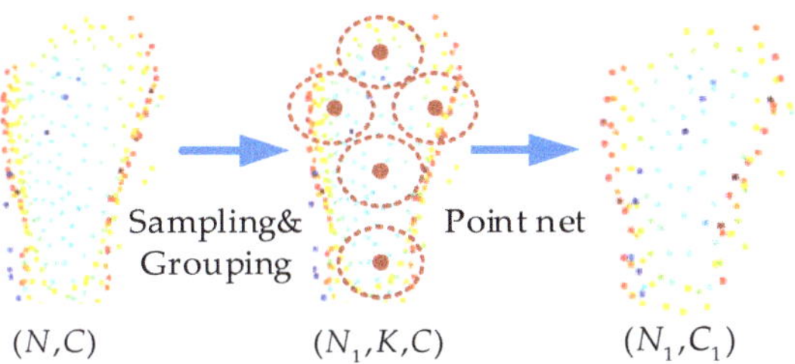

Figure 4. The structure of a set abstraction. The *farthest point sampling* method is used to sample feature points and group them, and a set of Pointnets is used to extract the features of each local region.

Secondly, we group these points in $\{p_{11}, p_{12}, p_{13}, \ldots, p_{1N_1}\}$; in this step, we regard each point as the center of a sphere with radius K (our network contains two set abstractions with a K of 0.2 and 0.4, respectively), and points in the same sphere are grouped into one group. After this step, we obtain a new grouping set $\{g_1, g_2, g_3, \cdots, g_{N_1}\}$, and each group represents a local region of its own central point.

Finally, we use a Pointnet, as shown in Figure 3, to extract features of each group, and obtain a set of local features $\{f_1, f_2, f_3, \ldots, f_{N_1}\}$ (the dimension of these features is C_1). We regard $\{f_1, f_2, f_3, \ldots, f_{N_1}\}$ as a new point set for the abstraction of the next step.

The process of our method is shown in Figure 5, and we use a pair of completely parallel branches to extract head features separately. Each branch contains two set abstractions. The first set abstraction adopts Pointnet1 to extract local features, which has three convolutional layers, and the filter of each layer is 64, 64, 128, respectively. The second set abstraction adopts Pointnet2 to extract local features, which also has three convolutional layers, and the filter of each layer is 128, 128, 256, respectively. After the second set abstraction, each branch employs Pointnet3 (The filters of three convolutional layers are 256, 512, 1024) to extract the local-to-global features of the object.

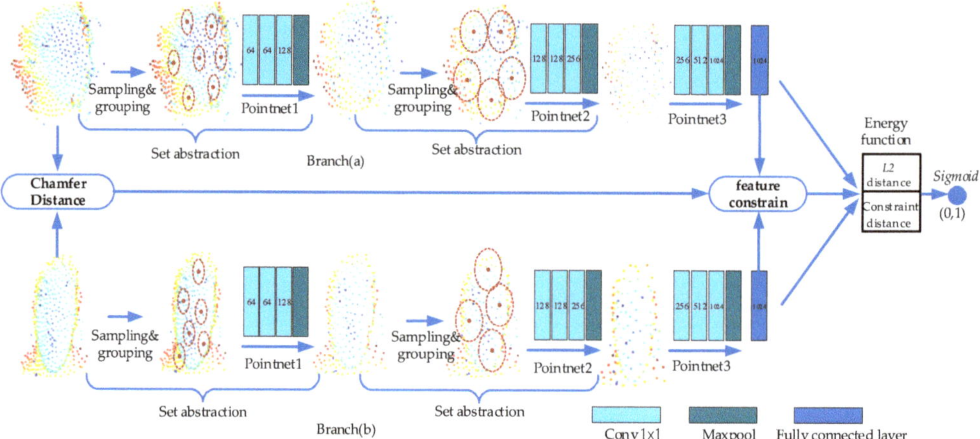

Figure 5. The whole structure of our method. Two identical branches are adopted to extract features separately, and the chamfer distance is used to constrain original point clouds. At the end of the network, a novel energy function is introduced to distinguish the similarity of objects.

In practice, although the *furthest point sampling* method samples uniformly, due to the unevenness of the point cloud, some groups have fewer points. During the grouping process, the density and sparseness of points will affect the feature extraction. Therefore, we use *multi-resolution grouping* to obtain the features of each layer.

As shown in Figure 6, the features of a set abstraction are composed of two vectors. The left vector is the features of each group in this set abstraction. The right vector is the features of the original points of the previous layer for groups with sparse points which makes the first vector less reliable. Therefore, the second vector learns a higher weight during training. On the other hand, for groups with dense points, the networks obtain finer feature information, and the first vector learns a higher weight. In the training process, the network adjusts the weights in the above way to find the optimal weights for different point densities [42].

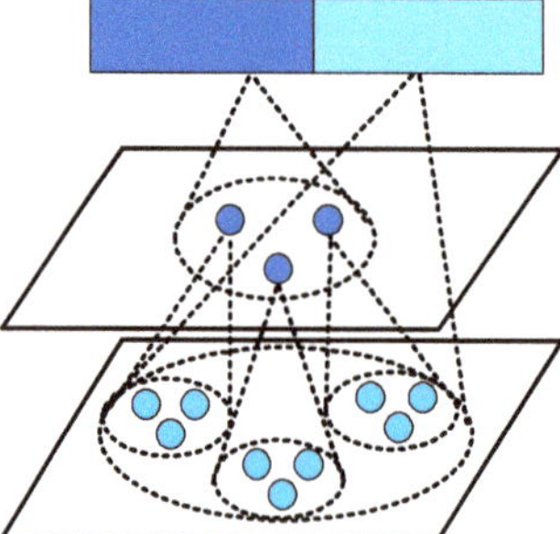

Figure 6. Hierarchical feature extraction schematic.

3.4. Feature Constraint

Siamese networks measure the difference between high-dimensional features but lack a description of the difference between original point clouds. The chamfer distance can represent the original differences of point clouds and is widely used in point cloud reconstruction [4]. In order to reduce the influence of the self-generated point clouds, we adopt chamfer distance (CD) to constrain their features. It is defined as follows:

$$d_{CD}(S_1, S_2) = \sum_{p \in S_1} \min_{q \in S_2} d(p, q) + \sum_{p \in S_2} \min_{q \in S_1} d(p, q) \tag{3}$$

where $S_1, S_2 \in \mathbb{R}^3$ represent two sets of point clouds. $d(p, q)$ measures the $L2$ distance between points p and q. The first term represents the sum of the minimum distances from any points in S_1 to S_2, whereas the second term represents the sum of the minimum distances from any points in S_2 to S_1. If the chamfer distance is greater, two sets of point clouds are more distinct, and vice versa.

As mentioned above, the ends of the Siamese network are connected by an energy function which measures the difference between a pair of objects. Based on the chamfer distance, we design a new energy function to measure features, which is as follows:

$$E_{constrain} = \begin{cases} D^2(f_i, f_j), (i, j) \in C \\ \max(0, m - D(f_i, f_j)), (i, j) \in \tilde{C} \end{cases} \tag{4}$$

where C is a set of correspondence point clouds, which has a low chamfer distance (the threshold is 0.02 in our method). f_i is the feature extracted by our network, and D is the Euclidean distance. m is the margin value (the threshold is 0.7 in our method). The first term constrains the same objects closer in the feature space, and the second term leads different objects to have a large distance (greater than the margin value).

In face verification tasks, the $L2$ distance is commonly used to measure the difference between two features. The whole energy function of our network is shown below:

$$E_{total} = \lambda E_{constrain} + (1 - \lambda) E_{L_2} \tag{5}$$

where E_{L_2} is the $L2$ distance between two objects. λ is the ratio of the contribution of the $E_{constrain}$.

According to Equation (6), we adopt *sigmoid* to map the value of energy function to probability distribution between $(0, 1)$.

$$S(E_{total}) = \frac{1}{1 + e^{-E_{total}}} \tag{6}$$

Face verification can be regarded as a classification task; our network uses cross-entropy as the loss function:

$$H(p, q) = -\sum_{x} p(x) \log q(x) \tag{7}$$

where $p(x)$ represents the ground truth, when $p(x)$ is 1, the pair of objects belong to the same object, and when $p(x)$ is 0, the pair of objects belong to different objects. The $q(x)$ represents the predicted value. The whole structure of our network is shown in Figure 5; the chamfer distance is used to constrain the features of original point clouds, and a new energy function is used to measure the difference between objects.

In the selection of hyperparameters, the batch size is 64, the learning rate is 0.001, the decay rate is 0.99, and the decay step size is 500.

4. Experiments

In this section, we first introduce two public datasets, the Pandora dataset [23] and the Curtin Faces dataset [43], for our experiments. Secondly, we conduct an experiment to investigate the similarity threshold of our Siamese network, which determines whether a pair of objects belong to the same object or not. Thirdly, we conduct ablation experiments to verify the effect of the set abstractions and chamfer distance and analyze the parameter λ in Equation (5). Fourthly, we explore the influence of the input numbers of points. Finally, we conduct comparison experiments with current methods and divide the Pandora dataset into a series of subsets to validate the performance of our network under pose interference and partial occlusion.

4.1. Dataset

Pandora dataset: Borghi et al. [23] created this dataset for head and shoulder pose estimation. This dataset collected upper body information of 22 subjects (10 males and 12 females) with Microsoft Kinect One. There are 110 sequences with over 250,000 images. Each depth map corresponds to an RGB image and has the ground truth of head center and pose angles. Interference is generated by glasses, scarves, mobile phones, and various postures.

Curtin Faces dataset: Li et al. [43] collected this dataset with the Microsoft Kinect Sensor. This dataset is created specifically for face verification and contains 5000 samples from 52 subjects. Each subject has 97 images, which contain varying head poses, facial expressions, occlusion, and illumination.

In our experiments, we only focus on face verification and not face detection and head center localization; we directly use ground truth to obtain face information.

4.2. Similarity Threshold

Ideally, in our method, the similarity threshold is close to 1 for the same objects and close to 0 for different objects, but due to the influence of head pose and partial occlusion, etc., the network cannot reach the optimal condition. As a result, the value of the similarity threshold directly affects the result. We conduct an experiment with the Pandora dataset to determine the similarity threshold. In order to reflect the initial performance of our network, we remove the feature constraint part and only use the *L2* distance as the energy function to investigate the similarity threshold. The results are reported in Table 1.

Table 1. Similarity threshold of our network for face verification.

Threshold	Acc	Threshold	Acc
0.1	82.62%	**0.6**	**85.52%**
0.2	83.97%	0.7	84.33%
0.3	84.85%	0.8	78.56%
0.4	85.14%	0.9	71.78%
0.5	84.79%	\	\

As shown in Table 1, when the threshold is selected as 0.1, our network has good performance, but it is difficult to distinguish between different objects with a similar appearance, and when the threshold is 0.9, a wrong prediction is often taken from the same objects. When the threshold is selected as 0.6, our network has the best performance because the network has good compatibility with the entire dataset under this setting and can minimize the influence of posture and partial occlusion. According to Table 1, we set the threshold to 0.6 for the subsequent experiments.

4.3. Ablation Experiments

As described above, we adopt set abstractions to extract local-to-global features and use the chamfer distance to constrain original point clouds; in this section, we conduct ablation experiments to verify the performance of our method.

In the first step, we conduct experiments on the Pandora dataset to verify the effect of the set abstractions. Firstly, we only employ one Pointnet to extract global features for face verification. Secondly, we adopt one set abstraction to extract features, and finally, we use two set abstractions to extract local-to-global features hierarchically. The results are reported in Table 2.

Table 2. Performance evaluation with different structures of a Siamese network on the Pandora dataset.

Method	Pointnet	One Set Abstraction	Two Set Abstractions
Acc	72.9%	80.1%	85.2%
fps	650	355	215

According to Table 2, the set abstraction can significantly improve the accuracy of our network. This is because multi-layer feature extraction can better describe the details of the objects, but it consumes more time. Considering both the accuracy and real-time performance, we use two set abstractions for feature extraction (215 fps can meet the real-time requirements of most tasks).

As shown in Equation (5), the parameter λ determines the contribution of the constraint function. In order to confirm λ, we can keep λ as a fixed value throughout training or let the network learn the parameter. The second way is elegant and always improves the regular loss [44], but the parameter learned by the network provides it with greater freedom to fit the easy samples, which results in a relaxed chamfer distance constraint. Therefore, we fix the parameter for the ablation experiment on the Pandora dataset to investigate the λ.

The results are reported in Table 3. When $\lambda = 0$, the feature constraint function is not utilized, and with the constraint of the chamfer distance, the performance of our network improved considerably. However, the accuracy decreases when $\lambda > 0.4$ because the chamfer distance mainly acts as a feature constraint; when λ is too large, the metric of the energy function is reduced, which is not conducive to distinguishing facial features. When λ is too small, the constraint of the chamfer distance is limited, and smaller constraint ratios lead to a limited improvement of the network's performance. According to Table 3, the network performs best when $\lambda = 0.4$ because the feature constraint in Equation (5) reaches an equilibrium value under this setting.

Table 3. Performance evaluation of face verification with different λ on the Pandora dataset.

λ	Acc	λ	Acc
0	85.52%	0.6	86.01%
0.1	86.69%	0.7	81.21%
0.2	87.13%	0.8	75.79%
0.3	88.95%	0.9	74.32%
0.4	**90.4%**	1.0	74.03%
0.5	88.04%	\	\

4.4. Point Number for Network Performance

Point clouds represent the geometric shape of an object in a 3D space. As shown in Figure 7, the number of points determines the detailed information of the shape; when the number of points is higher, the geometric texture is clearer. According to the sampling process of our network, the number of input points affects the efficiency of our network. In this section, we investigate the effect of the input number of points. Table 4 lists the experimental results on the Pandora dataset with different input numbers of points, of 1024, 2048, and 4096. As shown in Table 4, when the input number of points is 1024, our network has the lowest accuracy but the fastest speed. When the input number of points is 4096, because more detailed information about the faces is presented, the network has the highest accuracy, but this is more time-consuming; however, in the above three cases, the accuracy is relatively close. This is because even in the case of 1024 points, the geometric shapes of the objects can also be well characterized.

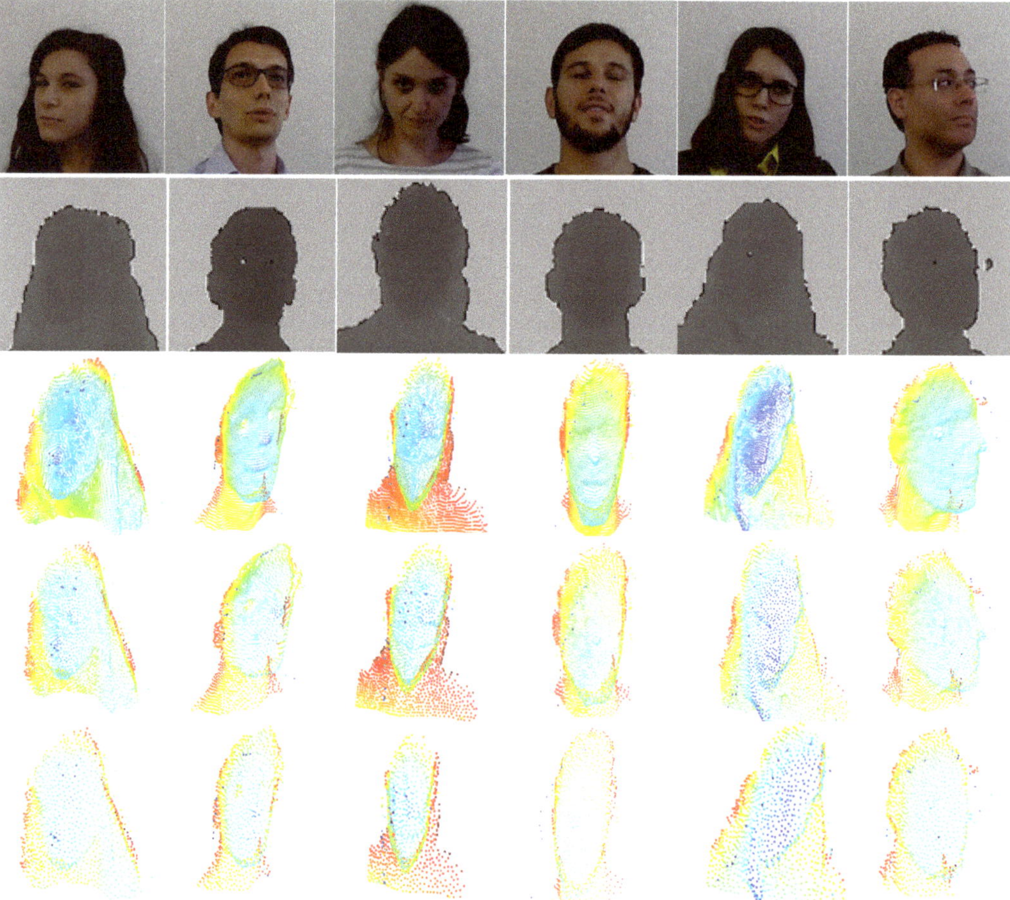

Figure 7. The first row shows the corresponding RGB images, and the second row shows the depth images. The third to fifth rows are point clouds transformed from the depth images, and they contain 4096, 2048, and 1024 points, respectively.

Table 4. Performance evaluation with different input numbers of points.

Input Number	Acc	fps
1024	88.2	420
2048	89.7	305
4096	**90.4**	**225**

As described above, the network performs best when the similarity threshold is 0.6, λ is 0.4, and the input number of points is 4096. We use our best result for the following comparison experiments. Figure 8 shows the loss and accuracy of our network during training under this setting.

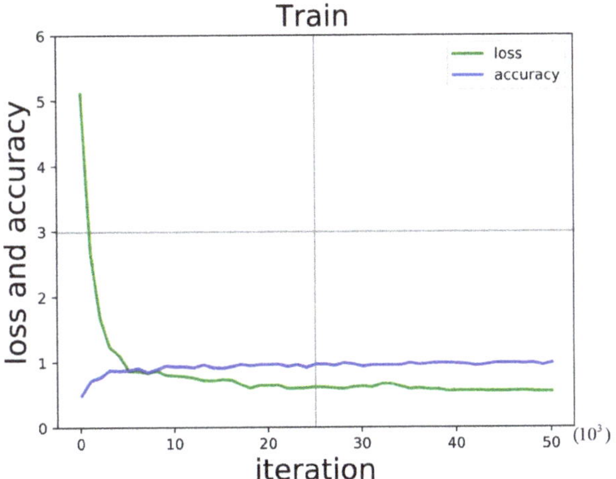

Figure 8. Curve of loss and accuracy when the threshold is 0.6, $\lambda = 0.4$, and the point number is 4096.

4.5. Comparison Experiments

The Pandora dataset and the Curtin Faces dataset contain two types of data, namely, RGB images and depth maps. The sensors of depth maps do not depend on lighting conditions but lack detailed contours compared with RGB images. The point clouds in our method are derived from depth maps; therefore, for a fair comparison, we compare this with other methods which only rely on depth maps. The experimental results are reported in Tables 5 and 6.

Table 5. Comparison of results achieved by different methods on the Pandora dataset.

Method	Input Images			Model	GPU:1080ti
	Train	Test	Input Size	Acc	fps
JanusNet [3]	RGB+Depth	Depth	100×100	81.4%	202
Siamese [2]	Depth	Depth	variable	85.3%	604
One-shot [25]	Depth	Depth	variable	89.2%	43
L2-Sia [28]	Depth	Depth	100×100	89.9%	148
Ours	**Depth**	**Depth**	variable	**90.5%**	225

Table 6. Comparison of results on the Curtin Faces dataset.

Methods	Siamese [2]	Ours
Acc	86%	89%

In the same experimental environment, comparison results with the current state-of-the-art methods on the Pandora dataset are reported in Table 5. The fully convolutional network method [2] has the fastest speed, but our accuracy improved by 5.2%. The method detailed in [28] explores an *L2-constraint* on pose features; although our accuracy is very close to the results of this experiment, with only a 0.6% increase, the efficiency of our method is significantly improved.

Table 6 lists the comparison results for the Curtin Faces dataset. We follow the evaluation procedure described in [2] with only 18 images per subject for the training phases and our accuracy increased by 3% under the same experimental conditions (this dataset is specifically used for face identification tasks which are rarely used for face verification and lacks other reference results).

According to Tables 5 and 6, our method achieves the highest accuracy and also has good real-time performance.

In order to further verify the performance of our network under the interference of head pose, according to [2,3,25,28], the Pandora dataset is split as follows:

$$A_1 = \left\{ s_{\rho\theta\sigma} \middle| \forall \gamma \in \{\rho, \theta, \sigma\} : -10^\circ \leq \gamma \leq 10^\circ \right\} \tag{8}$$

$$A_2 = \left\{ s_{\rho\theta\sigma} \middle| \exists \gamma \in \{\rho, \theta, \sigma\} : \gamma < -10^\circ \cup \gamma > 10^\circ \right\} \tag{9}$$

$$A_3 = \left\{ s_{\rho\theta\sigma} \middle| \forall \gamma \in \{\rho, \theta, \sigma\} : \gamma < -10^\circ \cup \gamma > 10^\circ \right\} \tag{10}$$

where ρ, θ, and σ are Euler angles, representing the yaw, pitch, and roll angles of the head pose. Figure 9 shows examples of group A_1, A_2 and A_3. In group A_1, all pose angles are within 10°, and less interference can be seen for the head pose. In group A_2, there exists at least one pose angle greater than 10° which has a little interference from the head pose, whereas in group A_3, three pose angles are greater than 10°, and head pose interferes the most. After the Pandora dataset is split, cross-training and testing are performed. The results are reported in Table 7. When A_1 is adopted as the training sequence, all methods achieve good results because the training samples are least disturbed by head pose. Our method achieves 91% accuracy. When A_3 is adopted as the training sequence, the samples are most affected by head pose, compared with the method in [28]; even in the A_3 testing sequence with the largest pose interference in both training and testing, our accuracy improved by 4%. When using the $\{A_1, A_2\}$ sequence for training, our network achieves the best results due to more abundant training samples.

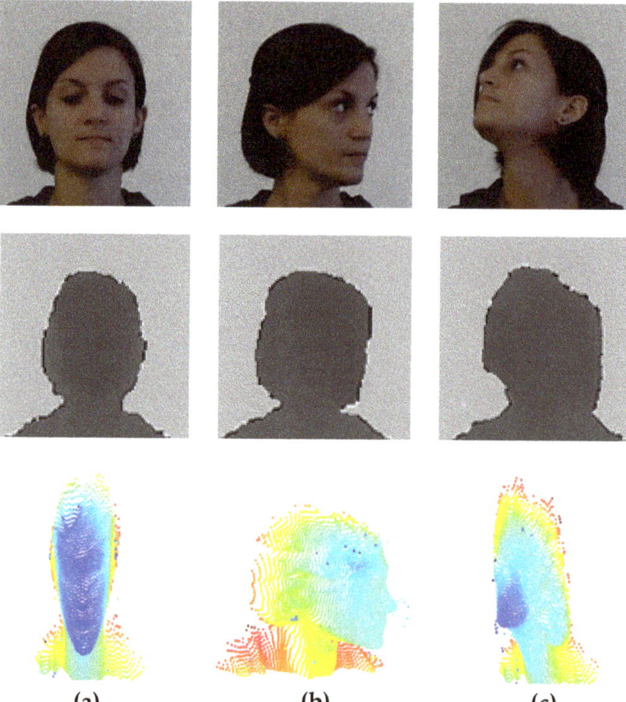

(a)   **(b)**   **(c)**

Figure 9. Example of Pandora split. Where (**a**) represents the A_1 sequence, with the least head pose interference; (**b**) represents the A_2 sequence with a little head pose interference; and (**c**) represents the A_3 sequence. Tree pose angles are greater than 10°.

Table 7. Comparison of results achieved by different methods on the dataset splits according to head pose.

Train	One-Shot [25]				L2-Sia [28]				Our			
						Test						
	A_1	A_2	A_3	$\{A_1,A_2\}$	A_1	A_2	A_3	$\{A_1,A_2\}$	A_1	A_2	A_3	$\{A_1,A_2\}$
A_1	0.90	0.81	0.78	0.82	0.90	0.83	0.78	0.82	**0.91**	**0.83**	**0.80**	**0.84**
A_2	0.91	0.88	0.87	0.90	0.90	0.88	0.87	0.89	**0.90**	**0.86**	**0.87**	**0.90**
A_3	0.81	0.77	0.67	0.73	0.82	0.78	0.73	0.75	**0.84**	**0.78**	**0.77**	**0.79**
$\{A_1, A_2\}$	0.90	0.84	0.85	0.89	0.91	0.88	0.87	0.90	**0.92**	**0.87**	**0.89**	**0.90**

According to Table 7, regardless of which sequence is chosen for training, our accuracy outperforms other methods, which proves that our method is more robust against pose interference.

In order to verify the performance of the network under movements and partial occlusion, the dataset is divided into five subsets S_1, S_2, S_3, S_4, and S_5. As shown in Figure 10, S_1, S_2, and S_3 only have limited movement (least pose interference from head and shoulder). S_4, S_5 contain complex and free movements, and the angles of the head and shoulder mainly vary one at a time and also contain partial occlusions. According to the methods of [2,3], the above five subsets are divided into three groups, where $G_1 = \{S_1, S_2, S_3\}$, $G_2 = \{S_4, S_5\}$, and $G_1 = \{S_1, S_2, S_3, S_4, S_5\}$; cross-training and testing are then performed. The results are reported in Table 8. When G_1 is used for both training and testing, all methods achieve good results due to the least interference from movements and partial occlusions; however, when G_2 is used for testing, due to the lack of corresponding training samples, the accuracy decreases. However, our accuracy is 83%, which also increased by 5% compared to other methods. When G_3 is used as the training sequence, the training samples are more abundant and include more common and complex samples; rich samples can effectively improve the generalization ability of our network and achieve the best results under all testing sequences.

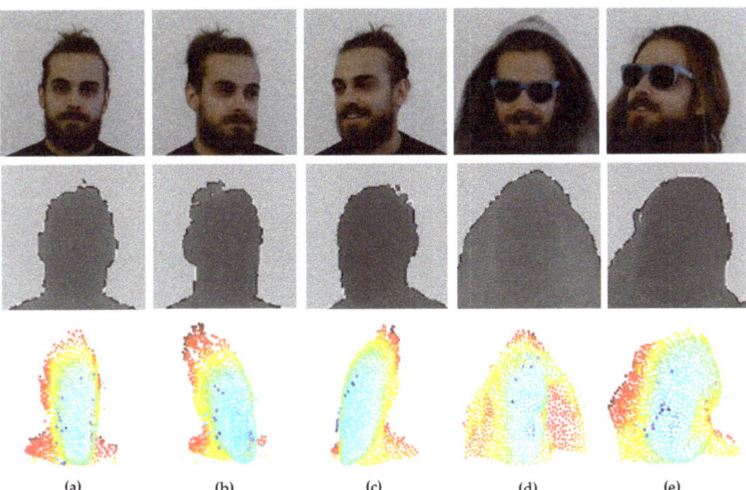

(a) (b) (c) (d) (e)

Figure 10. Example of Pandora split, where (**a**–**e**) represent the sequence S_1, S_2, S_3, S_4, and S_5, respectively, where the subset (**a**–**c**) contains constrained movements and subsets (**d**,**e**) contain complex movements and occlusions.

Table 8. Comparison of results achieved by different methods on the dataset splits according to head and shoulder movements and partial occlusions.

Train	Test								
	JanusNet [3]			Siamese [2]			Our Method		
	G_1	G_2	G_3	G_1	G_2	G_3	G_1	G_2	G_3
G_1	0.84	0.75	0.77	0.89	0.78	0.82	**0.90**	**0.83**	**0.84**
G_2	0.72	0.71	0.74	0.87	0.80	0.83	**0.87**	**0.84**	**0.86**
G_3	0.80	0.73	0.76	0.90	0.83	0.85	**0.91**	**0.87**	**0.88**

As shown in Table 8, under all the training and testing sequences, our network obtained better results than other methods, which proves that our network can cope well with the interference of movements and partial occlusions.

Combining the results of Tables 7 and 8, it is noticeable that our network can effectively solve face verification in the case of pose interference, movements, and partial occlusions, and obtain higher accuracy than other methods. Our experiments are implemented on a desktop computer with the Ubuntu16.04 operating system; the CPU is an Intel Core i7 (3.40GHz), and the GPU is an NVIDIA GTX1080ti

5. Conclusions

In this study, a novel Siamese network was developed for 3D face verification which employs two shared weight branches to extract features separately and calculate the similarity. For each branch, two set abstractions are adopted to group local regions and extract local-to-global features hierarchically. In order to reduce the influence of the self-generated point clouds, the chamfer distance is introduced to constrain the original point clouds and design a new energy function to distinguish features. The experimental results prove the effectiveness of the set abstraction and the chamfer distance for feature extraction. Comparison experiments on public datasets show that under large pose interference and partial occlusion, the accuracy is improved by 4% and 5%, respectively, and the whole accuracy also outperforms other methods. However, the network performs transformations from depth images and adopts a multi-layer structure to extract features which would lead to extra computational costs. In the case of large pose interference and partial occlusions, the accuracy is still not sufficient. In our future work, we will further optimize the network to improve efficiency and explore new algorithms to improve accuracy in more complex situations.

Author Contributions: Conceptualization, Q.W.; data curation, Q.W.; formal analysis, Q.W and W.-Z.Q.; investigation, W.-Z.Q.; methodology, Q.W.; project administration, H.L.; resources, Q.W. and L.C.; software, Q.W.; supervision, H.L.; visualization, Q.W. and L.C; Writing—Original draft, Q.W.; Writing—Review and editing, Q.W. All authors have read and agreed to the published version of the manuscript.

Funding: This research is funded by The National Natural Science Foundation of China (61802052).

Conflicts of Interest: The authors declare no conflict of interest.

References

1. Masi, I.; Wu, Y.; Hassner, T.; Natarajan, P. Deep Face Recognition: A survey. In Proceedings of the 2018 31st SIBGRAPI Conference on Graphics, Patterns and Images (SIBGRAPI), Parana, Brazil, 29 October–1 November 2018; pp. 471–478.
2. Borghi, G.; Pini, S.; Vezzani, R.; Cucchiara, R. Driver face verification with depth maps. *Sensors* **2019**, *19*, 3361. [CrossRef] [PubMed]
3. Borghi, G.; Pini, S.; Grazioli, F.; Vezzani, R.; Cucchiara, R. Face Verification from Depth Using Privileged Information. In Proceedings of the BMVC 2018 - 29th British Machine Vision Conference, Newcastle, Britain, 2–6 September 2018; p. 303.
4. Zhou, J.; Wang, M.J.; Mao, W.D.; Gong, M.L.; Liu, X.P. SiamesePointNet: A Siamese Point Network Architecture for Learning 3D Shape Descriptor. In *Computer Graphics Forum*; Wiley: Hoboken, NJ, USA, 2020; Volume 39, pp. 309–321.

5. Wang, Q.; Lei, H.; Ma, X.; Xiao, S.; Wang, X. CNN Network for Head Detection with Depth Images in cyber-physical systems. In Proceedings of the 2020 International Conferences on Internet of Things (iThings) and IEEE Green Computing and Communications (GreenCom) and IEEE Cyber, Physical and Social Computing (CPSCom) and IEEE Smart Data (SmartData) and IEEE Congress on Cybermatics (Cybermatics), Rhodes, Greece, 2–6 November 2020; pp. 544–549.
6. Qi, C.R.; Su, H.; Mo, K.; Guibas, L.J. Pointnet: Deep Learning on Point Sets for 3D Classification and Segmentation. In Proceedings of the IEEE Conference on Computer Vision and Pattern Recognition, Honolulu, HI, USA, 21–26 July 2017; pp. 652–660.
7. Ballotta, D.; Borghi, G.; Vezzani, R.; Cucchiara, R. Fully Convolutional Network for Head Detection with Depth Images. In Proceedings of the 2018 24th International Conference on Pattern Recognition (ICPR), Beijing, China, 20–24 August 2018; pp. 752–757.
8. Anith, S.; Vaithiyanathan, D.; Seshasayanan, R. Face Recognition System Based on Feature Extraction. In Proceedings of the 2013 International Conference on Information Communication and Embedded Systems (ICICES), Chennai, India, 21–22 February 2013; pp. 660–664.
9. Huang, G.B.; Mattar, M.; Berg, T.; Learned-Miller, E. Labeled Faces in the Wild: A Database Forstudying Face Recognition in Unconstrained Environments. In Proceedings of the Workshop on Faces in 'Real-Life' Images: Detection, Alignment, and Recognition, Marseille, France, 16–18 October 2008.
10. Schroff, F.; Kalenichenko, D.; Philbin, J. Facenet: A Unified Embedding for Face Recognition and Clustering. In Proceedings of the IEEE conference on computer vision and pattern recognition, Boston, MA, USA, 7–12 June 2015; pp. 815–823.
11. Phillips, P.J. A Cross Benchmark Assessment of a Deep Convolutional Neural Network for Face Recognition. In Proceedings of the 2017 12th IEEE International Conference on Automatic Face & Gesture Recognition (FG 2017), Washington, DC, USA, 30 May–3 June 2017; pp. 705–710.
12. Richardson, E.; Sela, M.; Or-El, R.; Kimmel, R. Learning Detailed Face Reconstruction from a Single Image. In Proceedings of the IEEE Conference on Computer Vision and Pattern Recognition, Honolulu, HI, USA, 21–26 July 2017; pp. 1259–1268.
13. Deng, W.; Hu, J.; Guo, J. Compressive binary patterns: Designing a robust binary face descriptor with random-field eigenfilters. *IEEE Trans. Pattern Anal. Mach. Intell.* **2018**, *41*, 758–767. [CrossRef] [PubMed]
14. Wu, Y.; Liu, H.; Li, J.; Fu, Y. Improving face representation learning with center invariant loss. *Image Vis. Comput.* **2018**, *79*, 123–132. [CrossRef]
15. Wang, F.; Cheng, J.; Liu, W.; Liu, H. Additive margin softmax for face verification. *IEEE Signal Process. Lett.* **2018**, *25*, 926–930. [CrossRef]
16. Ding, Z.; Guo, Y.; Zhang, L.; Fu, Y. One-Shot Face Recognition via Generative Learning. In Proceedings of the 2018 13th IEEE International Conference on Automatic Face & Gesture Recognition (FG 2018), Xi'an, China, 15–19 May 2018; pp. 1–7.
17. Liu, B.; Deng, W.; Zhong, Y.; Wang, M.; Hu, J.; Tao, X.; Huang, Y. Fair Loss: Margin-Aware Reinforcement Learning for Deep Face Recognition. In Proceedings of the IEEE/CVF International Conference on Computer Vision, Seoul, Republic of Korea, 27 October–2 November 2019; pp. 10052–10061.
18. Zhu, N.; Yu, Z.; Kou, C. A new deep neural architecture search pipeline for face recognition. *IEEE Access* **2020**, *8*, 91303–91310. [CrossRef]
19. Jiao, Q.; Li, R.; Cao, W.; Zhong, J.; Wu, S.; Wong, H.S. DDAT: Dual domain adaptive translation for low-resolution face verification in the wild. *Pattern Recognit.* **2021**, *120*, 108107. [CrossRef]
20. Lin, Y.S.; Liu, Z.Y.; Chen, Y.A.; Wang, Y.S.; Chang, Y.L.; Hsu, W.H. xCos: An explainable cosine metric for face verification task. *ACM Trans. Multimed. Comput. Commun. Appl.* **2021**, *17*, 1–16. [CrossRef]
21. Verma, R.; Bhardwaj, N.; Bhavsar, A.; Krishan, K. Towards facial recognition using likelihood ratio approach to facial landmark indices from images. *Forensic Sci. Int. Rep.* **2022**, *5*, 100254. [CrossRef]
22. Cao, Z.; Schmid, N.A.; Cao, S.; Pang, L. GMLM-CNN: A Hybrid Solution to SWIR-VIS Face Verification with Limited Imagery. *Sensors* **2022**, *22*, 9500. [CrossRef]
23. Borghi, G.; Fabbri, M.; Vezzani, R.; Calderara, S.; Cucchiara, R. Face-from-depth for head pose estimation on depth images. *IEEE Trans. Pattern Anal. Mach. Intell.* **2018**, *42*, 596–609. [CrossRef]
24. Ballotta, D.; Borghi, G.; Vezzani, R.; Cucchiara, R. Head detection with depth images in the wild. *arXiv* **2017**, arXiv:1707.06786.
25. Wang, Q.; Lei, H.; Wang, X. A Siamese Network for Face Verification with Depth Images. In Proceedings of the 2021 International Conference on Intelligent Technology and Embedded Systems (ICITES), Chengdu, China, 31 October–2 November 2021; pp. 138–143.
26. Zou, H.; Sun, X. 3D Face Recognition Based on an Attention Mechanism and Sparse Loss Function. *Electronics* **2021**, *10*, 2539. [CrossRef]
27. Rajagopal, S.D.; Ramachandran, B. 3D face expression recognition with ensemble deep learning exploring congruent features among expressions. *Comput. Intell.* **2022**, *38*, 345–365. [CrossRef]
28. Wang, Q.; Lei, H.; Wang, X. Deep face verification under posture interference. *J. Comput. Appl.* **2022**. [CrossRef]
29. Chen, G.; Han, K.; Wong, K.Y.K. PS-FCN: A flexible learning framework for photometric stereo. In Proceedings of the European Conference on Computer Vision (ECCV), Munich, Germany, 8–14 September 2018; pp. 3–18.
30. Ju, Y.; Peng, Y.; Jian, M.; Gao, F.; Dong, J. Learning conditional photometric stereo with high-resolution features. *Comput. Vis. Media* **2022**, *8*, 105–118. [CrossRef]

31. Woodham, R.J. Photometric method for determining surface orientation from multiple images. *Opt. Eng.* **1980**, *19*, 139–144. [CrossRef]
32. Ju, Y.; Shi, B.; Jian, M.; Qi, L.; Dong, J.; Lam, K.M. Normattention-psn: A high-frequency region enhanced photometric stereo network with normalized attention. *Int. J. Comput. Vis.* **2022**, *130*, 3014–3034. [CrossRef]
33. Qi, C.R.; Yi, L.; Su, H.; Guibas, L.J. Pointnet++: Deep hierarchical feature learning on point sets in a metric space. In Proceedings of the Conference on Neural Information Processing Systems (NIPS), Long Beach, CA, USA, 4–9 December 2017; pp. 5099–5108.
34. Li, Y.; Bu, R.; Sun, M.; Wu, W.; Di, X.; Chen, B. Pointcnn: Convolution on x-transformed points. In Proceedings of the Conference and Workshop on Neural Information Processing Systems (NIPS), Montreal, QC, Canada, 2–8 December 2018; pp. 820–830.
35. Guerrero, P.; Kleiman, Y.; Ovsjanikov, M.; Mitra, N.J. Pcpnet Learning Local Shape Properties from Raw Point Clouds. In *Computer Graphics Forum*; Wiley: Hoboken, NJ, USA, 2018; Volume 37, pp. 75–85.
36. Deng, H.; Birdal, T.; Ilic, S. Ppfnet: Global Context Aware Local Features for Robust 3D Point Matching. In Proceedings of the IEEE Conference on Computer Vision and Pattern Recognition, Salt Lake City, UT, USA, 18–22 June 2018; pp. 195–205.
37. Xiao, S.; Sang, N.; Wang, X.; Ma, X. Leveraging Ordinal Regression with Soft Labels for 3D Head Pose Estimation from Point Sets. In Proceedings of the ICASSP 2020—2020 IEEE International Conference on Acoustics, Speech and Signal Processing (ICASSP), Barcelona, Spain, 4–8 May 2020; pp. 1883–1887.
38. Ma, X.; Sang, N.; Xiao, S.; Wang, X. Learning a deep regression forest for head pose estimation from a single depth image. *J. Circuits Syst. Comput.* **2021**, *30*, 2150139. [CrossRef]
39. Cao, Y.; Liu, S. RP-Net: A PointNet++ 3D face recognition algorithm integrating RoPS local descriptor. *IEEE Access* **2022**, *10*, 91245–91252. [CrossRef]
40. Bromley, J.; Guyon, I.; LeCun, Y.; Säckinger, E.; Shah, R. Signature verification using a "siamese" time delay neural network. In , Denver, USA, 7-, pp 737-744. In Proceedings of the Conference and Workshop on Neural Information Processing Systems (NIPS), Denver, CO, USA, 7–11 December 1994; pp. 737–744.
41. Koch, G.; Zemel, R.; Salakhutdinov, R. Siamese Neural Networks for One-Shot Image Recognition. Ph.D. Thesis, University of Toronto, Toronto, ON, Canada, 2015.
42. Xiao, S.; Sang, N.; Wang, X. 3D point cloud head pose estimation based on deep learning. *J. Comput. Appl.* **2020**, *40*, 996.
43. Li, B.Y.L.; Mian, A.S.; Liu, W.; Krishna, A. Using Kinect for Face Recognition under Varying Poses, Expressions, Illumination and Disguise. In Proceedings of the 2013 IEEE Workshop on Applications of Computer Vision (WACV), Clearwater Beach, FL, USA, 15–17 January 2013; pp. 186–192.
44. Ranjan, R.; Castillo, C.D.; Chellappa, R. L2-constrained softmax loss for discriminative face verification. *arXiv* **2017**, arXiv:1703.09507.

Article

Siamese PointNet: 3D Head Pose Estimation with Local Feature Descriptor

Qi Wang, Hang Lei and Weizhong Qian *

School of Information and Software Engineering, University of Electronic Science and Technology of China, Chengdu 610054, China
* Correspondence: wzqian@uestc.edu.cn

Abstract: Head pose estimation is an important part of the field of face analysis technology. It can be applied to driver attention monitoring, passenger monitoring, effective information screening, etc. However, illumination changes and partial occlusion interfere with the task, and due to the non-stationary characteristic of the head pose change process, normal regression networks are unable to achieve very accurate results on large-scale synthetic training data. To address the above problems, a Siamese network based on 3D point clouds was proposed, which adopts a share weight network with similar pose samples to constrain the regression process of the pose's angles; meanwhile, a local feature descriptor was introduced to describe the local geometric features of the objects. In order to verify the performance of our method, we conducted experiments on two public datasets: the Biwi Kinect Head Pose dataset and Pandora. The results show that compared with the latest methods, our standard deviation was reduced by 0.4, and the mean error was reduced by 0.1; meanwhile, our network also maintained a good real-time performance.

Keywords: head pose estimation; convolutional neural network; Siamese network

1. Introduction

Citation: Wang, Q.; Lei, H.; Qian, W. Siamese PointNet: 3D Head Pose Estimation with Local Feature Descriptor. *Electronics* **2023**, *12*, 1194. https://doi.org/10.3390/ electronics12051194

Academic Editor: George A. Tsihrintzis

Received: 9 February 2023
Revised: 23 February 2023
Accepted: 27 February 2023
Published: 1 March 2023

Head pose estimation is an important part of the field of computer vision and also an important indicator for studying human behavior and attention. It can provide key information for many facial analysis tasks, such as face recognition, facial expression recognition, and driving concentration prediction [1]. The essence of the task is to predict the three pose angles (roll pitch yaw) of the object relative to the sensors. An effective algorithm should include the following main factors: a high accuracy, real-time performance, and the ability to cope with partial occlusion and large pose variations [2]. With respect to the above factors, many RGB-based head pose estimation algorithms have been proposed and achieved a very good performance [2]. However, the imaging quality of ordinary RGB sensors depends on light conditions, making them difficult to apply in some scenarios where light is weak or variable, such as night driving concentration detection, expression recognition in weak light environments, etc. [3]. With the development of depth sensors, it is more convenient to obtain high-quality depth images (also known as 2.5D images) [4]. Compared with ordinary RGB sensors, depth cameras have the following two main advantages. One advantage is that their infrared-based imaging principle—where each pixel represents the distance from the target to the sensors—makes the imaging quality mainly related to distances and is stable against variations in the light conditions; thus, it can be safely applied to human daily life [3]. The other advantage is that it can easily achieve background separation based on distance information, which can reduce the interference of the background and enable the task to focus on the object itself [1]. Depth maps can be easily converted into 3D point clouds by a simple coordinate transformation, which enables point clouds to inherit the above advantages of depth maps. Meanwhile, point clouds can better describe the spatial geometric feature of objects in 3D space, and the contours are stretched to have a more hierarchical appearance and clearer outlines; some important information around the outlines can be well retained [5].

Recently, many 3D methods based on different data types were proposed for face analysis, such as mesh, voxel grid, octree, and surface normal map. Compared with these four data types, the mathematical expression of a point cloud is more concise and can directly represent the spatial geometry information of an object. However, the disorder of the point clouds makes them difficult to apply to deep learning. Pioneers Qi et al. [6] relied on the idea of symmetric functions to solve the disorder of 3D point clouds and constructed PointNet. Many point clouds deep learning networks were proposed, such as those by Qi et al. [7], who optimized PointNet and proposed PointNet++. Deng et al. [8] introduced a local region representation to extract local features. Many point cloud methods were proposed for 3D computer vision. However, in the field of head pose estimation, due to the lack of the detailed textures of point clouds, the current pose estimation methods have not focused much attention on the local features of original point clouds, which leads to larger errors under large pose variations. Meanwhile, due to the non-stationary characteristic of the pose change process, previous regression networks were unable to achieve very good results on large-scale synthetic training data [9]. In order to deal with the above problems, we introduce a local feature descriptor coupled with a Siamese regression network for 3D head pose estimation. In our method, we first employ a local feature descriptor to describe the spatial geometric features of the objects; then, a group of PointNets is adopted to extract the local features, and three fully connected layers are used to map the head features to pose angles. Second, we utilize a share weight regression network with similar pose samples to guide the regression process of the pose angles. Finally, a novel loss function is introduced to constrain the difference between two similar features. In order to verify the effectiveness of the proposed method, we conduct experiments on two public datasets: the Biwi Kinect Head Pose dataset and Pandora.

The main contributions of this paper are summarized as follows:

1. We introduce a local feature descriptor to describe the detailed features of the point clouds to reduce the impact of their lack of detailed texture.
2. We present a new Siamese network to constrain the regression process of 3D head pose angles, which significantly reduced the errors of the original regression network. To the best of our knowledge, this is the first work to estimate 3D head poses by using a Siamese network.
3. The experimental results on public datasets show that our accuracy outperforms the latest approaches and also exhibits a good real-time performance.

2. Related Works

In recent years, the most widely used head pose estimation methods have mainly been proposed on RGB images. Drouard et al. [10] extracted HOG-based descriptors from face bounding boxes and mapped them to the corresponding head poses. Patacchiola et al. [11] proposed a convolutional neural network (CNN) supplemented with adaptive gradient methods to make the method robust for real-world applications. Hsu et al. [9] adopted a classification network to supervise the regression process of pose angles, which significantly improved the accuracy of the head pose estimation. Ruiz et al. [12] jointly combined pose classification and regression training with a multi-loss convolutional neural network on a large synthetically expanded dataset, which reduced the dependence on landmarks and enhanced the robustness of the network. Recently, Huang et al. [13] introduced a head pose estimation method using two-stage ensembles with average top-k regression, which combined the two subtasks by considering the task-dependent weights instead of setting coefficients by using grid search. Based on the driver's head pose and multi-head attention, Mercat et al. [14] proposed a vehicle motion forecasting method. In order to cope with complex situations, Liu et al. [15] proposed a robust three-branch model with a triplet module and matrix Fisher distribution module. Considering the discontinuity of Euler angles or quaternions and the observation that MAE may not reflect the actual behavior,

Cao et al. [16] proposed an annotation method which uses three vectors to describe the head poses and measurements using the mean absolute error of the vectors (MAEV) to assess the performance. Relying on head poses, Jha et al. [17] proposed a formulation based on probabilistic models to create salient regions describing the driver's visual attention. In order to bridge the gap between better predictions and incorrectly labeled pose images, Liu et al. [18] introduced probability values to encode labels, which took advantage of the adjacent pose's information and achieved a very good performance.

Compared to RGB images, depth maps cope well with dramatic light changes but lack texture detail [5], and very few studies only rely on depth maps [3]. Ballotta et al. [4] constructed a fully convolutional network to predict the location of the head's center. Wang et al. [19] combined the perception of deep learning and the decision-making power of machine learning to propose a convolutional neural network for multi-target head center localization. Borghi et al. [1] converted the depth maps into gray-level images and motion images via the GAN network, and they combined them to predict the head pose; this method relies on three types of training samples and greatly improved the head pose's prediction accuracy. Lei et al. [20] only relied on depth maps and constructed a one-shot network for face verification, which achieved a high accuracy with a small training sample. Recently, Wang et al. [21] employed an L2 norm to constrain head features in order to reduce the interference of partial occlusions for face verification.

As mentioned above, based on point clouds, many methods have been proposed and made breakthrough progress. Xiao et al. [2] utilized PointNet++ to extract the global features of the head and constructed a regression network for pose estimation. Xu et al. [22] presented a statistical and articulated a 3D human shape modeling pipeline, which captured various poses together with additional closeups of the individual's head and facial expressions. Then, Xiao et al. [23] adopted a classification network associated with soft labels to supervise the regression process of the pose angles. Hu et al. [24] leveraged the 3D spatial structure of the face and combined it with bidirectional long short-term memory (BLSTM) layers to estimate head poses in naturalistic driving conditions. Considering that the point clouds lack texture, Zou et al. [25] combined gray images and proposed a sparse loss function for 3D face recognition. Recently, Ma et al. [26] combined PointNet and deep regression forests to construct a new deep learning method in order to improve the efficiency of the head pose estimations. Cao et al. [27] proposed the RoPS local descriptor to map local features to three different planes and leveraged FaceNet to achieve 3D face recognition with a high accuracy. Based on a multi-layer perceptron (MLP), Xu et al. [28] constructed a classification network to predict the probability of each angle, and they also combined it with a graph convolutional neural network to reduce computation and memory costs.

In our method, we employ a Siamese network to supervise the regression process of the pose angles. The Siamese network was first proposed by Bromley et al. [29]; they applied this network to signature and verification certificate tasks. Based on the Siamese network, many methods have been proposed for computer vision. Melekhov et al. [30] used a Siamese network to extract a pair of features and calculated the similarity to determine whether the images matched. Varga et al. [31] introduced a deep multi-instance learning approach for person re-identification. Considering the local patterns of the target and their structural relationships, Zhang et al. [32] proposed a local structure learning method, which provides more accurate target tracking. Recently, Wang et al. [33] conducted a formal study on the importance of asymmetry by explicitly distinguishing the two encoders within the network and exploiting the asymmetry for Siamese representation learning.

3. Methods

In this section, we first introduce PointNet for point cloud feature extraction, and we propose a local feature descriptor to describe the local regions. Second, we construct a head pose regression network for the pose estimation. Finally, a Siamese network with similar samples is introduced to guide the training process of the pose regression network.

3.1. Introduction of Point Clouds and Feature Extraction

A point cloud is a series of points in a 3D space, and it is expressed as matrix $n \times 3$, where n is the number of points and 3 represents the coordinate (x, y, z) of a point in the world coordinate system, but the sequence of the points of the same object is not necessarily consistent [5]; moreover, due to the disorder of the point clouds, they cannot have an index sequence similar to regular 2D images or 3D voxels to achieve weight sharing for convolution operations [34]. Solving the disorder of the point clouds and performing an effective feature extraction is the key factor for facial analysis based on point clouds [2]. According to Theorem 1, Qi et al. [6] utilized the idea of a symmetric function to construct a deep learning network in order to deal with the disorder of the point clouds.

Theorem 1. *Suppose* $f : \chi \to \mathbb{R}$ *is a continuous set function w.r.t Hausdorff distance* $d_H(\cdot, \cdot)$ $\forall \varepsilon > 0, \exists$ *a continuous function* h, *and a symmetric function* $g(x_1, x_2, x_3, \cdots, x_n) = \gamma \circ MAX$ *such that for any* $S \in \chi$,

$$\left| f(S) - \gamma \left(\underset{x_i \in S}{MAX} \{ h(x_i) \} \right) \right| < \varepsilon \tag{1}$$

where $x_1, x_2, x_3, \cdots, x_n$ *is the full list of elements in* S *ordered arbitrarily,* γ *is a continuous function, and MAX is a vector max operator that takes* n *vectors as the input and returns a new vector of the element-wise maximum.*

Theorem 1 shows that if there are enough feature dimensions in the MAX operator, function f can be arbitrarily approximated by $\gamma \left(\underset{x_i \in S}{MAX} \{ h(x_i) \} \right)$.

Inspired by Theorem 1, a multilayer perceptron (MLP) is adopted to construct the right side of Equation (2) in order to approximate the left side:

$$f(x_1, x_2, x_3, \cdots, x_n) \approx \gamma \circ g(h(x_1), h(x_2), h(x_3), \cdots, h(x_n)) \tag{2}$$

where f and h are different general functions that map independent variables $(x_1, x_2, x_3, \cdots, x_n)$ and x_i to different feature spaces $\mathbb{R}^m$ and $\mathbb{R}^l$, respectively. G is a symmetric function (approximates the MAX operator in Theorem 1, and the function result is independent of the input order of the arguments). R is another general function $\mathbb{R}^l \to \mathbb{R}^m$ which maps the result of the symmetric function g to feature space $\mathbb{R}^m$ [5]. For a disordered point cloud, Qi et al. [6] employed a convolutional neural network as the MLP and a Max pooling layer as the symmetric function to extract the global feature of the object for classification and segmentation tasks. However, head pose estimation is a regression task, and it has difficulty achieving accurate results when only using global features. In this step, we adopt a shallow network structure, which deletes the transform net of PointNet, and we adjust the dimensions of each layer to make it suitable for local feature extractions in the next step. The structure of our proposed network is shown in Figure 1.

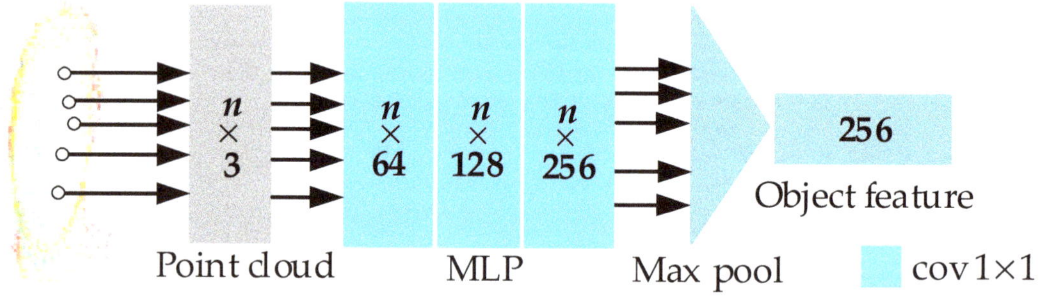

Figure 1. The structure of PointNet for extracting features of the point object. The MLP consists of three convolutional layers with filters 64, 128, and 256. The convolution kernel is 1×1.

As shown in Figure 1, for an input point cloud object with n points, we use three convolutional layers with 64, 128, and 256 filters to map every point to a high-dimensional feature space: $\mathbb{R}^3 \to \mathbb{R}^{64} \to \mathbb{R}^{128} \to \mathbb{R}^{256}$. Moreover, inspired by [6,35–37], a Max pooling layer was utilized as the symmetric function (the *MAX* operator in Theorem 1) to solve the disorder of the points set and to extract the feature in $\mathbb{R}^{256}$.

In order to ensure that the network has the same feature input dimension and can evenly sample the points, the farthest point sampling method is adopted to sample a fixed number of points for each object (each object samples 4096 points) before PointNet.

3.2. Local Feature Descriptor

Compared with RGB images, point clouds lack detailed textures, which results in difficulty in effectively characterizing objects by only using global features [2], and the position information of the points cannot directly reflect the geometric relationship between the points [8]. In order to enhance the description of the geometric details of the local region, in this step we adopt a local feature descriptor to describe the geometric characteristics of the local region.

For a pair of points (p_i, p_j) in a local region, in order to describe the geometric relationship between two points, a four-dimensional descriptor is introduced:

$$\psi_{ij} = (\|d\|, \angle(n_i, d), \angle(n_j, d), \angle(n_i, n_j)) \tag{3}$$

where d is the vector, which represents the difference between two points in the feature space, and $\|\cdot\|$ is the Euclidean distance. n_i and n_j are the normal vectors of p_i and p_j in the local region, respectively. As shown in Equation (4), $\angle$ is the angle between two normal vectors.

$$\angle(n_i, n_j) = a \tan 2(\|n_i \times n_j\|, n_i \cdot n_j) \tag{4}$$

The four-dimensional descriptor describes the spatial geometric characteristics of the points pair. For all points $\{p_1, p_2, p_3, \ldots, p_j\}$ in a local region, with p_i as the center and k as the radius (k is 0.4 in our method), we contain j point pairs with center point p_i. Then, the encoding method of this local region is expressed as Equation (5):

$$F_i = [p_1, n_1, p_2, n_2, \ldots, p_j, n_j, \psi_{i1}, \psi_{i2} \ldots, \psi_{ij}] \tag{5}$$

where n_j is the normal vector of point p_j, and ψ_{ij} is the four-dimensional feature descriptor between point p_j and center point p_i. As shown in Figure 2, F_i describes the spatial geometric characteristics of the local region via the local feature descriptor between all points with center point p_i in this local region.

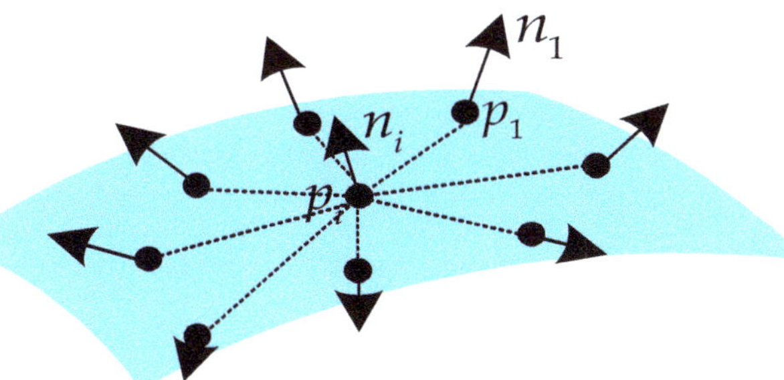

Figure 2. Schematic diagram of spatial geometric characteristics with center point p_i in a local region.

3.3. Pose Prediction Network

In this section, we utilize the PointNet with the local feature descriptor to construct a prediction network for head pose estimations; the structure of the head pose prediction network is shown in Figure 3.

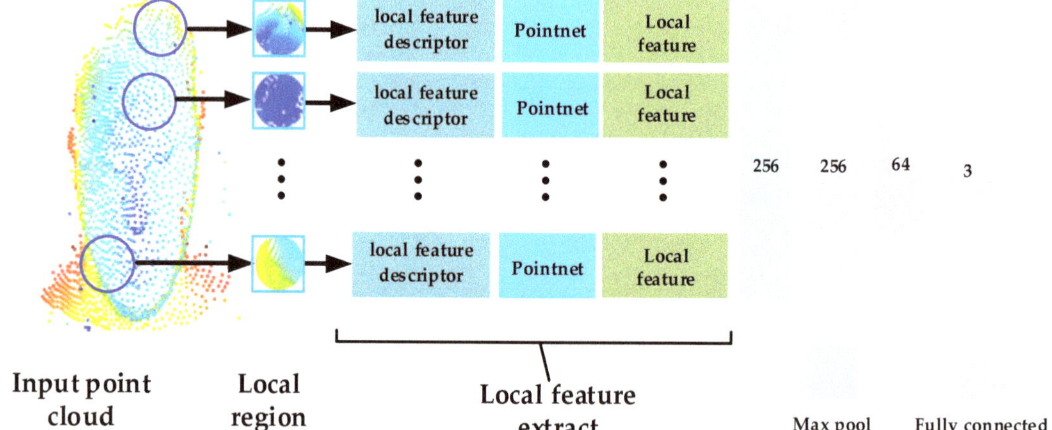

Figure 3. The structure of the head pose prediction network with local feature descriptors. In the figure, 3 fully connected layers are used to map head features to pose angles, where the last layer has 3 filters, which represent 3 pose angles (roll, pitch, and yaw).

As shown in Figure 3, for an input object with 4096 points $\{p_1, p_2, p_3, \cdots, p_{4096}\}$, we select every point as the center of a sphere with radius k (k is 0.4 in our method), and the points in the same sphere are regarded as being in the same local region $\{L_1, L_2, L_3, \cdots, L_{4096}\}$. For each L_i, we adopt the local feature descriptor to describe the geometric characteristics of the local region: $\{\psi_1, \psi_2, \psi_3, \cdots, \psi_{4096}\}$ (ψ_i represents a local geometric characteristic of local region L_i). Then, PointNet, as shown in Figure 1, is utilized to extract the features of each ψ_i. After the above steps, we obtain a set of local features in high-dimensional feature space $\{f_1, f_2, f_3, \cdots, f_{4096}\}$. Subsequently, a Max pooling layer is used to extract the entirety of feature F_w of all the local features $\{f_1, f_2, f_3, \cdots, f_{4096}\}$. Finally, three fully connected layers with 256, 64, and 3 filters are adopted to map head feature F_w to three pose angles.

The loss function of our head pose prediction network is defined as follows:

$$L_{predict} = \sum_j^n \|G_j - P_j\|_2^2 \tag{6}$$

where G_i represents the ground truth of three pose angles (expressed in Euler angles: roll, pitch, and yaw), and P_i is the prediction value of our head pose prediction network.

3.4. Siamese Network for Pose Constraint

As described above, a regression network is constructed to predict head poses, but due to the non-stationary characteristic of the head pose change process, it is difficult for a single regression network to cope with large-scale synthetic training data [23], which will result in a large prediction error. In order to deal with the above problem, a Siamese network with similar samples was proposed to constrain the prediction values and guide the regression process of the pose prediction network.

The structure of the proposed Siamese network is shown in Figure 4. The network consists of two identical branches, which accept similar pose samples as the inputs and extract features. The ends of the two branches are connected by an energy function to compute the difference between the two features:

$$L_{energy} = \sum_j^n \|D_{net}(x_i) - D_{gt}(x_i)\|_2^2 \tag{7}$$

$$D_{net}(x_j) = P_{1j} - P_{2j} \tag{8}$$

$$D_{gt}(x_j) = G_{1j} - G_{2j} \tag{9}$$

where D_{net} is the difference between the two predicted pose angles extracted by their own branch, and D_{gt} represents the difference in their ground truth [38].

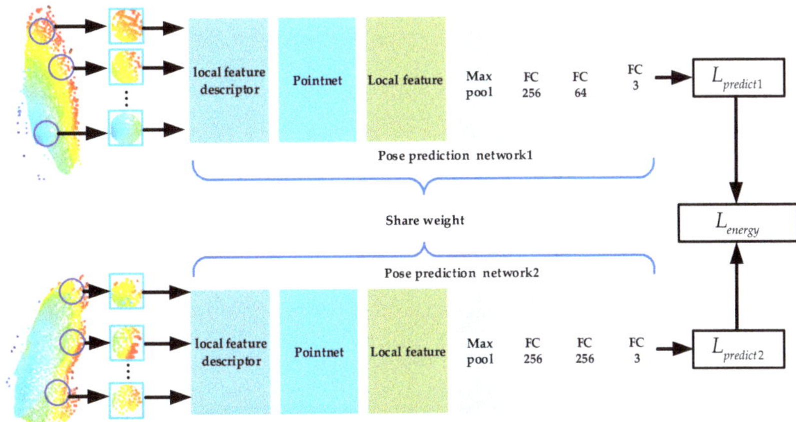

Figure 4. The structure of the Siamese network for head pose estimations. Two shared weight networks extract similar pose objects, and an energy function (the loss function of the Siamese network) concatenates the prediction results of two branches to constrain the prediction values and guide the regression process.

Considering that the training dataset has a total of N samples, a large number of N/2 possible pairs can be used, and for a specific pair of samples (S_i, S_j), only those with at least γ degrees of the total difference between all the pose angles (ground truth value) are selected:

$$\left| G_{S_i} - G_{S_j} \right| < \gamma \tag{10}$$

where γ determines the similarity of the pair of samples. In the training process, the energy function L_{energy} is also regarded as the loss function of the Siamese network.

Compared with a single-branch network, the proposed Siamese network has two main advantages. First, the parameters between the identical networks are shared, which can guarantee that a pair of very similar samples is not mapped to very different locations in a feature space by the respective networks. Second, as the loss function (L_{energy}) converges during training, similar pose samples within γ are extracted by their own network, which enables two regression networks to supervise each other and prevents the other side from being mapped to a more distant area in the feature space. In the testing stage, we only employ one pose prediction network to estimate the head pose (the parameters of the two networks are tied).

The hyperparameters of our Siamese network are as follows: the learning rate is 0.001, the decay rate is 0.99, the batch size is 64, and the decay step size is 500.

4. Experiments

In this section, we first introduce two public datasets for experiments: the Biwi Kinect Head Pose dataset and Pandora. Second, in order to verify the effect of the local feature descriptor and investigate similarity γ in Equation (10), we conduct ablation experiments on the Biwi Kinect Head Pose dataset. Third, we investigate the influence of the input number of points. Finally, we use our best results for comparison experiments with the latest methods and analyze the results.

4.1. Datasets

With respect to the Biwi Kinect Head Pose dataset, Fanelli et al. [39] utilized Kinect to collect this dataset. This dataset has a total of more than 15,000 head pose images, each object contains depth maps and the corresponding RGB images, and the resolution is 640 × 480. Biwi records 24 sequences of 20 different objects (6 females and 14 males, some of them are recorded twice). It is a challenging dataset with various head poses and partial occlusion. The test set includes sequences 11 and 12, which contain around 1304 images, and the training set contains the remaining 22 sequences, which contain around 14,000 images.

With respect to the Pandora dataset, Borghi et al. [1] collected this dataset specifically for head and shoulder pose estimations. Pandora has a total of more than 250,000 images, and each object contains depth maps (the resolution is 512 × 424) and corresponding RGB images (the resolution is 1920 × 1080). The dataset records 110 sequences of 10 male and 12 female objects. The recorded objects belonging to the upper body contain various postures, hairstyles, glasses, scarf, etc.

The above two datasets only provide RGB and depth images. We should transform depth images to point clouds before sending them into the Siamese network. First, we directly use the ground truth of the head center H_c with its depth value D_c to obtain the head areas (head detection is not the focus of our method), and we remove the background: we set the depth value as greater than $D_c + 300$ to 0 (300 is the general amount of space for a real head and expressed in mm). Second, we transform the depth map from an image coordinate system to a world coordinate system.

$$
\begin{bmatrix} x \\ y \\ z \end{bmatrix} = D_c \begin{bmatrix} \frac{1}{f_x} & 0 & 0 \\ 0 & \frac{1}{f_y} & 0 \\ 0 & 0 & 1 \end{bmatrix} \begin{bmatrix} x^i \\ y^i \\ 1 \end{bmatrix}
\tag{11}
$$

(x^i, y^i) denotes the pixel in the image coordinate system, and f_x and f_y represent the horizontal and vertical focal length of the internal parameters of the depth sensors. (x, y, z) is the position of the point converted from the pixel. Figure 5 shows examples of RGB images, depth maps, and point clouds from the Biwi Kinect Head Pose and Pandora datasets.

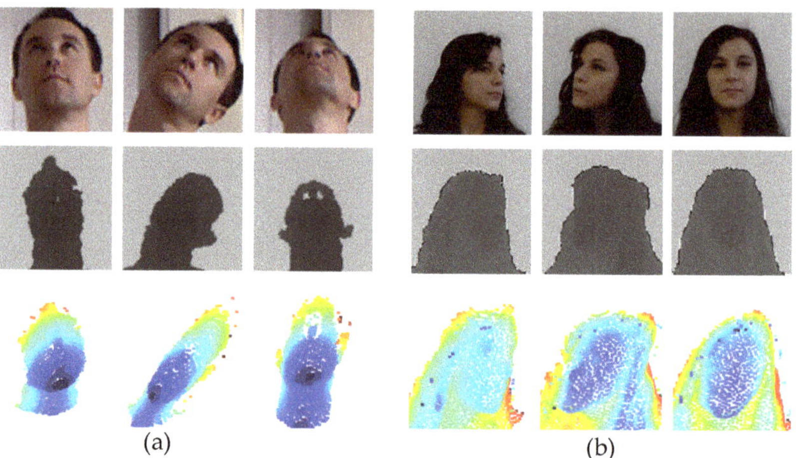

(a)   (b)

Figure 5. Examples of Biwi Kinect Head Pose dataset (**a**) and Pandora dataset (**b**). The first line is the RGB images, and the second and third lines are the corresponding depth maps and point clouds, respectively.

Both Biwi Kinect Head Pose and Pandora datasets provide ground truth pose angles (roll, pitch, and yaw). In our experiments, according to previous methods [1,2,23,26,38], we use the mean of the absolute values and the standard deviation to quantitatively evaluate the accuracy:

$$S = \delta \pm \beta \tag{12}$$

δ denotes the mean of the absolute values (MAE) of the difference between all ground truth and predicted values, and β is the standard deviation of the absolute values (SD) of the difference between all ground truth and predicted values.

4.2. Ablation Experiments

As described in Section 3.2, we introduced a local feature descriptor to describe the local region. In order to verify the effect of our method, according to the method in [7], we replace the local feature descriptor and only use the position information of the points to describe the local region.

In this section, to intuitively demonstrate the effect of the descriptor, we only use a single branch, as shown in Figure 3, to conduct the ablation experiment. The results are reported in Table 1.

Table 1. Performance evaluation with different local region expressions on the Biwi Kinect Head Pose dataset.

Local Region	Position	Local Feature Descriptor
Roll	2.2 ± 2.6	1.7 ± 2.0
Pitch	2.4 ± 2.1	2.0 ± 2.2
Yaw	2.4 ± 2.2	2.4 ± 2.1
Avg	2.3 ± 2.3	2.0 ± 2.1
fps	385	288

As shown in Table 1, the accuracy of the head pose prediction network greatly improved with the local feature descriptor, where the MAE is reduced by 0.3 and the SD is reduced by 0.2. This is because the descriptor provides the network with detailed local geometric features, which are more conducive to the extraction of the pose characteristics. On the other hand, our method would lead to extra computational costs, but it still maintains a good real-time performance. The results in Table 1 prove the effectiveness of the proposed local feature descriptor.

According to Equation (10) in Section 3.4, γ represents the similarity of the pair of samples. For a deep learning network, training samples are a key factor for the performance. In this step, we conduct comparison experiments on Biwi to decide the best γ for the Siamese network; the results are reported in Table 2.

Table 2. Performance evaluation with different γ on the Biwi Kinect Head Pose dataset.

γ	Roll	Pitch	Yaw	Avg
0	1.7 ± 2.0	2.0 ± 2.2	2.4 ± 2.1	2.0 ± 2.1
5	1.7 ± 2.0	1.9 ± 2.1	2.2 ± 2.1	2.0 ± 2.1
10	1.5 ± 1.9	1.7 ± 2.0	2.2 ± 1.9	1.8 ± 1.9
15	1.3 ± 1.7	1.5 ± 1.8	2.2 ± 1.7	1.6 ± 1.7
20	1.3 ± 1.7	1.6 ± 1.8	2.3 ± 1.8	1.7 ± 1.8
25	1.4 ± 1.7	1.7 ± 2.0	2.3 ± 1.8	1.8 ± 1.8
30	1.5 ± 1.8	1.8 ± 2.0	2.4 ± 1.9	1.9 ± 1.9
35	1.6 ± 1.9	1.9 ± 2.2	2.4 ± 2.1	2.0 ± 2.0
40	1.9 ± 2.2	2.2 ± 2.3	2.4 ± 2.2	2.2 ± 2.2
45	2.3 ± 2.4	2.4 ± 2.5	2.5 ± 2.3	2.4 ± 2.4
50	2.5 ± 2.8	2.6 ± 2.7	2.5 ± 2.5	2.5 ± 2.7

As shown in Table 2, when $\gamma = 0$, the inputted pair of samples has the same pose angles (same sample), and when the loss function is $L_{energy} = 0$, the constraint of the Siamese network is not utilized. As γ increases, the two branches of the Siamese network start to constrain each other. When $\gamma = 15$, our network achieves the best results. As γ continues to increase, the accuracy begins to decline. This is because obtaining more similar pose samples is more conducive to constraining pose angles within a smaller range. However, when γ is too small, the Siamese network also cannot achieve the best results because the pose features of the samples are too close, which makes it difficult for the Siamese network to distinguish the difference.

Figure 6 shows the prediction accuracy with different θ metrics. For each pose angle, if the absolute value of the difference between the prediction value and the ground truth is less than θ, the pose angle is regarded as accurately predicted. According to Figure 6, when θ is too small, the accuracy is obviously low. When the total difference, γ, is 5 and 10, the difference in the head pose is quite small, especially for a certain angle.

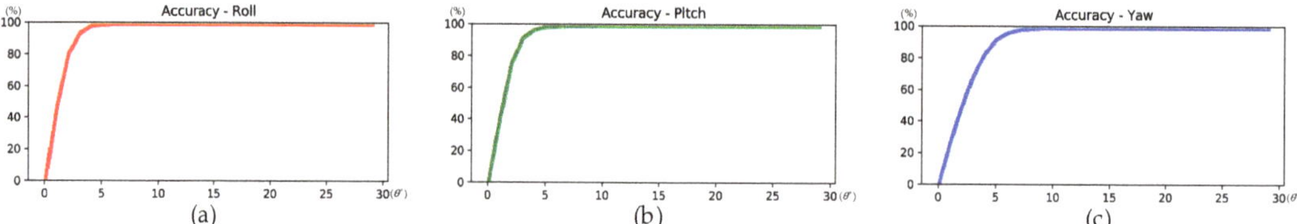

Figure 6. (**a**–**c**) represent the Curves of prediction accuracies with different metrics θ for Roll, Pitch, Yaw respectively.

According to Table 2, we set $\gamma = 15$ as our best result for comparison experiments. Figure 7 shows the curves of the loss function and the accuracy of the Siamese network during training when $\gamma = 15$.

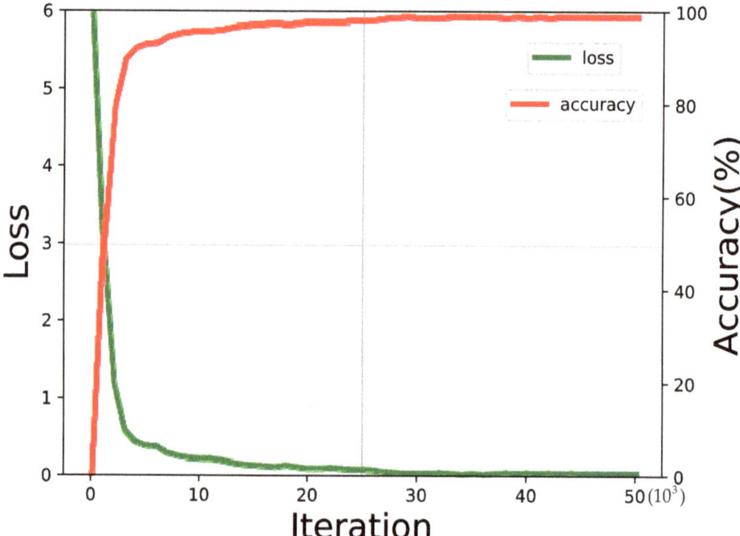

Figure 7. Curves of accuracy and loss when training our network at $\gamma = 15$.

4.3. Input Number of Points

As mentioned above, we sampled 4096 points for each object, but the number of input points affects the performance of the network. As shown in Figure 8, this is because the number of points affects the detailed information of the object and also determines the number of local regions. In this section, we investigate the input number of the points by losing half points for each step (we also adopt the farthest point sampling method to the sample points). The results are reported in Table 3.

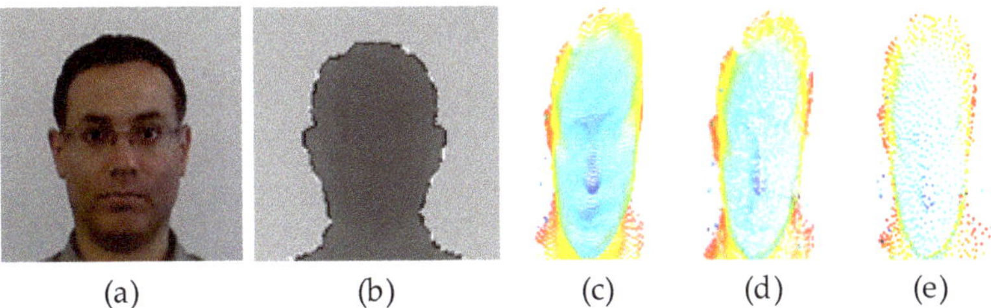

(a)   (b)   (c)   (d)   (e)

Figure 8. Examples of different input numbers of points, where (**a**,**b**) are corresponding RGB images and depth maps, and (**c–e**) represent input point clouds with 4096, 2048, and 1024 points, respectively.

Table 3. Results of the different input numbers of points on the Biwi Kinect Head Pose dataset.

Input Number	Acc	fps
4096	1.6 ± 1.7	288
2048	1.7 ± 1.9	398
1024	2.0 ± 2.2	558

As reported in Table 3, our network has a higher accuracy when there is an increased input number of points, which indicates that more points are beneficial for describing the more detailed features of the object and can significantly improve the accuracy of the network, but time consumption increases. However, 4096 points with 288 fps also maintained a good real-time performance for most applications.

4.4. Comparison Experiments

In this section, we conduct comparison experiments on two public datasets, and we analyze the results. Table 4 reports the comparison of the results with the latest methods on the Biwi Kinect Head Pose dataset.

Table 4. Comparison of results achieved by different methods on the Biwi Kinect Head Pose dataset.

Methods	Input	Roll	Pitch	Yaw	Avg
Venturelli et al. [38]	Depth	2.1 ± 2.2	2.3 ± 2.7	2.8 ± 3.3	2.4 ± 2.7
Borghi et al. [1]	Depth	1.8 ± 1.8	1.6 ± 1.7	1.7 ± 1.5	1.7 ± 1.7
Xiao et al. [2]	Point cloud	1.5 ± 1.4	2.3 ± 1.7	2.4 ± 1.8	2.1 ± 1.6
Huang et al. [13]	RGB	3.1	5.2	4.6	4.3
Ma et al. [26]	Point cloud	1.4 ± 2.0	1.5 ± 2.3	1.5 ± 2.1	1.5 ± 2.1
Cao et al. [16]	RGB	4.1	4.8	3.0	4.0
Liu et al. [18]	RGB	2.6	4.7	3.4	3.6
Ours	**Point cloud**	1.3 ± 1.7	1.5 ± 1.8	2.2 ± 1.7	1.6 ± 1.7

Table 4 lists a comparison of the experimental results on the Biwi Kinect Head Pose dataset. The methods in [13,16,18] only report their MAE, and other methods report $MAE \pm SD$. As shown in Table 4, the accuracy of the depth and point cloud methods is obviously higher than the RGB methods. This is because geometric information is more conducive to the extraction of the pose features, especially under partial occlusion and large pose interferences. Compared with depth maps, point clouds have more abundant geometric information and clearer contours, which are more beneficial to pose feature extraction. Although Borghi et al. [1] achieved a very high accuracy and only relied on depth maps, they used two Gan networks to generate gray and motion images, which leveraged three types of images to jointly predict the head pose, and the entire network structure is too complex.

As per the results reported in Table 4, compared with the methods in [1], our *MAE* was reduced by 0.1, and compared with the methods in [26], although their *MAE* is lower, our *SD* was reduced by 0.4. Overall, the accuracy of our method is higher than that of the other methods.

In order to intuitively show the test results on the Biwi Kinect Head Pose dataset, Figure 9a shows the ground truth and the prediction values of all the test samples, and Figure 9b shows the error distributions for each pose angle. As shown in Figure 9, the prediction results are very close to the ground truth, and the error distribution is convergent.

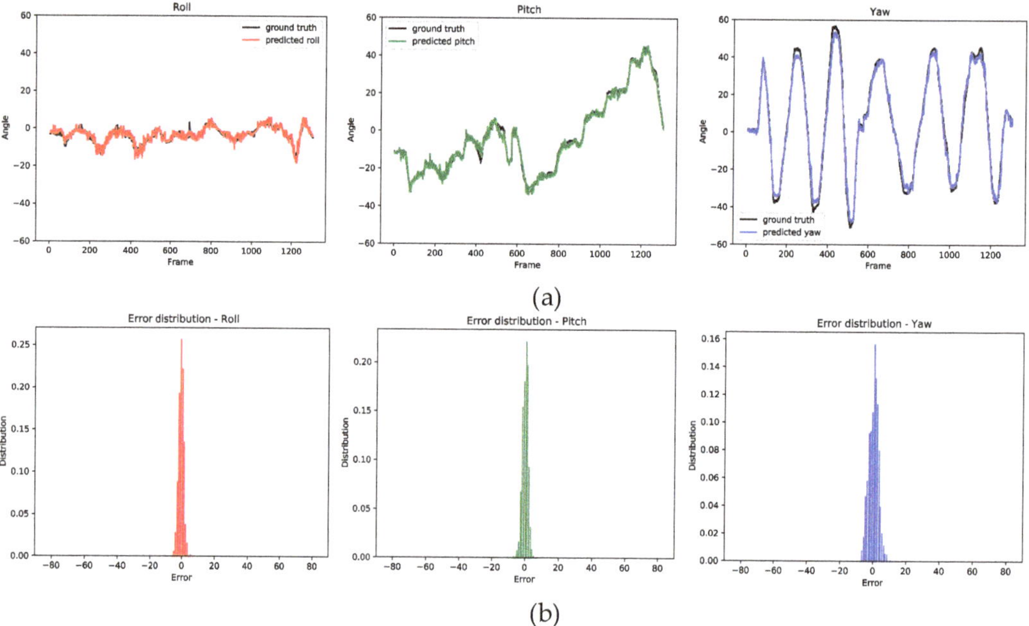

Figure 9. Results of the Biwi dataset: (**a**) reports the comparison between the ground truth and the predicted value for each frame (ground truth is the black line). (**b**) reports the error distributions for each angle.

Table 5 lists a comparison of the experimental results on the Pandora dataset, which contains more abundant samples with a series of large body gestures and partial occlusion. As reported in Table 5, our accuracy outperforms the latest methods. Compared with Xiao et al. [23], our accuracy is very close to theirs, and only the *MAE* was reduced by 0.1, but for each pose angle, our *MAE* and *SD* were better than theirs, except for the *SD* of the roll angle. Figure 10 shows the examples of our method on Pandora.

Table 5. Comparison of results achieved by different methods on the Pandora dataset.

Methods	Input	Roll	Pitch	Yaw	Avg
Borghi et al. [1]	Depth	5.4 ± 5.1	6.5 ± 6.6	10.4 ± 11.8	7.4 ± 7.8
Xiao et al. [23]	Point cloud	4.3 ± 4.5	6.1 ± 5.6	8.6 ± 9.8	6.3 ± 6.6
Ma et al. [26]	Point cloud	4.9 ± 7.4	6.4 ± 10.5	9.6 ± 15.3	7.0 ± 11.0
Ours	**Point cloud**	4.3 ± 4.7	6.0 ± 5.2	8.3 ± 9.8	6.2 ± 6.6

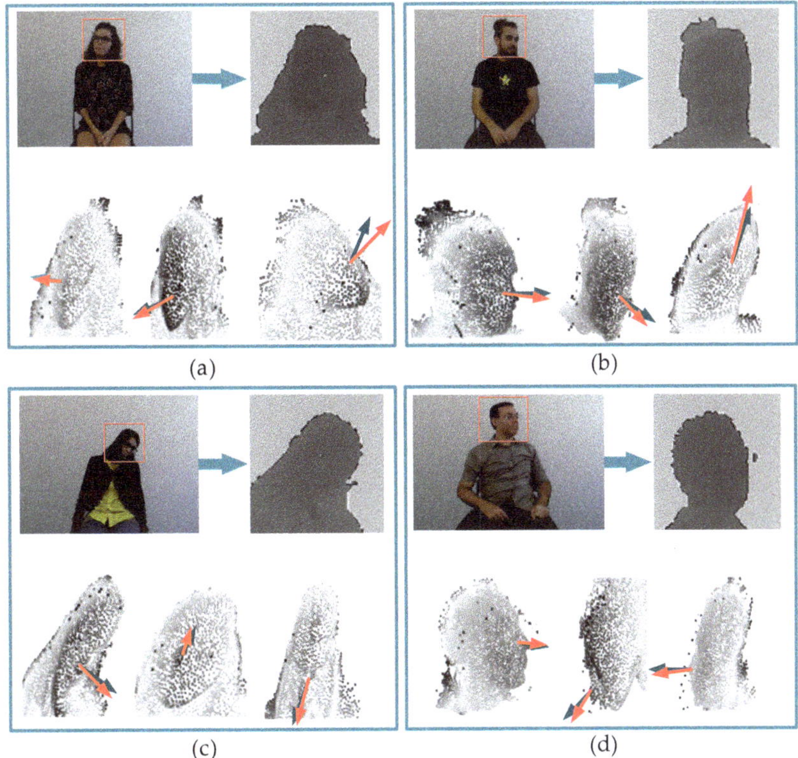

Figure 10. Examples on the Pandora dataset. (**a**–**d**) are different objects with variable head poses. The first rows show the RGB images and the corresponding depth maps of head regions. The second rows show the point clouds of the objects and the pose prediction results, where the red arrows are the ground truth, and the dark blue arrows are the prediction values.

As shown in Figure 10, our method can cope well with pose predictions with respect to various pose changes and provide an accurate pose angle estimation.

For the head pose estimation task, except for the accuracy, the time cost is also an important indicator for measuring performance, which determines whether the method can be applied to real application scenarios. Table 6 lists a comparison of different methods in terms of time costs. Because different data types are processed in different ways, for a fair comparison, we only conducted comparisons with point cloud methods.

Table 6. Comparison of different methods in terms of time costs.

Methods	fms
Xiao et al., 2020 [2]	125
Xiao et al., 2020 [23]	117
Wang et al., 2022 [21]	148
Wang et al., 2023 [5]	225
Ours	**288**

As shown in Table 6, compared with recent head feature extraction methods, our method is faster. This is because the local feature descriptor described the spatial geometric features of the local regions in detail before the deep learning network, which allows us to adopt a shallow network to extract the features and enables the network to maintain a good real-time performance.

Combining Tables 4 and 5, it is noticeable that our accuracy outperforms the latest methods, and Table 6 proves that our network also has a good real-time performance.

We conducted our experiments on the following operating system: Ubuntu16.04. The used hardware is listed as follows: the GPU is NVIDIA GTX1080ti, the CPU is Intel Core i7 (3.40 GHz), the display is SAMSUNG S27R350FHC (75 Hz, resolution: 1920×1080), and the depth cameras are Kinect v2 (resolution: 640×480) for the Biwi Kinect Head Pose dataset and the Kinect one (resolution: 512×424) for the Pandora dataset.

5. Conclusions

In this study, in order to cope with the non-stationary characteristic of the head pose change process, a new Siamese network with a local feature descriptor was constructed for 3D head pose estimations. In the feature extraction stage, a four-dimensional descriptor is introduced to describe the geometrical relationship between a pair of points, which can describe the geometric characteristics of the local regions in detail. In the head pose estimation stage, similar pose samples were used to constrain the regression process of the pose angles. Ablation experiments proved the effectiveness of the local feature descriptor, and the results of the experiments on public datasets show that compared with the latest methods, our accuracy outperformed the other methods (where *SD* was reduced by 0.4 and *MAE* was reduced by 0.1). Simultaneously, the proposed method also maintained real-time performance and can be applied to real application scenarios. However, in the case of partial occlusions, the accuracy is still not sufficient. In future studies, we will further explore algorithms and optimize the network and explore new methods for other 3D face analysis technologies.

Author Contributions: Conceptualization, Q.W.; data curation, Q.W.; formal analysis, Q.W. and W.Q.; investigation, Q.W.; methodology, Q.W.; project administration, H.L.; resources, Q.W.; software, Q.W.; supervision, H.L.; visualization, Q.W.; writing—original draft, Q.W.; writing—review and editing W.Q. All authors have read and agreed to the published version of the manuscript.

Funding: This research is funded by the National Natural Science Foundation of China (61802052).

Conflicts of Interest: The authors declare no conflict of interest.

References

1. Borghi, G.; Fabbri, M.; Vezzani, R.; Calderara, S.; Cucchiara, R. Face-from-Depth for Head Pose Estimation on Depth Images. *IEEE Trans. Pattern Anal. Mach. Intell.* **2018**, *42*, 596–609. [CrossRef] [PubMed]
2. Xiao, S.; Sang, N.; Wang, X. 3D point cloud head pose estimation based on deep learning. *J. Comput. Appl.* **2020**, *40*, 996.
3. Ballotta, D.; Borghi, G.; Vezzani, R.; Cucchiara, R. Head detection with depth images in the wild. *arXiv* **2017**, arXiv:1707.06786.
4. Ballotta, D.; Borghi, G.; Vezzani, R.; Cucchiara, R. Fully convolutional network for head detection with depth images. In Proceedings of the 2018 24th International Conference on Pattern Recognition (ICPR), Beijing, China, 20–24 August 2018; pp. 752–757.
5. Wang, Q.; Qian, W.Z.; Lei, H.; Chen, L. Siamese Neural Pointnet: 3D Face Verification under Pose Interference and Partial Occlusion. *Electronics* **2023**, *12*, 620. [CrossRef]
6. Qi, C.R.; Su, H.; Mo, K.; Guibas, L.J. Pointnet: Deep Learning on Point Sets for 3D Classification and Segmentation. In Proceedings of the IEEE Conference on Computer Vision and Pattern Recognition, Honolulu, HI, USA, 21–26 July 2017; pp. 652–660.
7. Qi, C.R.; Yi, L.; Su, H.; Guibas, L.J. Pointnet++: Deep hierarchical feature learning on point sets in a metric space. In Proceedings of the Conference on Neural Information Processing Systems (NIPS), Long Beach, CA, USA, 4–9 December 2017; pp. 5099–5108.
8. Deng, H.; Birdal, T.; Ilic, S. Ppfnet: Global Context Aware Local Features for Robust 3D Point Matching. In Proceedings of the IEEE Conference on Computer Vision and Pattern Recognition, Salt Lake City, UT, USA, 18–22 June 2018; pp. 195–205.
9. Hsu, H.W.; Wu, T.Y.; Wan, S.; Wong, W.H.; Lee, C.-Y. Quatnet: Quaternion-based head pose estimation with multiregression loss. *IEEE Trans. Multimed.* **2018**, *21*, 1035–1046. [CrossRef]
10. Drouard, V.; Ba, S.; Evangelidis, G.; Deleforgr, A.; Horaud, R. Head pose estimation via probabilistic high-dimensional regression. In Proceedings of the 2015 IEEE International Conference on Image Processing (ICIP), Quebec City, QC, Canada, 27–30 September 2015; pp. 4624–4628.
11. Patacchiola, M.; Cangelosi, A. Head pose estimation in the wild using convolutional neural networks and adaptive gradient methods. *Pattern Recognit.* **2017**, *71*, 132–143. [CrossRef]
12. Ruiz, N.; Chong, E.; Rehg, J.M. Fine-grained head pose estimation without keypoints. In Proceedings of the IEEE Conference on Computer Vision and Pattern Recognition Workshops, Salt Lake City, UT, USA, 18–22 June 2018; pp. 2074–2083.

13. Huang, B.; Chen, R.; Xu, W.; Zhou, Q. Improving head pose estimation using two-stage ensembles with top-k regression. *Image Vis. Comput.* **2020**, *93*, 103827. [CrossRef]
14. Mercat, J.; Gilles, T.; El Zoghby, N.; Sandou, G. Multi-head attention for multi-modal joint vehicle motion forecasting. In Proceedings of the 2020 IEEE International Conference on Robotics and Automation (ICRA), Paris, France, 31 May–4 June 2020; pp. 9638–9644.
15. Liu, H.; Fang, S.; Zhang, Z.; Li, D.; Lin, K. MFDNet: Collaborative poses perception and matrix Fisher distribution for head pose estimation. *IEEE Trans. Multimed.* **2021**, *24*, 2449–2460. [CrossRef]
16. Cao, Z.; Chu, Z.; Liu, D.; Chen, Y. A vector-based representation to enhance head pose estimation. In Proceedings of the IEEE/CVF Winter Conference on Applications of Computer Vision, Online, 5–9 January 2021; pp. 1188–1197.
17. Jha, S.; Busso, C. Estimation of Driver's Gaze Region from Head Position and Orientation Using Probabilistic Confidence Regions. *IEEE Trans. Intell. Veh.* **2022**, *8*, 59–72. [CrossRef]
18. Liu, H.; Liu, T.; Zhang, Z.; Arun Kumar, S.; Yang, B.; Li, Y. ARHPE: Asymmetric relation-aware representation learning for head pose estimation in industrial human–computer interaction. *IEEE Trans. Ind. Inform.* **2022**, *18*, 7107–7117. [CrossRef]
19. Wang, Q.; Lei, H.; Ma, X.; Xiao, S.; Wang, X. CNN Network for Head Detection with Depth Images in cyber-physical systems. In Proceedings of the 2020 International Conferences on Internet of Things (iThings) and IEEE Green Computing and Communications (GreenCom) and IEEE Cyber, Physical and Social Computing (CPSCom) and IEEE Smart Data (SmartData) and IEEE Congress on Cybermatics (Cybermatics), Rhodes, Greece, 2–6 November 2020; pp. 544–549.
20. Wang, Q.; Lei, H.; Wang, X. A Siamese Network for Face Verification with Depth Images. In Proceedings of the 2021 International Conference on Intelligent Technology and Embedded Systems (ICITES), Chengdu, China, 31 October–2 November 2021; pp. 138–143.
21. Wang, Q.; Lei, H.; Wang, X. Deep face verification under posture interference. *J. Comput. Appl.* **2022**, *43*, 595–600. [CrossRef]
22. Xu, H.; Bazavan, E.G.; Zanfir, A.; Freeman, W.T.; Sukthankar, R.; Sminchisescu, C. Ghum & ghuml: Generative 3d human shape and articulated pose models. In Proceedings of the IEEE/CVF Conference on Computer Vision and Pattern Recognition, Seattle, WA, USA, 14–19 June 2020; pp. 6184–6193.
23. Xiao, S.; Sang, N.; Wang, X.; Ma, X. Leveraging Ordinal Regression with Soft Labels for 3D Head Pose Estimation from Point Sets. In Proceedings of the ICASSP 2020—2020 IEEE International Conference on Acoustics, Speech and Signal Processing (ICASSP), Barcelona, Spain, 4–8 May 2020; pp. 1883–1887.
24. Hu, T.; Jha, S.; Busso, C. Temporal head pose estimation from point cloud in naturalistic driving conditions. *IEEE Trans. Intell. Transp. Syst.* **2021**, *23*, 8063–8076. [CrossRef]
25. Zou, H.; Sun, X. 3D Face Recognition Based on an Attention Mechanism and Sparse Loss Function. *J. Electron.* **2021**, *10*, 2539. [CrossRef]
26. Ma, X.; Sang, N.; Xiao, S.; Wang, X. Learning a Deep Regression Forest for Head Pose Estimation from a Single Depth Image. *J. Circuits Syst. Comput.* **2021**, *30*, 2150139. [CrossRef]
27. Cao, Y.; Liu, S. RP-Net: A PointNet++ 3D face recognition algorithm integrating RoPS local descriptor. *IEEE Access* **2022**, *10*, 91245–91252. [CrossRef]
28. Xu, Y.; Jung, C.; Chang, Y. Head pose estimation using deep neural networks and 3D point clouds. *Pattern Recognit.* **2022**, *121*, 108210. [CrossRef]
29. Bromley, J.; Guyon, I.; LeCun, Y.; Säckinger, E.; Shah, R. Signature verification using a "siamese" time delay neural network. In Proceedings of the Conference and Workshop on Neural Information Processing Systems (NIPS), Denver, CO, USA, 7–11 December 1994; pp. 737–744.
30. Melekhov, I.; Kannala, J.; Rahtu, E. Siamese network features for image matching. In Proceedings of the 2016 23rd International Conference on Pattern Recognition (ICPR), Cancun, Mexico, 4–8 December 2016; pp. 378–383.
31. Varga, D.; Szirányi, T. Person re-identification based on deep multi-instance learning. In Proceedings of the 2017 25th European Signal Processing Conference (EUSIPCO), Kos, Greece, 28 August–2 September 2017; pp. 1559–1563.
32. Zhang, Y.; Wang, L.; Qi, J.; Wang, D.; Feng, M.; Liu, H. Structured siamese network for real-time visual tracking. In Proceedings of the European Conference on Computer Vision (ECCV), Munich, Germany, 8–14 September 2018; pp. 351–366.
33. Wang, X.; Fan, H.; Tian, Y.; Kihara, D.; Chen, X. On the importance of asymmetry for siamese representation learning. In Proceedings of the IEEE/CVF Conference on Computer Vision and Pattern Recognition, New Orleans, LA, USA, 18–24 June 2022; pp. 16570–16579.
34. Li, Y.; Bu, R.; Sun, M.; Wu, W.; Di, X.; Chen, B. Pointcnn: Convolution on x-transformed points. In Proceedings of the Conference and Workshop on Neural Information Processing Systems (NIPS), Montreal, QC, Canada, 2–8 December 2018; pp. 820–830.
35. Guerrero, P.; Kleiman, Y.; Ovsjanikov, M.; Mitra, N.J. Pcpnet learning local shape properties from raw point clouds. In *Computer Graphics Forum*; Wiley: Hoboken, NJ, USA, 2018; Volume 37, pp. 75–85.
36. Ju, Y.; Peng, Y.; Jian, M.; Gao, F.; Dong, J. Learning conditional photometric stereo with high-resolution features. *Comput. Vis. Media* **2022**, *8*, 105–118. [CrossRef]
37. Chen, G.; Han, K.; Wong, K.Y.K. PS-FCN: A flexible learning framework for photometric stereo. In Proceedings of the Proceedings of the European Conference on Computer Vision (ECCV), Munich, Germany, 8–14 September 2018; pp. 3–18.

38. Venturelli, M.; Borghi, G.; Vezzani, R.; Cucchiara, R. From depth data to head pose estimation: A siamese approach. *arXiv* **2017**, arXiv:1703.03624.
39. Fanelli, G.; Gall, J.; Van Gool, L. Real time head pose estimation with random regression forests. In Proceedings of the 2011 IEEE Conference on Computer Vision and Pattern Recognition, Colorado Springs, CO, USA, 21–23 June 2011; pp. 617–624.

Article

Deep Learning Architecture Improvement Based on Dynamic Pruning and Layer Fusion

Qi Li [†], Hengyi Li [†] and Lin Meng [*,†]

Department of Electronic and Computer Engineering, Ritsumeikan University, Kusatsu 525-8577, Japan
* Correspondence: menglin@fc.ritsumei.ac.jp
† Current address: College of Science and Engineering, Ritsumeikan University, 1-1-1, Nojihigashi, Kusatsu 525-8577, Japan.

Abstract: The heavy workload of current deep learning architectures significantly impedes the application of deep learning, especially on resource-constrained devices. Pruning has provided a promising solution to compressing the bloated deep learning models by removing the redundancies of the networks. However, existing pruning methods mainly focus on compressing the superfluous channels without considering layer-level redundancies, which results in the channel-pruned models still suffering from serious redundancies. To mitigate this problem, we propose an effective compression algorithm for deep learning models that uses both the channel-level and layer-level compression techniques to optimize the enormous deep learning models. In detail, the channels are dynamically pruned first, and then the model is further optimized by fusing the redundant layers. Only a minor performance loss results. The experimental results show that the computations of ResNet-110 are reduced by 80.05%, yet the accuracy is only decreased by 0.72%. Forty-eight convolutional layers could be discarded from ResNet-110 with no loss of performance, which fully demonstrates the efficiency of the proposal.

Keywords: convolutional neural network; architecture improvement; dynamic channel pruning; memory access improvement

Citation: Li, Q.; Li, H.; Meng, L. Deep Learning Architecture Improvement Based on Dynamic Pruning and Layer Fusion. *Electronics* **2023**, *12*, 1208. https://doi.org/10.3390/electronics12051208

Academic Editor: Donghyeon Cho

Received: 1 February 2023
Revised: 22 February 2023
Accepted: 1 March 2023
Published: 2 March 2023

1. Introduction

Convolution neural networks (CNNs) have been proven to be effective in various applications [1–3]: object detection [4,5], cultural heritage protection [6], environment monitoring [7], robotics [8–10] and healthcare [11].

CNNs are designed to extract features from the input, which are used to reflect whether a region of the input has certain properties [12]. Based on these features, CNNs can accomplish tasks such as classification or detection. For example, in drone-based disaster management applications, CNNs techniques are used to quickly and accurately extract features of disasters, such as forest fires, landslides, and volcanic eruptions, from images captured by drone camera [13].

However, the enormous numbers of computations and parameters of CNNs hinder further development. Thus, it is not practical to deploy heavy CNNs on resource-constrained computing devices, such as embedded systems and mobile devices [14–16]. To address the problems, substantial research efforts have been devoted to compression techniques: channel pruning [17–20], low-rank decomposition [21–23], and weight quantization [24,25]. Channel pruning is performed by locating and removing redundant channels to reduce the numbers of floating-point operations (FLOPs) and parameters. In addition, the pruned model is intact in parallelism, which contributes to the efficient utilization of hardware resources [26].

After the model is compressed by channel pruning, many convolutional layers are equipped with only a few channels. These layers are defined as thin layers. Channel pruning is designed to remove unimportant channels and keep relatively important ones, so the remaining channels have valuable contributions. However, since the residual connection is effective, mainstream models [27] (such as ResNet [28], DenseNet [29], and MobileNet [30]) adopt design patterns where multiple layers form the bottleneck structure and residual connection are applied to the input and output of the bottleneck structure. To maintain the functional integrity of the residual connection, most pruning strategies [31,32] do not modify the input channels of the first layer and the output channels of the last layer in the bottleneck structure. Hence, even though the thin layer is important to the model, there is a lot of redundancy in the corresponding bottleneck structure. It is feasible to further compress the model by utilizing the redundancy in the bottleneck structure.

On the other hand, models with fewer layers have more benefits for hardware. In recent years, the use of CNNs on resource-constrained devices has gained attention, e.g., the field-programmable gate array (FPGA). The combination of high performance and high power efficiency is leading to the adoption of FPGAs in a variety of CNN-based applications. However, since CNN models are designed to be bloated, a large number of weights need to be stored in external memory and transferred to the FPGAs during computation [33,34]. This process requires additional energy and time. The energy cost due to the increased memory accesses and data movement even exceeds the energy cost of computation [35–38]. As a result, the implementation of deep learning models on FPGAs or other lightweight devices should be accompanied by optimizations, such as model compression and weight quantization. Removing layers could further contribute to solving this problem by reducing the load of the layer weights and the feature maps. Thus, a method to remove thin layers and the corresponding bottleneck structure while preserving the feature extraction capability is urgently needed [39].

Therefore, this paper proposes an architecture improvement approach for CNNs that aims to improve the performance of a model on resource-constrained devices by optimizing the model at the channel level and layer level. Specifically, first, the model is compressed by dynamic pruning, where highly sparse channels are dynamically removed. Then the channel-level compressed model is further optimized by layer fusion, the redundant structure is removed, and other layers substitute its function. Moreover, knowledge distillation and short–long fine-tuning are introduced to layer fusion to reduce performance loss. As layer fusion proceeds, the optimal architecture for the current task is obtained. The proposal was applied to various models, and experimental results show that the improved models can achieve high performance with fewer computational resources.

The main contributions of the paper are as follows:

- A method for layer-level compression of CNNs is proposed. By introducing knowledge distillation and short–long fine-tuning, redundant layers are removed with lower accuracy loss.
- The proposal may provide an idea for applications that desire to reduce memory access more than reduce computational complexity.

The rest of the paper is organized as follows: Section 2 introduces related works. Section 3 details the methodology. Section 4 shows the experimental results. Section 5 concludes the paper. In addition, all the abbreviations and definitions are listed in the Appendix A.

2. Related Work

This section reviews channel pruning and knowledge distillation, and then gives a short introduction to the related work.

Channel pruning is an efficient approach to compressing CNN models. The challenge of channel pruning is to remove channels with the minimal performance loss. He et al. [40] proposed a new channel-pruning method. Inspired by tensor factorization improvement based on feature-map reconstruction, the proposal fully exploits the redundancy of feature maps between channels. Specifically, for the trained model, it aims to reduce the dimensions of the input feature maps of the layer, while minimizing the reconstruction error of the output feature maps, to achieve pruning of the layer. The minimization problem is solved by two key steps: in the first step, the most representative channels are selected based on lasso regression, and the redundant channels are pruned. In the other step, the output of the remaining channels are reconstructed with linear least squares. Experiments showed that a 2× speed-up is achieved with a 1% accuracy loss.

As a representative method for model compression and acceleration, knowledge distillation effectively can learn small student models from large teacher models [41]. Huang et al. [42] proposed a new type of knowledge from the teacher model and transferred it to the student model. Specifically, the selective knowledge of neurons was exploited. Each neuron essentially extracts a certain pattern from the raw input. If a neuron is activated, that suggests some common property in the corresponding region that is relevant to the target task. Such information is valuable for the student model, as it provides an explanation for the prediction results of the teacher model. Hence, they proposed to align the distribution of neuron selectivity pattern between the student model and the teacher model. The maximum mean difference was introduced as a loss function to measure the discrepancy between the output feature maps of the teacher and student intermediate layers. The experimental results indicate that the proposal improves the performance of the student model significantly.

In some studies, knowledge distillation and channel pruning are combined. Aghli et al. [43] proposed a compression method for CNNs by combining knowledge distillation and weight pruning based on activation analysis. In detail, a select number of the layers in ResNet are pruned to avoid breaking the network's structure. Then, a new knowledge distillation architecture and loss function are used to compress the layers that were untouched in the previous step. The proposal was applied to the image classification task of head pose. Experimental results show that the model was significantly compressed while maintaining accuracy close to the baseline.

3. Methods

This paper intends to improve the architecture of CNNs. The proposals include dynamic pruning and layer fusion. First, the unimportant channels in the trained model are removed by dynamic pruning. Then, the redundant layers in the pruned model are further removed by layer fusion. The flow of the proposed method is described in Figure 1. In the next section, the details of each part are explained.

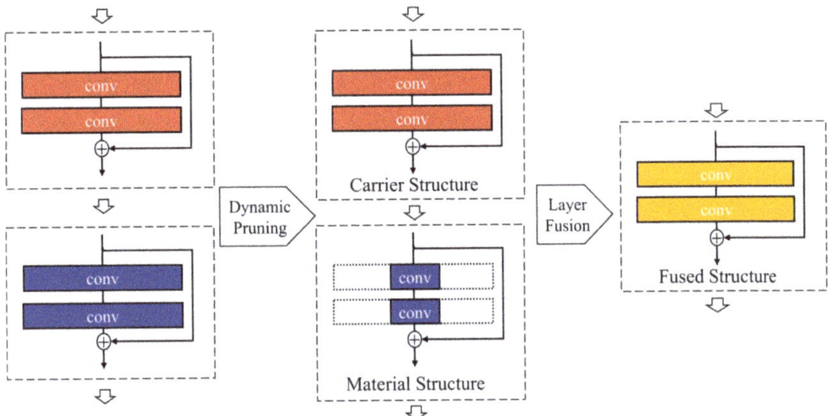

Figure 1. A diagram about the flow of proposed architecture-improvement strategy. The part enclosed by the dotted line indicates the compressed channel.

3.1. Preliminary

First, the convolution operation is introduced. $\mathbf{F}_{ij}$ indicates the filter that connects the ith input channel to the jth output channel. With the batch size set to 1, the feature map is a two-dimensional matrix that propagates between layers. If $\mathbf{M}_i$ denotes the input feature map on the ith channel, $\circledast$ and b denote the convolution operator and the bias, and then the jth output feature map $\mathbf{O}_j$ is generated as

$$\sum_{i=1}^{C} \mathbf{M}_i \circledast \mathbf{F}_{ij} + b = \mathbf{O}_j. \tag{1}$$

The sparsity is introduced to describe the percentage of redundant data (i.e., zero elements) in the feature map. Since the output feature map is the sum of the convolution results of the input feature map and the filter, if the input feature map has high sparsity, the convolution result of the corresponding element is close to zero and has no effect on the output, which means the redundancy of input channels can be measured by the sparsity of the input feature map and removes redundant channels while having a minor impact on the model.

In addition, in CNNs, layers are tightly connected to each other by channels. Removing an input channel is also removing the corresponding output channel from the previous layer. As the highly sparse feature maps are generated from the corresponding output channel of the previous layer, the loss of the entire output channel is acceptable to the model. In detail, when the channel is removed, the corresponding filter is pruned and the model is compressed.

3.2. Dynamic Pruning

In the previous section, the sparsity and redundancy were explained. This part proposes a method to dynamically determine the pruning target based on the sparsity of the feature map. Let the model infer the entire validation set, and calculate the average sparsity of each feature map in each input channel. The average sparsity is defined as channel sparsity, which is used to evaluate the importance of the channel. Here, the pruning threshold is introduced to distinguish high sparsity from low sparsity. When the value of channel sparsity is greater than the pruning threshold, the corresponding channel is considered to be the pruning target. After pruning the model, fine-tune the remaining parts of the model to restore accuracy. If a low pruning threshold is set, major parts of the model are removed, which leads to difficulties in recovering the accuracy. Thus, determining an appropriate pruning threshold is critical.

The investigation of different thresholds on pruning results is conducted. Figure 2 shows the results. It should be noted that the accuracy of the model after fine-tuning is given in the figure. It can be seen that fine-tuning accuracy increases as the pruning threshold increases. When the accuracy is close to the baseline, the effect of increasing the pruning threshold is slight. In addition, according to Figure 2b, it can be concluded that the smaller the pruning threshold, the greater the compression ratio.

We empirically summarize the following: (1) When the fine-tuning accuracy is lower than the target accuracy, increasing the pruning threshold improves the fine-tuning accuracy. (2) When the fine-tuning accuracy is higher than the target accuracy, the compression ratio could be further enhanced by lowering the threshold slightly while the fine-tuning accuracy remains at the same level. Note that the target accuracy is not the baseline. Considering that the fine-tuning accuracy is unstable, target accuracy was set to slightly below the baseline in the experiment.

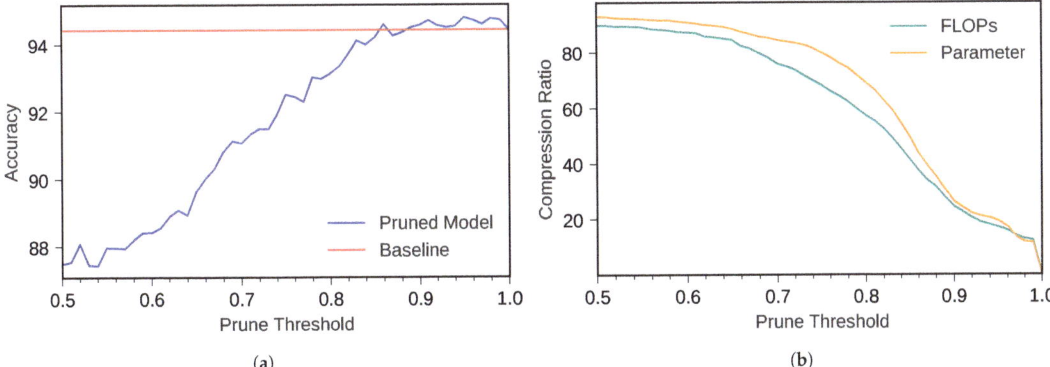

Figure 2. Results of pruning experiments on CIFAR10 with pruning thresholds from 0.5 to 1.0. ResNet-56 model was pruned in the experiment. (**a**) Accuracy. (**b**) Compression ratio of FLOPs and parameters.

Therefore, the binary search algorithm is introduced to adjust the pruning threshold based on feedback from fine-tuning. Algorithm 1 describes the proposal in detail. P_u and P_l are the two endpoints of the search interval for the optimal pruning threshold and are initialized to 1 and 0.5, respectively. P_c is the current pruning threshold and is initialized to the midpoint of the search interval. The pruning process is simplified into the following steps: (1) Load the original trained model, or the compressed model, from the previous iteration. (2) One pruning attempt is performed based on P_c, and then the model is fine-tuned to get the corresponding accuracy. (3) If the fine-tuning accuracy is higher than the target accuracy, the upper endpoint, P_u, is updated to P_c. Then, P_c is reduced by a quarter of the search interval. If the fine-tuning accuracy is lower than the target accuracy, the lower endpoint P_l is updated to the value of P_c. Then, P_c is increased by half of the search interval. The above steps repeat until the gap between P_l and P_u is less than 0.03, an empirically determined termination condition. When the loop ends, the compression result of this iteration is obtained.

Algorithm 1: Algorithm for dynamic pruning.

Data: Pre-trained network
Result: The compressed network
initialization: the current pruning threshold P_c;
upper endpoint of target interval $P_u = 1.0$;
lower endpoint of target interval $P_l = 0.5$;
$P_c \leftarrow (1/2P_u + 1/2P_l)$;
while $P_u - P_l > 0.03$ **do**
    load trained network;
    select and delete channels based on P_c;
    fine-tune the pruned network and measure accuracy;
    if *accuracy $\leq$ target accuracy* **then**
        $P_u \leftarrow P_c$;
        $P_c \leftarrow (3/4P_u + 1/4P_l)$;
        save current network as the result;
    else
        $P_l \leftarrow P_c$;
        $P_c \leftarrow (1/2P_u + 1/2P_l)$;
    end
end

In addition, dynamic pruning is iterative to get better compression results. The compressed model of the previous iteration is the original model of this iteration. By analyzing the experimental data, too many pruning iterations make little improvement on the compression result; thus, the iterations were set to 3 by us.

3.3. Layer Fusion

After dynamic pruning, the pruned model has multiple thin layers. This part intends to deprecate these thin layers by layer fusion, causing slight performance loss.

The bottleneck structure where the thin layer lies is defined as the material structure. Another bottleneck structure in the fusion operation is defined as the carrier structure (described in Figure 1). The challenge of layer fusion is that the impact of losing an entire layer on feature propagation is serious; thus, the carrier structure should undertake the function of the material structure. Therefore, the nearest bottleneck structure of the material structure is chosen as the carrier structure. Normally, carrier structure is the layer before the material structure. Then, the model is fine-tuned to adjust the output of the carrier structure to be similar to that of the material structure, so the carrier structure is functionally equivalent to the two layers before fusion. If x represents the input, and $\mathcal{F}()$ and $\mathcal{G}()$ represent the carrier structure and the material structure, respectively, then the fused layer $\mathcal{H}()$ should function as:

$$\mathcal{H}(x) \; = \; \mathcal{G}(\mathcal{F}(x)). \tag{2}$$

There are two key points in layer-fusion fine-tuning: knowledge distillation and short–long fine-tuning.

3.3.1. Knowledge Distillation

Knowledge distillation [44] is a method for transferring knowledge from a complex teacher network to a simple student network. A critical part of knowledge distillation is the soft label, which is a learning objective obtained from the output of the teacher network. The soft label is defined as:

$$q_i \; = \; \frac{\exp\left(z_i/T\right)}{\sum_j \exp\left(z_j/T\right)}. \tag{3}$$

Here, z_i is the probability of the ith class and T is the temperature of knowledge distillation. In knowledge distillation, the student network is optimized according to the soft labels and the ground-truth labels to get better training results. Benefiting from knowledge distillation, the student network is trained to generalize in the same way as the teacher network, and the training difficulty of the student network is reduced.

Therefore, in layer fusion, knowledge distillation is introduced as the fine-tuning method. After a material structure is removed, a sub-network with a simpler structure is obtained. The model compressed by dynamic pruning is considered as the teacher network, and the sub-network is considered as the student network. The sub-network is fine-tuned with knowledge distillation after removing one material structure. Fine-tuning with knowledge distillation could optimize the output of the carrier structure to close to the material structure, which means the loss of layer fusion is minimal.

3.3.2. Short–Long Fine-Tuning

In general, training the model optimizes all of the parameters. However, after the material structure is removed, only the carrier structure needs to be fine-tuned. Optimizing the entire network would increase the difficulty of searching for the optimal solution. This part is intended to keep the fine-tuning focused on the output of the carrier structure. Thus, after removing the material structure, only the weights of the carrier structure are tuned, and other layers are frozen. As a result, it is enough to take a small number of epochs for fine-tuning only the carrier structure of the network, the process of which is denoted as short fine-tuning. The short fine-tuning is done after each material structure is removed.

In addition, it is difficult to fine-tune the output of the carrier structure to be exactly the same as that of the material structure. A tiny offset remains in the fused layers after short fine-tuning. When the offsets accumulate too much, the performance of the model is severely degraded. Thus, after four bottleneck structures are fused, the model is fine-tuned without freezing. This process requires more iterations and is defined as long fine-tuning. Introducing long–short fine-tuning contributes to providing layer fusion with a low performance loss.

3.3.3. Iterative Layer Fusion

Layer fusion is an iterative process. After one bottleneck structure is fused, the next set of material and carrier structures is searched by the new model.

Algorithm 2 describes the flow of layer fusion. In detail, the duplicate of the pruned model always serves as the teacher model for knowledge distillation. First, find the bottleneck structure with the lowest number of channels as the material structure, and select the previous bottleneck structure as the carrier structure. Then, remove the material structure and short fine-tune the model. After repeating these two steps four times, a long fine-tuning is conducted. The above steps are repeated until the drop in accuracy is greater than 3%. Although the architecture is compressed by layer fusion, the model performance does not keep decreasing. When the model is modified to the appropriate architecture, the model's performance increases. Both increases and decreases in model accuracy are possible after each long fine-tuning, so a loose termination condition is adopted. The results after each long fine-tuning are kept, and we evaluate them in terms of compression ratio and performance.

Algorithm 2: Flow of layer fusion.

Data: The pruned model
Result: The model with improved architecture
duplicate the pruned model as the teacher model;
while *accuracy drop less than 3%* **do**
    for *iterations* **to** *4* **do**
        select the bottleneck structure with the least number of channels as the
          material structure;
        select the previous layer of the element layer as the carrier structure;
        remove the material structure from model;
        freeze weights except for the carrier structure;
        short fine-tuning with knowledge distillation;
    end
    long fine-tuning with knowledge distillation;
    save model;
end

---

4. Experimental

4.1. Experimental Configuration

The proposal was applied to the ResNet and DenseNet models to evaluate the improvement effect. CIFAR10 and ImageNet50 [45] were adopted as experimental datasets. CIFAR10 contains 50k training images and 10k test images, all of which are 32×32. ImageNet50 consists of 50 random classes chosen from the ILSVRC2012 dataset. It contains 51,614 training images, 6490 validation images, and 6440 test images, all of which are 224×224. Common ResNet models, such as ResNet-50 and ResNet-101, are designed for ImageNet, and their architectures are too complex for CIFAR10. ResNet-56 and ResNet-110 have been designed for CIFAR10 with a simpler architecture, and they expect an input size of 32×32. Therefore, ResNet-56 and ResNet-110 have been adopted as the base model for the CIFAR10 compression experiment. For the same reason, DenseNet-40 was adopted as the base model for the CIFAR10, and DenseNet-121 was compressed in the experiment on ImageNet50.

Each base model was trained on the datasets with a 5-epoch warm-up and 320 epochs, from scratch. The momentum was set to 0.9, the weight decay factor was 10^{-4}, and the batch size was 64. Experiments were conducted on the Nvidia GeForce GTX 3080 Ti GPU and Intel i9-10900 CPU, and the models were implemented by pytorch.

In the dynamic pruning phase, the optimizer SGD with a learning rate initialized to 0.01 was adopted. The learning rate was decayed by cosine annealing [46] with a period of 320 epochs and restarted at epoch 160. The network was fine-tuned on the training set, and the number of epochs was set to 320. When the best accuracy was not updated for more than 20 epochs, the fine-tuning was stopped. About the pruning strategy, only the input channels of second convolutional layers in bottleneck structures were selected as pruning targets. In DenseNet40, since a more dense structure than the bottleneck structure which contains one convolutional layer is adopted, the highly sparse output channels of each convolutional layer and all the corresponding input channels are removed.

In the layer-fusion phase, normally, the carrier structure is the previous bottleneck structure of the material structure. However, layer fusion should not fuse two layers with different sizes of output feature maps. When the output size of the previous bottleneck structure is different, the following one is picked as the carrier structure. The temperature of the knowledge distillation was set to 4. The number of epochs for short fine-tuning was set to 50, and 200 for long fine-tuning. The other settings of fine-tuning were the same as in dynamic pruning.

4.2. Experiments on CIFAR10

The experimental results of dynamic pruning on CIFAR10 are shown in Table 1. Top-1 accuracy and FLOPs are the focuses. The compression effect is noticeable on the ResNet series, especially ResNet110, which compresses 75.75% of FLOPs with a 0.42% drop in precision. In addition, layers with less than six output channels are considered thin layers, and the number of such layers is listed in the table. It can be seen that the number of thin layers after pruning was considerable, especially for ResNet110, which had 31 thin layers. In the next phase, layer fusion was mainly focused on these thin layers.

Table 1. Results of dynamic pruning on CIFAR10. "Acc." indicates accuracy. "Acc. ↓" and "FLOPs ↓" denote reductions compared to the base models. The other tables and figures follow the same conventions.

	Baseline (%)	Pruned Acc. (%)	Acc. ↓ (%)	FLOPs (M)	FLOPs ↓ (%)	Thin Layers
ResNet-56	93.52	93.1	0.42	50.92	60.10	10
ResNet-110	93.76	93.34	0.42	62.34	75.75	31
DenseNet-40	94.53	94.07	0.46	210.56	28.03	10

Table 2 details the experimental results of layer fusion. The effect of layer fusion was most significant on ResNet-56 and DenseNet-40. Compared to the results of dynamic pruning, the compression ratio was improved by 12.01% on ResNet-56, and the accuracy was further reduced by 0.35%. Additionally, FLOPs were further reduced in number by 11.78% in DenseNet-40, along with a further loss of 0.41% in accuracy. For ResNet-110, although only 4.3% of FLOPs were eliminated by layer fusion, up to 64 convolutional layers were fused, and the accuracy reduction was 0.3%. The results of the layer fusion, in order to prioritize the performance, are also listed in the table. It indicates that multiple layers were removed from the models with less than 0.07% in accuracy degradation. Specifically, after fusing the 48 convolutional layers in ResNet-110, the accuracy rose by 0.04% compared to the pruned model.

Table 2. Results of layer fusion on CIFAR10. The depth and fused layers refer to numbers of convolutional layers. ResNet-56* indicates the layer fusion results of the priority selection according to the precision, and same for ResNet-110* and DenseNet-40*.

	Fused Acc. (%)	Acc. $\downarrow$ (%)	FLOPs (M)	FLOPs $\downarrow$ (%)	Depth	Fused Layers
ResNet-56	92.75	0.77	35.59	72.11	24	32
ResNet-110	93.04	0.72	51.28	80.05	46	64
DenseNet-40	93.66	0.87	176.10	39.81	24	16
ResNet-56*	93.04	0.48	46.85	63.29	40	16
ResNet-110*	93.38	0.38	57.14	77.77	62	48
DenseNet-40*	94.07	0.46	187.54	35.90	28	12

4.3. Experiments on ImageNet50

The proposal achieved satisfactory results on CIFAR10. However, mainstream models are considered too complex for CIFAR10, which means there are plenty of redundant structures that can be easily removed from the model. Therefore, as a complement, the experiments were performed on ImageNet50, which consists of large input images. Since ImageNet50 is more complex compared to CIFAR10, it is challenging to compress the models without accuracy loss. Table 3 provides the results obtained in the experiment. As can be seen in the table, 34.54% of the FLOPs of DenseNet121 were compressed, and there was a 0.65% accuracy reduction. Then, layer fusion improved the compression of FLOPs to 36.48% while resulting in an overall accuracy loss of 0.92%. In detail, 32 convolutional layers of DenseNet121 were fused in the layer fusion.

Table 3. Compression results of DenseNet-121 on ImageNet50. "After layer fusion" indicates the result of layer fusion after dynamic pruning.

	Acc. (%)	Acc. $\downarrow$ (%)	FLOPs (M)	FLOPs $\downarrow$ (%)
Baseline	90.21	-	59.2	-
After dynamic pruning	89.56	0.65	38.75	34.54
After layer fusion	89.29	0.92	37.60	36.48

4.4. Analysis

The results of all experiments are summarized in Table 4. In addition, the intermediate results of layer fusion are analyzed. Figure 3 presents the layer fusion details of the experiments on CIFAR10. The baseline is the model's accuracy after dynamic pruning. It can be noticed that the accuracy is not continuously decreasing as the layers are fused. In the layer-fusion experiments on ResNet-110, the accuracy was higher than the baseline four times. Additionally, for DenseNet-40, there was a significant increase in accuracy of 0.38% after the first four layers were fused. Afterward, when the 12 convolutional layers were discarded, the accuracy was the same as the baseline. These data suggest that it is feasible to improve the compressed model's performance by layer fusion. Moreover, the variations in FLOPs are also shown in the figure. It can be found that since the bottleneck structure with the fewest channels was fused, the reduction in computational resources by layer fusion was not significant. However, considering that it was a further reduction of FLOPs from a compressed model, the enhanced compression ratio is valuable.

Table 4. All compression results by our proposed method.

Dataset	Model	Original Acc. (%)	Compressed Acc. (%)	Acc. ↓ (%)	Original FLOPs (M)	Compressed FLOPs (M)	FLOPs ↓ (%)
CIFAR10	ResNet-56	93.52	92.75	0.77	127.62	35.59	72.11
	ResNet-110	93.76	93.04	0.72	257.09	51.28	80.05
	DenseNet-40	94.53	93.66	0.87	292.56	176.1	39.80
ImageNet50	DenseNet-121	90.21	89.29	0.92	59.20	37.60	36.48

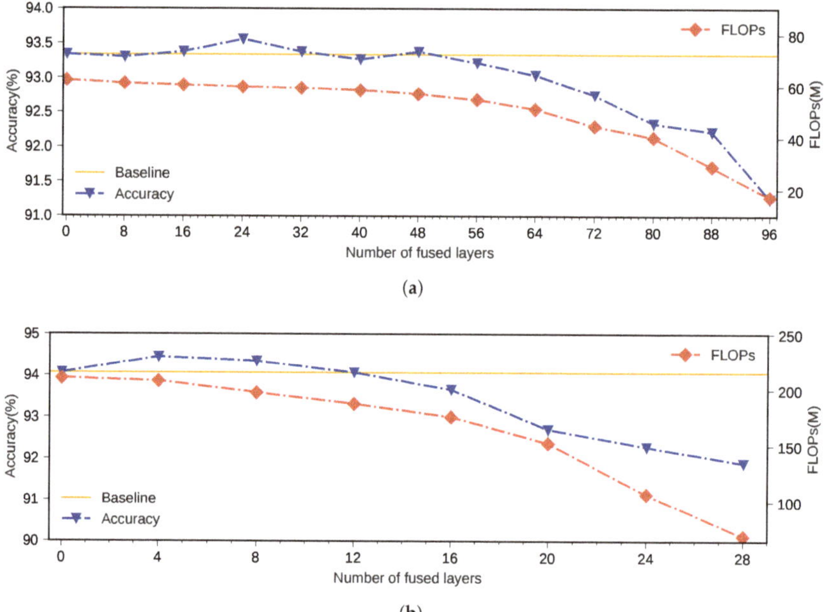

Figure 3. Layer fusion details of ResNet-110 and DenseNet-40. The accuracy and FLOPs of the model after all four layers are fused are shown in the figure. (**a**) ResNet-110. (**b**) DenseNet-40.

Moreover, comparison experiments were conducted to analyze the proposal:

- Knowledge distillation was replaced by cross-entropy loss in the fine-tuning.
- No short fine-tuning was performed after each structure was fused; only long fine-tuning was conducted after four layers were fused.
- We trained models from scratch with the optimized architectures.

Figure 4 presents the experiment without knowledge distillation on DenseNet-40. From the figure, it can be seen that without the benefit of knowledge distillation, it is difficult to recover the model accuracy to a satisfactory level. After 20 layers are fused, the model's accuracy degrades more seriously. Additionally, Figure 5 presents the experiment with no short fine-tuning on ResNet-110. After removing short fine-tuning, seven fine-tuning results were worse than before, and the accuracy dropped more severely after 44 layers were fused.

Since the complexity of the model architectures was reduced without a significant drop in performance, we can say that they nearly retain the performance of the complex models even though the architectures are relatively simple. To demonstrate that, models with architectures the same as those of the compressed models were built and then trained from scratch with the same training settings as the base models. The results are listed in Table 5. Acc. improved indicates the accuracy improved by the proposal compared to training from scratch. These data show that with similar complexity, the models compressed by the proposed method have higher performance.

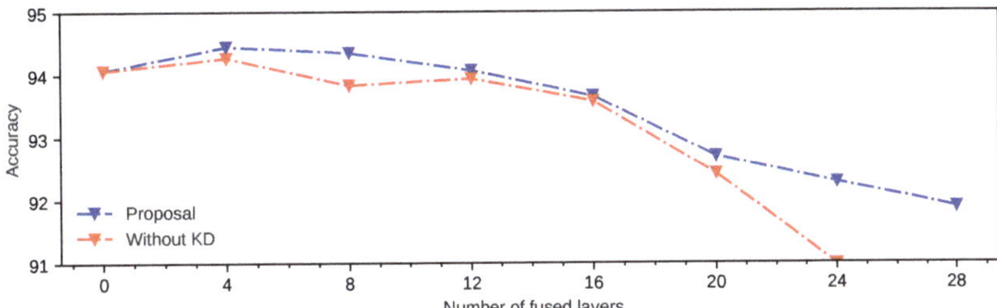

Figure 4. Details of the layer-fusion experiment on DenseNet-40 without knowledge distillation.

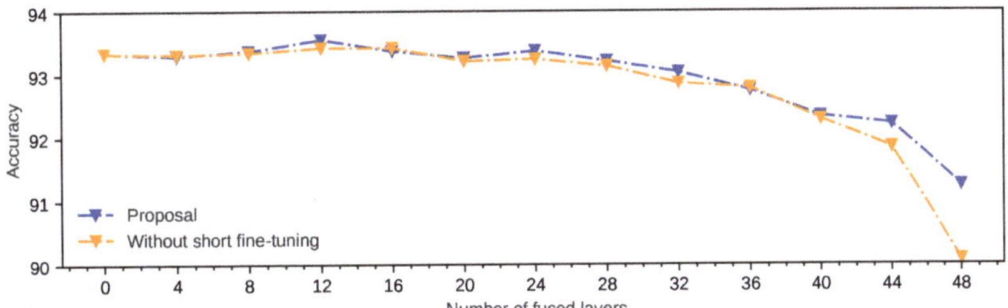

Figure 5. Details of the layer-fusion experiment on ResNet-110 without short fine-tuning.

Table 5. Accuracy of the compressed models and models trained from scratch with compressed architectures.

	Acc. by Proposal (%)	Acc. from Scratch (%)	Acc. Improved (%)
ResNet-56	92.75	91.65	1.10
ResNet-110	93.04	92.31	0.73
DenseNet-40	93.66	93.25	0.41

5. Conclusions

In this paper, we proposed a CNN architecture-improvement approach to optimize redundant models at the channel level and the layer level. First, the binary search method is used to dynamically determine the appropriate pruning threshold, and then redundant channels are removed based on the threshold. Then, bottleneck structures with only a few channels are eliminated by layer fusion to compress the model at the layer level. Knowledge distillation and short–long fine-tuning were introduced to layer fusion to enhance the performance of the fused models. The experimental results show the efficiency of the proposal: in terms of ResNet-56, 72.11% of FLOPs were eliminated, and there was a 0.77% drop in accuracy; as for ResNet-110, 80.05% of FLOPs were eliminated, and there was a drop in accuracy of 0.72%. In detail, the data demonstrate that there are 48 convolutional layers that could be removed from ResNet110 by our method without harming the model. We focused on the analysis of the proposal in the classification task. In future work, the effects of detection and segmentation tasks will be analyzed, since compression is more difficult for these tasks. Furthermore, the models with the optimized architecture are planned to be implemented on FPGAs to evaluate the compression effect on resource-constrained devices.

Author Contributions: Conceptualization, L.M. and H.L.; methodology, Q.L.; software, Q.L.; validation, Q.L. and H.L; formal analysis, Q.L.; investigation, Q.L.; resources, L.M.; data curation, Q.L.; writing—original draft preparation, Q.L.; writing—review and editing, H.L. and L.M.; visualization, Q.L.; supervision, L.M.; project administration, L.M.; funding acquisition, L.M. All authors have read and agreed to the published version of the manuscript.

Funding: This research received no external funding.

Data Availability Statement: No new data were created or analyzed in this study. Data sharing is not applicable to this article.

Conflicts of Interest: The authors declare no conflict of interest.

Appendix A

Table A1. List of abbreviations and definitions.

Abbreviation	Explanation
FLOP	Floating-point operation
CNN	Convolution neural network
FPGA	Field-programmable gate array
lasso	Least absolute shrinkage and selection operator
$\mathbf{F}_{ij}$	Filter that connects the ith input channel to the jth output channel
$\mathbf{M}_i$	Input feature map on the ith channel
$\mathbf{O}_i$	Output feature map on the ith channel
$\circledast$	Convolution operator
b	Bias
$\mathbf{P}_c$	Current demarcation point of En-sparsity
$\mathbf{P}_u$	Upper endpoint of target interval
$\mathbf{P}_l$	Lower endpoint of target interval
x	Input of bottleneck structure
$\mathcal{F}()$	Carrier structure
$\mathcal{G}()$	Material structure
$\mathcal{H}()$	Fused layer
z_i	The probability of the ith class
T	The temperature of knowledge distillation
q_i	The soft label of the ith class
GPU	Graphics processing unit
CPU	Central Processing Unit
Acc.	Top-1 accuracy
Acc. $\downarrow$	Reduction in accuracy compared to the base model
FLOPs $\downarrow$	Reduction in FLOP compared to the base model
SGD	Stochastic gradient descent algorithm

References

1. Li, Z.; Meng, L. Research on Deep Learning-based Cross-disciplinary Applications. In Proceedings of the 2022 International Conference on Advanced Mechatronic Systems (ICAMechS), Toyama, Japan, 17–20 December 2022; pp. 221–224. [CrossRef]
2. Chen, X.; Liu, L.; Tan, X. Robust Pedestrian Detection Based on Multi-Spectral Image Fusion and Convolutional Neural Networks. *Electronics* **2022**, *11*, 1. [CrossRef]
3. Avazov, K.; Mukhiddinov, M.; Makhmudov, F.; Cho, Y.I. Fire Detection Method in Smart City Environments Using a Deep-Learning-Based Approach. *Electronics* **2022**, *11*, 73. [CrossRef]
4. Yue, X.; Li, H.; Shimizu, M.; Kawamura, S.; Meng, L. YOLO-GD: A Deep Learning-Based Object Detection Algorithm for Empty-Dish Recycling Robots. *Machines* **2022**, *10*, 294. [CrossRef]
5. Ge, Y.; Yue, X.; Meng, L. YOLO-GG: A slight object detection model for empty-dish recycling robot. In Proceedings of the 2022 International Conference on Advanced Mechatronic Systems (ICAMechS), Toyama, Japan, 17–20 December 2022; pp. 59–63.
6. Yue, X.; Li, H.; Fujikawa, Y.; Meng, L. Dynamic Dataset Augmentation for Deep Learning-Based Oracle Bone Inscriptions Recognition. *J. Comput. Cult. Herit.* **2022**, *15*, 76. [CrossRef]

7. Meng, L.; Hirayama, T.; Oyanagi, S. Underwater-drone with panoramic camera for automatic fish recognition based on deep learning. *IEEE Access* **2018**, *6*, 17880–17886. [CrossRef]
8. Deng, M.; Inoue, A.; Shibata, Y.; Sekiguchi, K.; Ueki, N. An obstacle avoidance method for two wheeled mobile robot. In Proceedings of the 2007 IEEE International Conference on Networking, Sensing and Control, London, UK, 15–17 April 2007; pp. 689–692. [CrossRef]
9. Wen, S.; Deng, M.; Inoue, A. Operator-based robust non-linear control for gantry crane system with soft measurement of swing angle. *Int. J. Model. Identif. Control* **2012**, *16*, 86–96. [CrossRef]
10. Bergerman, M.; van Henten, E.; Billingsley, J.; Reid, J.F.; Deng, M. IEEE Robotics and Automation Society Technical Committee on Agricultural Robotics and Automation. *IEEE Robot. Autom. Mag.* **2013**, *20*, 20–125. [CrossRef]
11. Yue, X.; Lyu, B.; Li, H.; Meng, L.; Furumoto, K. Real-time medicine packet recognition system in dispensing medicines for the elderly. *Meas. Sens.* **2021**, *18*, 100072. [CrossRef]
12. Gu, J.; Wang, Z.; Kuen, J.; Ma, L.; Shahroudy, A.; Shuai, B.; Liu, T.; Wang, X.; Wang, G.; Cai, J.; et al. Recent advances in convolutional neural networks. *Pattern Recognit.* **2018**, *77*, 354–377. [CrossRef]
13. Daud, S.M.S.M.; Yusof, M.Y.P.M.; Heo, C.C.; Khoo, L.S.; Singh, M.K.C.; Mahmood, M.S.; Nawawi, H. Applications of drone in disaster management: A scoping review. *Sci. Justice* **2022**, *62*, 30–42. [CrossRef]
14. Ghimire, D.; Kil, D.; Kim, S.H. A Survey on Efficient Convolutional Neural Networks and Hardware Acceleration. *Electronics* **2022**, *11*, 945. [CrossRef]
15. Ahamad, A.; Sun, C.C.; Kuo, W.K. Quantized Semantic Segmentation Deep Architecture for Deployment on an Edge Computing Device for Image Segmentation. *Electronics* **2022**, *11*, 3561. [CrossRef]
16. Zhao, M.; Li, M.; Peng, S.L.; Li, J. A Novel Deep Learning Model Compression Algorithm. *Electronics* **2022**, *11*, 1066. [CrossRef]
17. Zhou, H.; Alvarez, J.M.; Porikli, F. Less is More: Towards Compact CNNs. In Proceedings of the Computer Vision–ECCV 2016: 14th European Conference, Amsterdam, The Netherlands, 11–14 October 2016; Springer International Publishing: Berlin/Heidelberg, Germany, 2016; pp. 662–677.
18. Li, H.; Yue, X.; Wang, Z.; Chai, Z.; Wang, W.; Tomiyama, H.; Meng, L. Optimizing the deep neural networks by layer-wise refined pruning and the acceleration on FPGA. *Comput. Intell. Neurosci.* **2022**, *2022*, 8039281. [CrossRef] [PubMed]
19. Jordao, A.; Lie, M.; Schwartz, W.R. Discriminative Layer Pruning for Convolutional Neural Networks. *IEEE J. Sel. Top. Signal Process.* **2020**, *14*, 828–837. [CrossRef]
20. Yuan, S.; Du, Y.; Liu, M.; Yue, S.; Li, B.; Zhang, H. YOLOv5-Ytiny: A Miniature Aggregate Detection and Classification Model. *Electronics* **2022**, *11*, 1743. [CrossRef]
21. Lin, S.; Ji, R.; Chen, C.; Tao, D.; Luo, J. Holistic CNN Compression via Low-Rank Decomposition with Knowledge Transfer. *IEEE Trans. Pattern Anal. Mach. Intell.* **2019**, *41*, 2889–2905. [CrossRef]
22. Li, H.; Wang, Z.; Yue, X.; Wang, W.; Hiroyuki, T.; Meng, L. A Comprehensive Analysis of Low-Impact Computations in Deep Learning Workloads. In Proceedings of the 2021 on Great Lakes Symposium on VLSI, GLSVLSI '21, Virtual Event, 22–25 June 2021; Association for Computing Machinery: New York, NY, USA, 2021; pp. 385–390. [CrossRef]
23. Hao, Z.; Li, Z.; Dang, X.; Ma, Z.; Liu, G. MM-LMF: A Low-Rank Multimodal Fusion Dangerous Driving Behavior Recognition Method Based on FMCW Signals. *Electronics* **2022**, *11*, 3800. [CrossRef]
24. Gong, C.; Chen, Y.; Lu, Y.; Li, T.; Hao, C.; Chen, D. VecQ: Minimal Loss DNN Model Compression With Vectorized Weight Quantization. *IEEE Trans. Comput.* **2021**, *70*, 696–710. [CrossRef]
25. Husham Almukhtar, F.; Abbas Ajwad, A.; Kamil, A.S.; Jaleel, R.A.; Adil Kamil, R.; Jalal Mosa, S. Deep Learning Techniques for Pattern Recognition in EEG Audio Signal-Processing-Based Eye-Closed and Eye-Open Cases. *Electronics* **2022**, *11*, 4029. [CrossRef]
26. Guo, K.; Zeng, S.; Yu, J.; Wang, Y.; Yang, H. [DL] A survey of FPGA-based neural network inference accelerators. *ACM Trans. Reconfigurable Technol. Syst. TRETS* **2019**, *12*, 1–26. [CrossRef]
27. Li, H.; Yue, X.; Wang, Z.; Wang, W.; Tomiyama, H.; Meng, L. A survey of Convolutional Neural Networks —From software to hardware and the applications in measurement. *Meas. Sens.* **2021**, *18*, 100080. [CrossRef]
28. He, K.; Zhang, X.; Ren, S.; Sun, J. Deep residual learning for image recognition. In Proceedings of the IEEE Conference on Computer Vision and Pattern Recognition, Las Vegas, NV, USA, 26 June–1 July 2016; pp. 770–778.
29. Huang, G.; Liu, Z.; Van Der Maaten, L.; Weinberger, K.Q. Densely connected convolutional networks. In Proceedings of the IEEE Conference on Computer Vision and Pattern Recognition, Honolulu, HI, USA, 21–26 July 2017; pp. 4700–4708.
30. Sandler, M.; Howard, A.; Zhu, M.; Zhmoginov, A.; Chen, L.C. Mobilenetv2: Inverted residuals and linear bottlenecks. In Proceedings of the IEEE Conference on Computer Vision and Pattern Recognition, Salt Lake City, UT, USA, 18–23 June 2018; pp. 4510–4520.
31. Kuang, J.; Shao, M.; Wang, R.; Zuo, W.; Ding, W. Network pruning via probing the importance of filters. *Int. J. Mach. Learn. Cybern.* **2022**, *13*, 2403–2414. [CrossRef]
32. Li, Y.; Gu, S.; Mayer, C.; Gool, L.V.; Timofte, R. Group sparsity: The hinge between filter pruning and decomposition for network compression. In Proceedings of the IEEE/CVF Conference on Computer Vision and Pattern Recognition, Seattle, WA, USA, 13–19 June 2020; pp. 8018–8027.
33. Shawahna, A.; Sait, S.M.; El-Maleh, A. FPGA-based accelerators of deep learning networks for learning and classification: A review. *IEEE Access* **2018**, *7*, 7823–7859. [CrossRef]

34. Misra, J.; Saha, I. Artificial neural networks in hardware: A survey of two decades of progress. *Neurocomputing* **2010**, *74*, 239–255. [CrossRef]
35. Han, S.; Liu, X.; Mao, H.; Pu, J.; Pedram, A.; Horowitz, M.A.; Dally, W.J. EIE: Efficient inference engine on compressed deep neural network. *ACM SIGARCH Comput. Archit. News* **2016**, *44*, 243–254. [CrossRef]
36. Chen, Y.H.; Emer, J.; Sze, V. Eyeriss: A spatial architecture for energy-efficient dataflow for convolutional neural networks. *ACM SIGARCH Comput. Archit. News* **2016**, *44*, 367–379. [CrossRef]
37. Hameed, R.; Qadeer, W.; Wachs, M.; Azizi, O.; Solomatnikov, A.; Lee, B.C.; Richardson, S.; Kozyrakis, C.; Horowitz, M. Understanding sources of inefficiency in general-purpose chips. In Proceedings of the 37th Annual International Symposium on Computer Architecture, Saint-Malo, France, 19–23 June 2010; pp. 37–47.
38. Keckler, S.W.; Dally, W.J.; Khailany, B.; Garland, M.; Glasco, D. GPUs and the Future of Parallel Computing. *IEEE Micro* **2011**, *31*, 7–17. [CrossRef]
39. Chen, S.; Zhao, Q. Shallowing Deep Networks: Layer-Wise Pruning Based on Feature Representations. *IEEE Trans. Pattern Anal. Mach. Intell.* **2019**, *41*, 3048–3056. [CrossRef]
40. He, Y.; Zhang, X.; Sun, J. Channel Pruning for Accelerating Very Deep Neural Networks. In Proceedings of the IEEE International Conference on Computer Vision (ICCV), Venice, Italy, 22–29 October 2017.
41. Gou, J.; Yu, B.; Maybank, S.J.; Tao, D. Knowledge distillation: A survey. *Int. J. Comput. Vis.* **2021**, *129*, 1789–1819. [CrossRef]
42. Huang, Z.; Wang, N. Like what you like: Knowledge distill via neuron selectivity transfer. *arXiv* **2017**, arXiv:1707.01219.
43. Aghli, N.; Ribeiro, E. Combining weight pruning and knowledge distillation for cnn compression. In Proceedings of the IEEE/CVF Conference on Computer Vision and Pattern Recognition, Nashville, TN, USA, 20–25 June 2021; pp. 3191–3198.
44. Hinton, G.; Vinyals, O.; Dean, J. Distilling the knowledge in a neural network. *arXiv* **2015**, arXiv:1503.02531.
45. CIFAR-10 and CIFAR-100 Datasets. Available online: https://www.cs.toronto.edu/~kriz/cifar.html (accessed on 3 October 2021).
46. Loshchilov, I.; Hutter, F. SGDR: Stochastic Gradient Descent with Warm Restarts. *arXiv* **2016**, arXiv:1608.03983.

 *electronics*

Article

Generalized Zero-Shot Image Classification via Partially-Shared Multi-Task Representation Learning

Gerui Wang [1,2,]* and Sheng Tang [1,2]

[1] School of Computer Science and Engineering, Central South University, Changsha 410083, China; 204712137@csu.edu.cn
[2] Hunan Engineering Research Center of Machine Vision and Intelligent Medicine, Central South University, Changsha 410083, China
* Correspondence: wjhzxy@csu.edu.cn

Abstract: Generalized Zero-Shot Learning (GZSL) holds significant research importance as it enables the classification of samples from both seen and unseen classes. A prevailing approach for GZSL is learning transferable representations that can generalize well to both seen and unseen classes during testing. This approach encompasses two key concepts: discriminative representations and semantic-relevant representations. "Semantic-relevant" facilitates the transfer of semantic knowledge using pre-defined semantic descriptors, while "discriminative" is crucial for accurate category discrimination. However, these two concepts are arguably inherently conflicting, as semantic descriptors are not specifically designed for image classification. Existing methods often struggle with balancing these two aspects and neglect the conflict between them, leading to suboptimal representation generalization and transferability to unseen classes. To address this issue, we propose a novel partially-shared multi-task representation learning method, termed PS-GZSL, which jointly preserves complementary and sharable knowledge between these two concepts. Specifically, we first propose a novel perspective that treats the learning of discriminative and semantic-relevant representations as optimizing a discrimination task and a visual-semantic alignment task, respectively. Then, to learn more complete and generalizable representations, PS-GZSL explicitly factorizes visual features into task-shared and task-specific representations and introduces two advanced tasks: an instance-level contrastive discrimination task and a relation-based visual-semantic alignment task. Furthermore, PS-GZSL employs Mixture-of-Experts (MoE) with a dropout mechanism to prevent representation degeneration and integrates a conditional GAN (cGAN) to synthesize unseen features for estimating unseen visual features. Extensive experiments and more competitive results on five widely-used GZSL benchmark datasets validate the effectiveness of our PS-GZSL.

Keywords: Generalized Zero-Shot Learning; discriminative; semantic-relevant; image classification; partially-shared multi-task learning; transferable representation

Citation: Wang, G.; Tang, S. Generalized Zero-Shot Image Classification via Partially-Shared Multi-Task Representation Learning. *Electronics* **2023**, *12*, 2085. https://doi.org/10.3390/electronics12092085

Academic Editor: George A. Papakostas

Received: 29 March 2023
Revised: 27 April 2023
Accepted: 29 April 2023
Published: 3 May 2023

1. Introduction

Generalized Zero-Shot Learning (GZSL) [1] has attracted significant research interest due to its ability to transfer knowledge to unseen classes using additional class-level semantic descriptors, such as word vectors [2] or attributes [3]. As an extension of Zero-Shot Learning (ZSL) [3,4], GZSL aims to classify both seen and unseen classes simultaneously during testing. This capability is crucial in various real-world applications where the availability of labeled samples for all possible classes is limited or infeasible [5,6].

A key idea in GZSL is learning transferable representations, which encompass two essential concepts: *discriminative* and *semantic-relevant* features. *Discriminative* features are crucial for accurate category discrimination, possessing strong decision-making power and promoting the classification task of unseen classes. In contrast, *semantic-relevant* facilitates a shared semantic space between seen and unseen classes using pre-defined semantic descriptors, reflecting the semantic relationships between different classes as accurately as possible. GZSL can be viewed as a multi-task problem, where learning discriminative features optimizes a discrimination sub-task, and learning semantically-relevant features optimizes a visual-semantic alignment sub-task. By adopting a multi-task perspective, GZSL aims to obtain comprehensive representations between tasks that can generalize well to unseen classes during testing. However, since semantic descriptors are not specifically designed for image classification [1,7,8], two main challenges arise: (1) appropriately balancing these sub-tasks and resolving their conflict, and (2) ensuring the stability and expressiveness of learned representations.

Unfortunately, existing methods tend to bypass or ignore these challenges between *discriminative* and *semantic-relevant*, resulting in passable performance on unseen classes. Specifically: (1) some researchers focus solely on semantic-relevant representations through elaborate visual-semantic alignment [8–10], while others concentrate on advanced discrimination techniques to extract more generalizable discriminative representations [11,12]. (2) Furthermore, the conflict between discrimination and visual-semantic alignment is often neglected, as recent methods primarily focus on learning shared representations between these two sub-task [7,13,14]. As a result, their poor generalization can be attributed to the discarding of some task-specific information between sub-tasks, which can be viewed as the "diamond in the rough" for GZSL. Some works in domain generalization (DG) have shown that this specific information could enhance a model's generalization performance when classifying unseen classes [15,16]. For example, in the AWA1 dataset shown in Figure 1. attributes like "Strong, Big" that are not visually discriminative can still reduce the misclassification between tigers and cats. Similarly, visual cues like the ear and nose shape are salient for classifying image samples but not represented in the semantic descriptors.

To address the aforementioned challenges and limitations, we propose a novel partially-shared representation learning network, termed PS-GZSL, which jointly preserves complementary and transferable information between discriminative and semantic-relevant features. First, to resolve the conflict between tasks and avoid information loss, PS-GZSL proposes a partially-shared multi-task learning mechanism to explicitly model both task-shared and task-specific representations. As depicted in Figure 2, PS-GZSL utilizes three Mixture-of-Experts (MoE) [17,18] to factorize a visual feature into three latent representations: a task-shared discriminative and semantic representation h_{ds}, a task-specific discriminative representation h_d, and a task-specific semantic-relevant representation h_s. Each sub-task corresponds to a task-specific and a task-shared representation. Second, to ensure the stability and expressiveness of learned representations, PS-GZSL draws inspiration from the success of contrastive learning [19] and metric learning [20], proposing two effective sub-tasks: an instance-level contrastive discrimination task and a relation-based visual-semantic alignment task. These tasks have been proven to achieve better generalization performance, respectively. To avoid representation degeneration, PS-GZSL randomly drops out experts in each MoE. Furthermore, PS-GZSL is a hybrid GZSL framework that integrates with a feature generation component. In feature generation, PS-GZSL adopts a conditional generative adversarial network [21] with a feedback mechanism to mitigate the bias towards seen classes in the latent representation space.

In summary, the main contributions of our work can be summarized:

1. We describe a novel perspective grounded in multi-task learning, which reveals that existing methods exhibit an inherent generalization weakness of losing some transferable visual features.
2. We propose a novel GZSL method, termed partially-shared multi-task representation learning network (PS-GZSL), to jointly preserve complementary and transferable information between discriminative and semantic-relevant features
3. Extensive experiments on five widely-used GZSL benchmark datasets validate the effectiveness of our PS-GZSL and show that the joint contributions of the task-shared and task-specific representations result in more transferability representation.

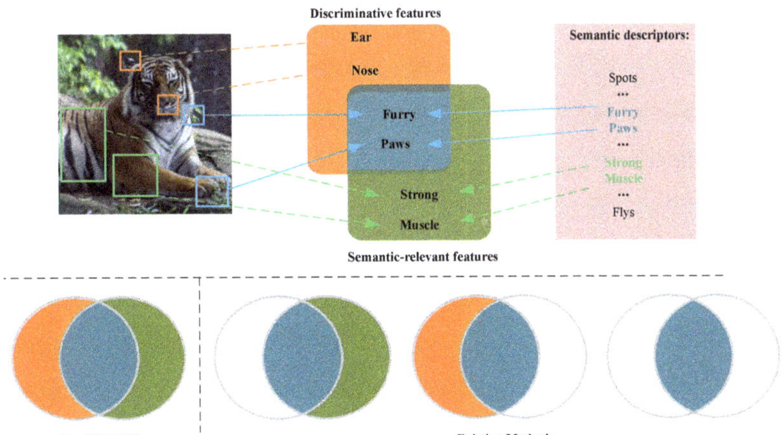

Figure 1. Existing GZSL methods either bypass or ignore the conflict between discriminative and semantic-relevant objectives, and may overlook some task-specific visual features (as indicated by the green and orange dashed lines). In contrast, PS-GZSL can preserve more complete sharable features.

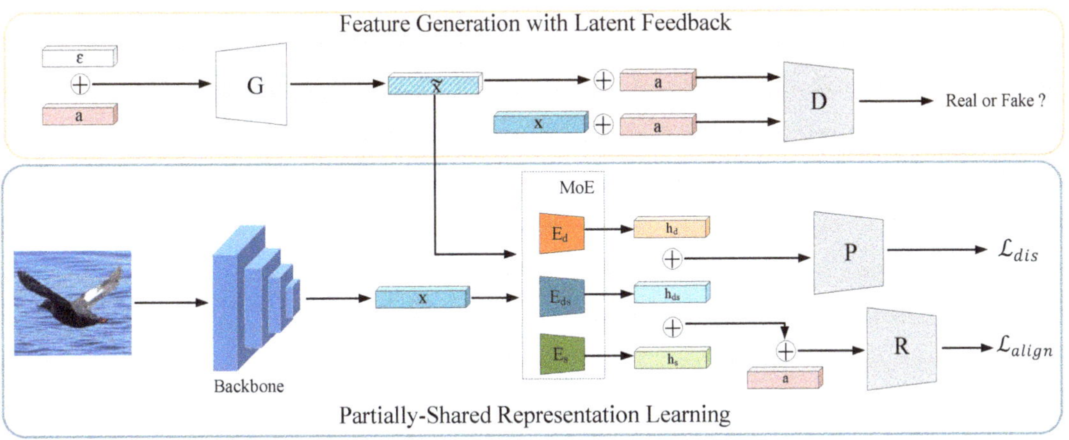

Figure 2. Illustration of our proposed PS-GZSL, which consists of (i) a conditional GAN network **D** and **G** with latent feedback mechanism; (ii) a multi-branch MoE network $E = [E_{ds}, E_d, E_s]$ for factorized latent representation learning. And two task modules **P** and **R** are extended to ensure the discriminative property and semantic property. Here, **a** denotes the semantic descriptors, and ϵ is a random gaussian noise.

2. Related Works

Early approaches for ZSL/GZSL can be broadly classified into two main groups: Embedding-based methods and Generative-based methods. The former group [22–27] learns an encoder to map the visual features of seen classes to their respective semantic descriptors. In contrast, the latter group [21,28–32] learns a conditional generator, such as cVAE [33] or cGAN [34], to synthesize virtual unseen features based on the seen samples and semantic descriptors of both classes.

Recent state-of-the-art methods typically graft an encoder on top of a conditional generator, with a focus on improving the transferability of visual representations. (1) Some methods emphasize preserving semantic-relevant information that corresponds to pre-defined descriptors. For example, CADA-VAE [9] employs two aligned Variational Autoencoders(VAEs) to learn shared latent representations between semantic descriptors and visual features. SDGZSL [10] integrates a disentanglement constraint and a Relation network [20] to ensure the semantic-consistency of the learned representation. SE-GZSL [35] uses two AutoEncoders and Mutual information maximization to capture semantic-relevant information. (2) Some others prioritize the preservation of more discriminative information. DLFZRL [11] adopts a hierarchical factorizing approach and adversarial learning to learn the discriminative latent representation, regardless of whether it is semantically relevant or not. DR-GZSL [7] utilizes an auxiliary classifier and a shuffling disentanglement mechanism to extract the discriminative part of the semantic-relevant representation. CE-GZSL [13] integrates the semantic-supervised learning module and label-supervised discrimination module in the latent space to learn discriminative visual representations. In summary, these methods differ in the transferable characteristics of the data they model for recognition.

In contrast to existing methods, we argue that both discriminative and semantic-relevant representations are important for recognizing test classes. However, due to the conflict between them, these methods implicitly discard some valuable features. We are thus motivated to adopt the soft-parameter sharing mechanism [17,36] in multi-task learning. This flexibility stems is derived from information routing between tasks, and its characteristics of seeking similarities while preserving differences have led to significant successes in multi-task learning domains such as recommendation systems. We are the first to apply this idea and revise it for representation learning in GZSL. A novel multi-task representation learning paradigm is proposed that models task-specific and task-shared representations in parallel, unlike existing paradigms [37,38] that use a single MoE for each sub-task and a hierarchical structure. For the sake of clear understanding, we highlight the distinctions between our approach and those counterparts in Table 1.

Table 1. Qualitative Model Comparison. The $\bigcirc$, $\square$, and $\triangle$ denote representations that are discriminative and semantic-relevant, only discriminative, and only semantic-relevant, respectively.

Model Comparison	Task-Shared	Task-Specific	
	$\bigcirc$	$\square$	$\triangle$
SP-AEN [8]	✓		✓
CADA-VAE [9]	✓		
SDGZSL [10]	✓		✓
DLFZRL [11]	✓	✓	
DR-GZSL [7]	✓		
CE-GZSL [13]	✓		
Our PS-GZSL	✓	✓	✓

3. Methods

To learn more transferable representations, in this section, we present our proposed PS-GZSL method, which combines MoE, a partially-shared mechanism, an instance contrastive discrimination module, and a relation-based visual-semantic alignment module. To alleviate the bias towards seen, we also adopt a feature generation module with latent feedback. The overall framework of our proposed PS-GZSL is shown in Figure 2, Then, the definition of the ZSL/GZSL problem and all the above modules are explained in detail.

3.1. Problem Definition

In Zero-Shot learning, we are given two disjoint sets of classes: $\{\mathcal{X}^s, \mathcal{Y}^s\}$ with $\mathcal{S}$ seen classes and $\{\mathcal{X}^u, \mathcal{Y}^u\}$ with $\mathcal{U}$ unseen classes, where we have $\mathcal{Y}^s \cap \mathcal{Y}^u = \varnothing$ and $\mathcal{Y}^{all} = \mathcal{Y}^s \cup \mathcal{Y}^u$. For the semantic descriptors $\mathcal{A} = \{a_1, \ldots, a_S, a_{S+1}, \ldots, a_{S+U}\}$, each class, whether seen or unseen, is associated with a semantic descriptor that can take the form of sentences or attributes. Under ZSL setting, we have $\{\mathcal{X}^s, \mathcal{Y}^s, \mathcal{A}^s\}$ and $\{\mathcal{Y}^u, \mathcal{A}^u\}$ available during training phase. Let $x \in \mathcal{X}$ denote the extracted feature instances of images. The goal of ZSL is to learn a model f to classify unseen samples during the test phase, which can be formulated as $f : x \rightarrow \mathcal{Y}^u$. GZSL is a more realistic and challenging problem that requires f to handle both seen and unseen samples: $f : x \rightarrow \mathcal{Y}^{all}$.

3.2. Task-Shared and Task-Specific Representations

To begin our PS-GZSL, we first provide definitions for three visual representations that are concerning discriminative and semantic-relevant concepts.
Discriminative and Semantic-relevant Representations. Firstly, we define task-shared discriminative and semantic-relevant representations h_{ds} to encode the discriminative features of images that are related to corresponding semantic descriptors. These visual features are used for the both discrimination task and the visual-semantic alignment task during the training phase.
Discriminative but Non-semantic Representations. Secondly, discriminative but non-semantic features are encoded in discrimination task-specific representations, denoted as h_d. These features are important for discrimination, but they may not contribute to the visual-semantic alignment task since not represented in the semantic descriptors.
Non-Discriminative but Semantic-relevant Representations. Finally, non-discriminative but semantic-relevant features are encoded in visual-semantic alignment task-specific representations, denoted as h_s. These features are not discriminative in seen classes but may be critical for recognizing unseen classes. Thus, these features only contribute to the visual-semantic alignment task during training.

3.3. Representation Learning

As shown in Figure 2, Our encoder module consists of three parallel Mixture-of-experts (MoE) modules ($E = [E_{ds}, E_d, E_s]$), which explicitly factorize a visual feature x into three latent representations: $h_{ds}, h_d,$ and h_s, i.e., $h_{ds} = E_{ds}(x), h_d = E_d(x)$ and $h_s = E_s(x)$.

3.3.1. Mixture-of-Experts

PS-GZSL adopts a gated MoE module to replace simple Multi-Layer Perceptrons (MLPs) in order to obtain more expressive representations, MoE is a neural network architecture that comprises several experts, each of which specializes in a specific part of the input space. The output of the network is then computed as a weighted combination of the outputs of the experts by a gating network, as shown in Figure 3.

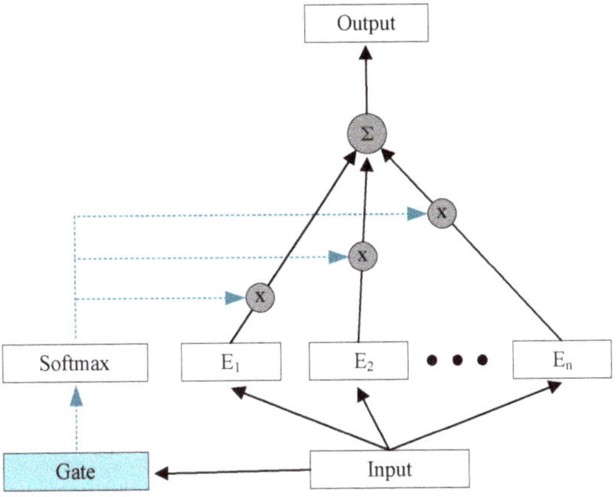

Figure 3. The architecture of MoE.

Given a visual feature as input, the MoE module can be formulated as:

$$E(x) = \sum_{i}^{n} g(x)_i e_i(x),$$

$$(1)$$

where, the gate network g combines the results of n expert networks, where $\sum_{i=1}^{n} g(x)_i = 1$ and $g(x)_i$ represents the ith logit of the output, indicating the weight assigned to expert e_i.

We denote the aforementioned three MoE modules as E_{ds}, E_d and E_s for the task-shared representation h_{ds} and two task-specific representation h_d and h_s, respectively. It's worth noting that we've incorporated the dropout technique in the gate network, which randomly discards some outputs of the experts. This technique helps prevent overfitting and also ensures that the representations (h_{ds}, h_d, and h_s) remain informative for subsequent sub-tasks.

3.3.2. Instance Contrastive Discrimination Task

According to the definition above, both h_{ds} and h_d are expected to capture the discriminative features. For convenience, we denote $w = h_{ds} \oplus h_d = E_{ds}(x) \oplus E_d(x)$. To compare the similarities and differences of visual representations w, an instance contrastive discrimination task is proposed, which assigns samples to different categories according to the comparison results. Specifically, PS-GZSL takes Supervised Contrastive Learning (SupCon) [19] loss as the objective function in this task since SupCon shows better generalization performance and stronger robustness in discriminative representation learning compared with other metric learning loss.

We follow the strategy proposed in [19] where the representation w is further propagated through a projection network P (as shown in Figure 4) to obtain a new representation denoted as $z = P(w)$. For every w_i encoded from a visual feature x_i, the SupCon loss of w_i is as follows:

$$\ell(z_i) = -\log\left\{ \frac{1}{P(i)} \sum_{p \in P_i} \frac{\exp\left(z_i^\top z_p / \tau_e\right)}{\sum_{k \in K(i)} \exp\left(z_i^\top z_k / \tau_e\right)} \right\},$$

$$(2)$$

where, $\tau_e > 0$ denotes the temperature parameter for stable training. $P(i) \equiv \{p \in K(i) : y_p = y_i\}$ represents the indices of all positives in the mini-batch that are distinct from i, and $|P(i)|$ is its cardinality.

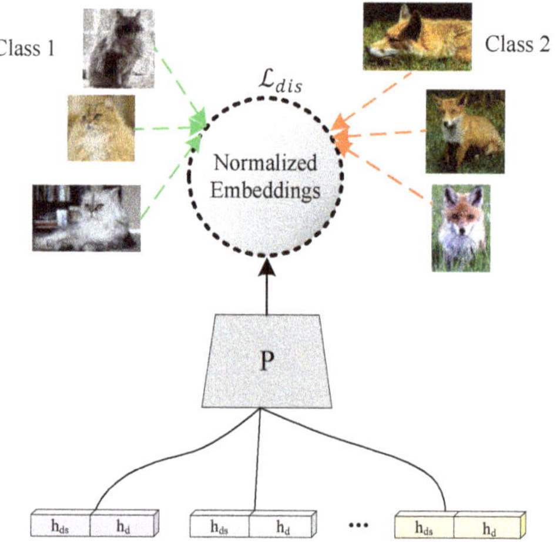

Figure 4. Illustration of Instance Contrastive Discrimination.

To simultaneously learn the MoE modules E_{ds}, E_d, and the projection network P, the loss function for this discrimination task is calculated as the sum of instance-level SupCon loss within a batch of samples I.

$$\mathcal{L}_{dis}(E_d, E_{ds}, P) = \sum_{i \in I} \ell(z_i). \tag{3}$$

Such a contrastive learning encourages E_{ds} and E_d to capture the strong inter-class discriminative features, and intra-class structure shared in the latent space, making both h_{ds} and h_d more discriminative and more transferable. Furthermore, we demonstrate the superiority of SupCon loss over softmax loss in ablation experiments.

3.3.3. Relation-Based Visual-Semantic Alignment Task

In the same way, both h_{ds} and h_s are devised to capture semantic-relevant information that corresponds to the annotated semantic descriptors A. For convenience, we denote $v = h_{ds} \oplus h_s = E_{ds}(x) \oplus E_s(x)$. In order to learn semantic-relevant representations v without directly mapping visual features into the semantic space, we adopt a Relation network in [20] as a visual-semantic alignment task. The goal is to maximize the similarity score (SS) between v and the corresponding semantic descriptor a through a deeper end-to-end architecture, which includes a learned nonlinear metric in the form of our alignment task. Thus, the objective of this task is to accurately measure the similarity score between pairs of v and a via a neural network. The similarity score SS of the matched pairs is set to 1, while mismatched pairs are assigned 0, which can be formulated as:

$$SS(v_t, a_c) = \begin{cases} 0, & y_t \neq y_c \\ 1, & y_t = y_c \end{cases}, \tag{4}$$

where t and c refer to the t-th visual sample's semantic-relevant representation and c-th class-level semantic descriptor from the seen classes, y_t and y_c denote the ground truth label of v_t and a_c.

In [20], they utilize mean square error(MSE) as a loss function while ignoring the class-imbalance problem in zero-shot learning. Moreover, as SupCon requires a large batch size, "Softmax + Cross Entropy" is a more efficient alternative than MSE in this scenario (as shown in Figure 5).

Denote the relation module as R. We can calculate the loss function of this task as:

$$\mathcal{L}_{align}(E_s, E_{ds}, R) = \sum_{i \in I} -\log \frac{\exp\left(R(v_i, a^+)/\tau_s\right)}{\sum_{s=1}^{S} \exp\left(R(v_i, a_s)/\tau_s\right)}, \tag{5}$$

where, S denotes the number of seen classes, and $\tau_s > 0$ denotes the scaling factor to stable the softmax activation for robust performance.

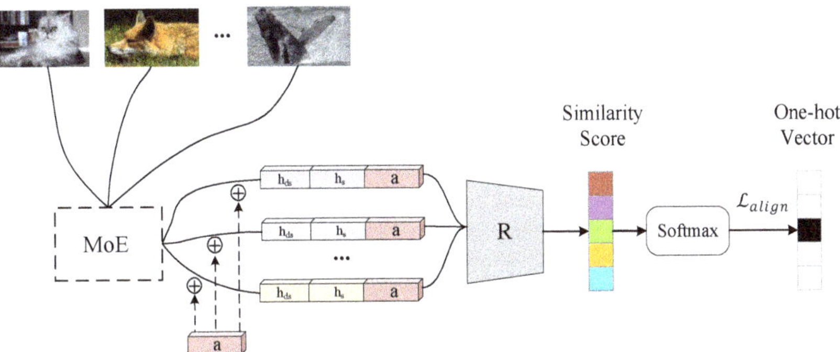

Figure 5. Illustration of Relation-Based Visual-Semantic Alignment.

3.4. Feature Generation with Latent Feedback

In order to alleviate the phenomenon that encoded representations are biased towards seen classes in GZSL, we integrate the proposed representation learning method on top of a conditional GAN (cGAN) [21]. Specifically, we adopt a conditional generator network G to generate virtual unseen features $\tilde{x} = G(a, \epsilon)$, here $\epsilon \sim \mathcal{N}(\mathbf{0}, \mathbf{I})$ represent a Gaussian noise. In the meanwhile, we train a discriminator D to distinguish between a real pair (x, a) and a generated pair $(\tilde{x}, a)$. The generator G and the discriminator D are jointly trained by minimizing the adversarial objective given as:

$$V(G, D) = \mathbb{E}_{p(x,a)}[\log D(x, a)] + \mathbb{E}_{p_G(\tilde{x},a)}[\log(1 - D(\tilde{x}, a))], \tag{6}$$

where $p_{(}x, a)$ and $p_G(\tilde{x}, a)$ represent the joint distribution of real/synthetic visual-semantic pairs, respectively.

However, the objective stated above does not guarantee that the generated features are discriminative or semantic-relevant. Drawing on the feedback mechanism in [13,21,39], we aim to improve the quality of generated features by passing them through the aforementioned multi-task network. Therefore, Equation (6) can be reformulated as:

$$\begin{aligned} V(G, D) =& \mathbb{E}_{p(x,a)}[\log D(x, a)] + \mathbb{E}_{p_G(\tilde{x},a)}[\log(1 - D(\tilde{x}, a))] \\ &+ \mathbb{E}_{p_G(\tilde{x},a)}[\delta_1 \mathcal{L}_{align} + \delta_2 \mathcal{L}_{dis}], \end{aligned} \tag{7}$$

3.5. Training and Inference

As a summary, the overall loss of our proposed method is formulated as:

$$\mathcal{L}_{total} = V(G, D) + \mathcal{L}_{dis}(E_d, E_{sh}, P) + \mathcal{L}_{align}(E_s, E_{sh}, R). \tag{8}$$

Given visual features and corresponding semantic descriptors from seen classes, PS-GZSL solves GZSL in four steps:

1. Training feature generation and representation learning models based on Equation (8).
2. These learned models are then used to synthesize and extract unseen class representations $\tilde{c}$.
3. Using real visual samples x from seen classes for training the partially-shared representation learning part and synthesized visual samples $\tilde{x}$ for tuning generator.
4. The final generalized zero-shot classifier is a single layer linear softmax classifier, learned on $\tilde{c}$ and c (extracted from real seen x and synthesized samples $\tilde{x}$), as depicted in Figure 6.

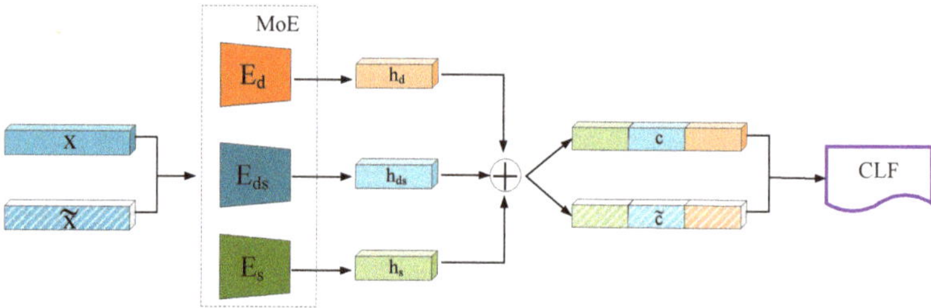

Figure 6. Using concatenated task-shared and task-specific representation for training classifier.

4. Experiments

4.1. Datasets

We perform our PS-GZSL on five widely used benchmark datasets for GZSL, including Animals with Attributes 1&2 (AWA1 [3] & AWA2 [1]), Caltech-UCSD Birds-200-2011 (CUB) [40], Oxford Flowers (FLO) [41]), and SUN Attribute (SUN) [42]. For visual features, we follow the standard GZSL practice of using ResNet101 [43] pre-trained on ImageNet-1k [44] without fine-tuning, resulting in 2048-dimensional features for each image. The semantic descriptors used for AWA1, AWA2, and SUN are their respective class-level attributes. For CUB and FLO, the semantic descriptors are generated from 10 textual descriptions by character-based CNN-RNN [45]. In addition, we employ the Proposed Split(PS) in [1] to split seen and unseen classes on each dataset. The statistics of the datasets and GZSL split settings are illustrated in Table 2.

Table 2. Statistics of the AWA1&2, CUB, and FLO, SUN datasets.

Dataset	AWA1	AWA2	CUB	FLO	SUN
#Seen Classes	40	40	150	82	645
#Unseen Classes	10	10	50	20	72
#Samples	30,475	37,322	11,788	8189	14,340
#Semantic Descriptors [1]	85	85	1024	1024	102
#Training Samples	19,832	23,527	7057	5394	10,320
#Test Seen Samples	4958	5882	1764	1640	2580
#Test Unseen Samples	5685	7913	2967	1155	1440

[1] #Semantic Descriptors indicate the dimensions of semantic descriptors per class.

4.2. Metrics

To assess the model performance in GZSL setting, we use the harmonic mean of per-class Top-1 accuracy on seen classes and unseen classes, formulated as $H = 2 \times S \times U/(S + U)$, where S and U represent seen accuracy and unseen accuracy, respectively. In addition, we adopt U as the evaluation metric for ZSL.

4.3. Implementation Details

In our PS-GZSL, all networks are implemented with Multi-Layer Perceptrons(MLPs). The architecture of the discriminator and generator of the feature generation architectures consist of single-layer MLPs with a 4096-unit hidden layer activated by LeakyReLU. In representation learning, each MoE module contains three experts and corresponds to a gate network. The dimension of task-specific representation (h_d & h_s) and task-shared representation (h_{sh}) are set to 1024 in all of the five datasets. For the projection network P, we set the size of the projection's output z to 256 for AWA2, FLO, and SUN and 512 for AWA1 and CUB. The relation network R contains two FC+ReLU layers, and we utilize 2048 hidden units for AWA1, AWA2, and CUB and 1024 units for FLO and SUN. The difference among datasets has motivated us to perform numerous experiments aimed at determining the optimal number of synthesized unseen visual instances in each dataset. Once PS-GZSL is trained, we use a fixed 400 per unseen class for CUB, 2400 for AWA1&2, 600 for FLO, and 100 for SUN. The weighting coefficients in Equation (7) are set to $\sigma_1 = 0.001$ and $\sigma_2 = 0.001$, and the value of temperature in Equations (2) and (5) are set to $\tau_e = 0.1$ and $\tau_s = 0.1$. We optimize the overall loss function (Equation (8)) with the Adam optimizer, using $\beta_1 = 0.5$, $\beta_2 = 0.999$. The mini-batch size is set to 512 for AWA1, AWA2, CUB, and SUN, and 3072 for FLO in our method. All experiments are implemented with PyTorch, and trained on a single NVIDIA RTX 2080Ti GPU.

4.4. Comparison with State-of-the-Arts

Recently, some methods have introduced transductive zero-shot learning on target datasets, where they use unlabeled unseen samples for training models, leading to significant performance increases. However, it is costly and even unrealistic in real-world zero-shot scenarios. Thus, we only present results under the inductive setting.

Our PS-GZSL is compared with other GZSL methods on five widely used datasets without fine-tuning the pre-trained backbone. Results of our method in GZSL are given in Table 3, which indicates that PS-GZSL is compatible with the state-of-the-art. Specifically, PS-GZSL attains the best harmonic mean **H** on four datasets, i.e., 70.6 on AWA1, 71.8 on AWA2, 67.4 on CUB, and 43.3 on SUN. Notably, on CUB, PS-GZSL is the first one that attains a performance > 70.0 on unseen accuracy, which is even higher than the seen accuracy. This is because PS-GZSL retains more information in the learned representations to enhance GZSL classification during testing. As a result, representations for seen classes contain some redundancy, which adversely affects their classification accuracy. On FLO, PS-GZSL achieves the second-best harmonic mean **H** with 73.8, only lower than FREE [14]. However, PS-GZSL outperforms FREE by a considerable margin on the other four datasets. These results show that PS-GZSL can acquire classification knowledge transferable to unseen classes by utilizing the partially-shared mechanism and MoE, thereby learning more transferable representations from the seen classes. Specifically, by explicitly preserving these task-specific representations, the three MoE modules can effectively reduce the loss of information caused by the conflict between discrimination and visual-semantic alignment, thus enabling the preservation of more useful features for the testing phase.

Table 3. Comparisons with the State-Of-The-Art GZSL Methods. The best results and the second-best results are respectively marked in red and blue.

Methods	AWA1			AWA2			CUB			FLO			SUN		
	U	S	H	U	S	H	U	S	H	U	S	H	U	S	H
DeViSE [22]	13.4	68.7	22.4	17.1	74.7	27.8	23.8	53.0	32.8	9.9	44.2	16.2	16.9	27.4	20.9
TCN [46]	49.4	76.5	60.0	61.2	65.8	63.4	52.6	52.0	52.3	-	-	-	31.2	37.3	34.0
DVBE [47]	-	-	-	63.6	70.8	67.0	53.2	60.2	56.5	-	-	-	45.0	37.2	40.7
f-CLSWGAN [21]	57.9	64.0	60.2	-	-	-	43.7	57.7	49.7	59.0	73.8	65.6	42.6	36.6	39.4
CADA-VAE [9]	57.3	72.8	64.1	55.8	75.0	63.9	51.6	53.5	52.4	-	-	-	47.2	35.7	40.6
SP-AEN [8]	-	-	-	23.3	90.9	37.1	34.7	70.6	46.6	-	-	-	24.9	38.6	30.3
LisGAN [28]	52.6	76.3	62.3	-	-	-	46.5	57.9	51.6	57.7	83.8	68.3	42.9	37.8	40.2
cycle-CLSWGAN [30]	56.9	64.0	60.2	-	-	-	45.7	61.0	52.3	59.2	72.5	65.1	49.4	33.6	40.0
DLFZRL [11]	-	-	61.2	-	-	60.9	-	-	51.9	-	-	-	-	-	42.5
cvcZSL [48]	62.7	77.0	69.1	56.4	81.4	66.7	47.4	47.6	47.5	-	-	-	36.3	42.8	39.3
f-VAEGAN-D2 [29]	57.9	61.4	59.6	-	-	-	43.7	57.7	49.7	59.0	73.8	65.6	42.6	36.6	39.4
LsrGAN [31]	54.6	74.6	63.0	-	-	-	48.1	59.1	53.0	-	-	-	44.8	37.7	40.9
TF-VAEGAN [39]	-	-	-	59.8	75.1	66.6	52.8	64.7	58.1	62.5	84.1	71.7	45.6	40.7	43.0
DR-GZSL [7]	60.7	72.9	66.2	56.9	80.2	66.6	51.1	58.2	54.4	-	-	-	36.6	47.6	41.4
SDGZSL [10]	-	-	-	64.6	73.6	68.8	59.9	66.4	63.0	62.2	79.3	69.8	48.2	36.1	41.3
CE-GZSL [13]	65.3	73.4	69.1	63.1	78.6	70.0	63.9	66.8	65.3	69.0	78.7	73.5	48.8	38.6	43.1
FREE [14]	62.9	69.4	66.0	60.4	75.4	67.1	55.7	59.9	57.7	67.4	84.5	75.0	47.4	37.2	41.7
Our PS-GZSL	67.5	74.1	70.6	66.4	78.1	71.8	70.6	64.5	67.4	66.8	82.5	73.8	50.1	38.1	43.3

Furthermore, we also report the performances of our PS-GZSL in the conventional ZSL scenario, as presented in Table 4. To provide a comprehensive comparison, we have selected both previous conventional ZSL methods and recent GZSL methods under the conventional zero-shot setting. PS-GZSL achieves the best performance on three datasets and the second-best on FLO and SUN. This shows its superiority over existing GZSL methods on unseen classes and its strong generalization ability. These results prove the effectiveness of our PS-GZSL in both GZSL and conventional ZSL.

Table 4. Results of conventional ZSL. The best and the second-best accuracy of unseen classes are respectively marked in red and blue.

Methods	AWA1	AWA2	CUB	FLO	SUN
DEVISE [22]	54.2	59.7	52.0	45.9	56.5
SJE [23]	65.6	61.9	53.9	53.4	53.7
ALE [24]	59.9	62.5	54.9	48.5	58.1
ESZSL [25]	58.2	58.6	53.9	51.0	54.5
DCN [26]	65.2	-	56.2	-	61.8
CADA-VAE [9]	-	64.0	60.4	65.2	61.8
SP-AEN [8]	58.5	-	55.4	-	59.2
cycle-CLSWGAN [30]	66.3	-	58.4	70.1	60.0
DLFZRL [11]	71.3	70.3	61.8	-	61.3
TCN [46]	70.3	71.2	59.5	-	61.5
f-CLSWGAN [21]	68.2	-	57.3	67.2	60.8
f-VAEGAN-D2 [29]	-	71.1	61.0	67.7	64.7
TF-VAEGAN [39]	-	72.2	64.9	70.8	66.0
AGZSL [12]	-	72.8	76.0	-	63.3
SDGZSL [10]	-	72.1	75.5	73.3	62.4
CE-GZSL [13]	71.0	70.4	77.5	70.6	63.3
Ours PS-GZSL	71.5	72.9	78.1	71.3	64.7

4.5. Ablation Studies

Ablation studies were conducted to gain further insight into our PS-GZSL, evaluating the effects of different model architectures and representation components.

4.5.1. t-SNE Visualization

To further validate the transferability of our PS-GZSL, we visualize the task-shared representation h_{ds} and the multi-task joint representation $h_{ds} \oplus h_d \oplus h_s$ from unseen visual samples in Figure 7. We choose 10 unseen categories of test unseen set on AWA2 and 50 unseen categories of test unseen set on CUB. These data are sufficient in quantity and explicitly show the model's learned representation for the class comparison in unseen classes. Clearly, as we expected, the multi-task joint representation is more discriminative than the individual task-shared representation. However, we can still see discriminative patterns from h_{ds}, which is consistent with the assumption of previous methods based on learning the shared parts. This demonstrates that these task-shared representations may help classify between these categories, but the discriminative knowledge transfer from known to unknown categories is impaired due to the loss of task-specific information.

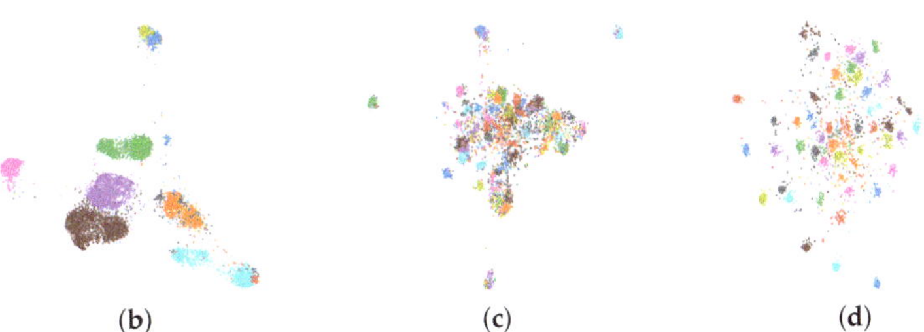

(a)   (b)   (c)   (d)

Figure 7. The t-SNE visualization:. (a) h_{ds} of unseen classes on AWA2, (b) $h_{ds} \oplus h_d \oplus h_s$ of unseen classes on AWA2, (c) h_{ds} of unseen classes on CUB and (d) $h_{ds} \oplus h_d \oplus h_s$ of unseen classes on CUB.

4.5.2. Effectiveness of Task-Shared & Task-Specific Representations

In order to validate our key motivation for the partially-shared mechanism of PS-GZSL: In addition to task-shared *discriminative and semantic-relevant* representations, task-specific *only discriminative* representations and *only semantic-relevant* representations are both useful in GZSL. We studied the performance of different combinations among h_{ds}, h_d and h_s. The results are presented in Figure 8, where we observe that using h_{ds} alone achieves comparable poor performance. However, when h_{ds} is concatenated with either h_d or h_s, the performance is improved, which demonstrates that both the h_d and h_s are helpful in GZSL. The best performance is achieved when we concatenate h_{ds}, h_d, and h_s together. This reveals that task joint representation $h_{ds} \oplus h_d \oplus h_s$ can capture complete correlation information among categories and their semantic descriptors, resulting in more informative and transferable representations for the test phase. Thus, both the task-shared and task-specific representations between discrimination and visual-semantic alignment are crucial to improve the classification performance in GZSL.

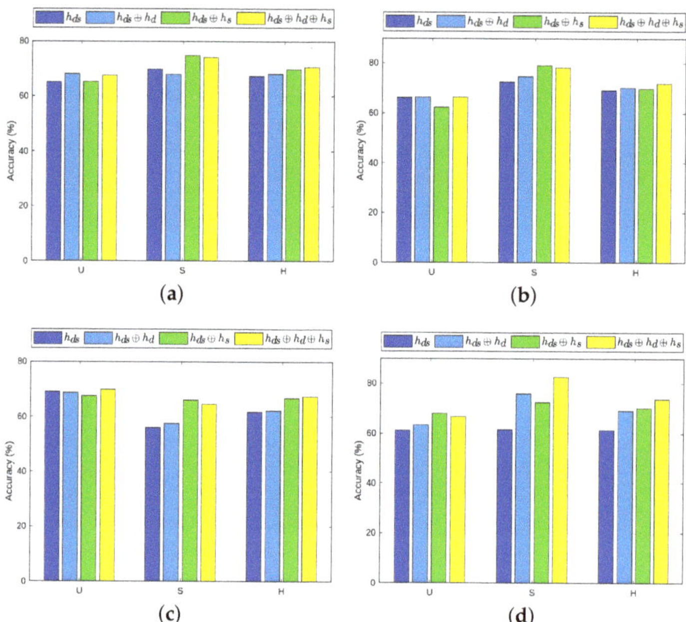

Figure 8. The effectiveness of various latent representations: (**a**) AWA1, (**b**) AWA2, (**c**) CUB and (**d**) FLO.

4.5.3. Analysis of Model Components

To assess the contributions of each component in PS-GZSL, different stripped-down architectures of we proposed methods were evaluated. The GZSL performance of each version on the AWA2 and CUB is represented in Table 5.

We observe that PS-GZSL outperforms PS-GZSL w/o MoE which validates that the MoE can improve the transferability of representation in GZSL. More importantly, we observe that PS-GZSL w/o MoE&PS outperforms PS-GZSL w/o MoE. This reveals the fact that simply splitting the visual encoder into three branches is not sufficient for learning the ideal transferable representations. Because any arbitrary mutually exclusive information decomposition can satisfy the regularizer, even if the h_{ds} encodes total information and h_d, h_s are non-informative for both tasks. This further demonstrates the superiority of our MoE module and expert dropout mechanism, which avoids the inexpressive issue among h_{ds}, h_d, and h_s. The above results indicate that our partially-shared mechanism and MoE module are mutually complementary in our method and prove that jointly preserving shared and specific representations between discriminative features and semantic features can preserve more complete and transferable information.

Table 5. Ablation study for different stripped-down architectures of PS-GZSL on the AWA2 and CUB dataset. PS is the partially-shared mechanism, $\mathcal{L}_{dis}$ is the adopted SupCon loss, $\mathcal{L}_{clf}$ is a classification loss of an auxiliary classifier for our discrimination task, and $\mathcal{L}_{mse}$ is the MSE version of our visual-semantic alignment task. The best and the second-best accuracy of unseen classes are respectively marked in red and blue.

Version	AWA2			CUB		
	U	**S**	**H**	**U**	**S**	**H**
PS-GZSL w/o MoE&PS	65.7	74.8	69.9	71.5	61.3	66.0
PS-GZSL w/o PS	66.9	74.8	70.7	67.0	66.8	66.9
PS-GZSL w/o MoE	61.4	79.8	69.4	68.4	63.1	65.6
PS-GZSL w/o $\mathcal{L}_{align}$ w/ $\mathcal{L}_{mse}$	66.0	75.5	70.5	66.9	66.2	66.5
PS-GZSL w/o $\mathcal{L}_{dis}$ w/ $\mathcal{L}_{clf}$	65.7	77.8	71.2	67.5	66.8	67.2
PS-GZSL	66.4	78.1	71.8	70.1	64.5	67.4

4.6. Hyper-Parameter Analysis

In our PS-GZSL approach, the hyperparameters that exert the greatest influence are the number of synthesized samples per class, the number of experts in each branch, and the dimensions of h_{ds}, h_d, and h_s.

Visualization of Different Number of Synthesized Samples. The number of synthesized samples per class was varied, as shown in Figure 9. The results show that the performance on all four datasets increased with an increasing number of synthesized examples. This demonstrated that the bias towards seen problems was relieved by the feature generation in our PS-GZSL. However, generating too many samples will impair the accuracy of seen classes (**S**) and eventually hamper the harmonic mean **H**. Therefore, selecting an appropriate value to achieve the balance between **S** and **U** is important.

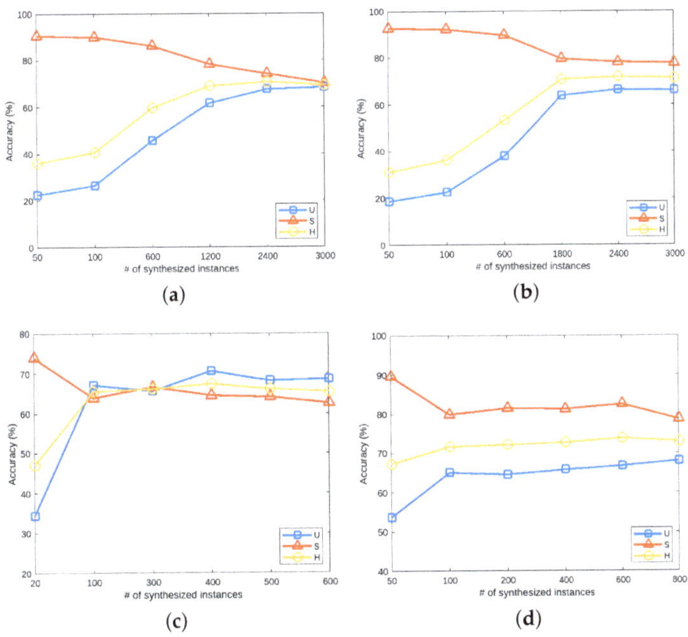

Figure 9. The influence of the number of synthesized visual instances in each unseen class. (**a**) AWA1, (**b**) AWA2, (**c**) CUB, and (**d**) FLO.

Visualization of Different Number of Experts. Since we use MoE modules for each branch, the architecture of the expert network is very important for our method. As shown in Figure 10, we study different numbers of experts for task-specific and task-shared, noted as **num_sp** and **num_sh**, respectively. As the numbers of task-specific experts and task-shared experts increase, the harmonic mean is boosted and then drops, which achieves the peek performance when num_sp = 3 and num_sh = 3. Thus, for convenience, both num_sp and num_sh are set to 3 in order to achieve a considerable performance in all of the remaining datasets.

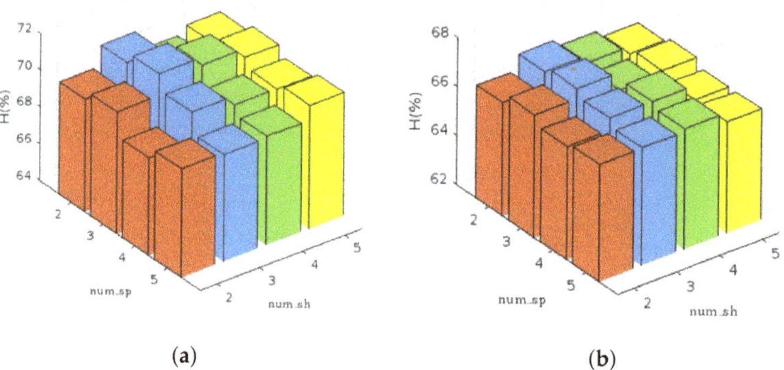

(a)   (b)

Figure 10. The effect of the number of task-specific and task-shared experts (denoted as **num_sp** and **num_sh**, respectively) : (**a**) AWA2 and (**b**) CUB.

Visualization of Different Representations Dimensions. Intuitively, the dimensions h_{ds}, h_d, and h_s will have a significant impact on the optimization of these two sub-tasks. This will ultimately affect the transferability and expressiveness of the concatenated final representations. To explore the sensitivity of our PS-GZSL to the dimensionality in the latent space. As shown in Figure 11, the harmonic mean accuracy of PS-GZSL for different latent dimensions on AWA2 and CUB, i.e., 256, 512, 1024, and 2048 for both task-specific and task-shared representations(denoted as **spSize** and **shSize**, respectively) are represented. As **spSize** and **shSize** are both set to 1024, PS-GZSL consistently performs better than all others on AWA2 and CUB. Therefore, both **spSize** and **shSize** are set to 1024 in all of the remaining datasets.

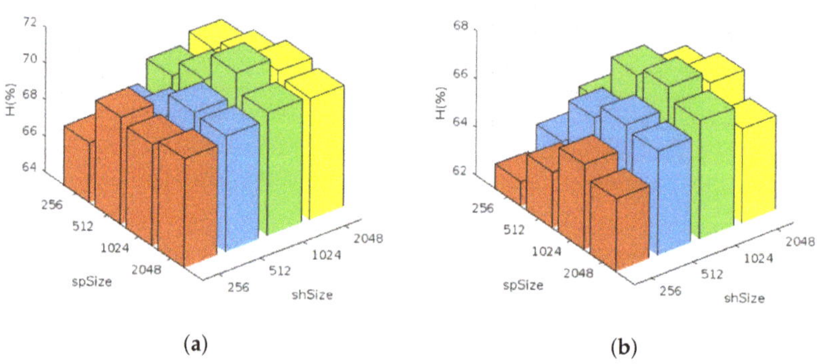

(a)   (b)

Figure 11. The effect of the dimension of task-specific representation and task-shared representation (denoted as **spSize** and **shSize**, respectively). (**a**) AWA2 and (**b**) CUB.

5. Conclusions

In this paper, we propose a new way of learning the composite method by accounting for all the features based on multi-task representation learning. Specifically, the recent representation learning method in GZSL discards some specific information between two tasks (i.e., classification task and visual semantic alignment task). As explained in the introduction, this specific information can be either discriminative or semantic-relevant, depending on their contribution to the testing phase.

Further on, we believe that jointly preserving task-specific and task-shared features leads to a more complete and more transferable representation in GZSL. To support this claim, a novel representation learning method termed PS-GZSL is proposed. Unlike most existing methods, PS-GZSL explicitly factorizes visual features into one task-shared and two task-specific representations through the partially-shared mechanism between the discrimination and visual semantic alignment task. This flexibility enables PS-GZSL to preserve more complete knowledge. Furthermore, PS-GZSL carefully designs the mixture of experts and gate networks for learning informative representations for each branch. As evaluated in extensive experiments, the good transferability of PS-GZSL has been demonstrated.

As a starting point, this study shows the potential ability of the partially-shared mechanism in learning transferable representation in GZSL. There is still a large research space in this direction. First, the relative loss weight ratio of each sub-task is set to 1, but future work could investigate the use of adaptive weights to balance the two tasks during optimization. Second, ideally, the encoding information of task-shared and task-specific representations should be no redundancy. It is also important to devise a regularizer to accomplish this. In the future, we will investigate these potential directions.

Author Contributions: Conceptualization, G.W. and S.T.; methodology, G.W.; software, G.W.; validation, S.T.; formal analysis, G.W. and S.T.; investigation, G.W.; resources, G.W.; data curation, G.W.; writing—original draft preparation, G.W.; writing—review and editing, S.T.; visualization, G.W.; supervision, S.T.; project administration, G.W.; funding acquisition, S.T. All authors have read and agreed to the published version of the manuscript.

Funding: This research received no external funding.

Data Availability Statement: The data presented in this study are available upon request from the first author.

Acknowledgments: We are grateful for resources from the High-Performance Computing Center of Central South University.

Conflicts of Interest: The authors declare no conflict of interest.

Abbreviations

The following abbreviations are used in this manuscript:

ZSL	Zero-Shot Learning
GZSL	Generalized Zero-Shot Learning
SupCon	Supervised Contrastive
MoE	Mixture-of-Experts
PS	Partially-Shared mechanism

References

1. Xian, Y.; Schiele, B.; Akata, Z. Zero-shot learning-the good, the bad and the ugly. In Proceedings of the IEEE Conference on Computer Vision and Pattern Recognition, Honolulu, HI, USA, 21–26 July 2017; pp. 4582–4591.
2. Mikolov, T.; Chen, K.; Corrado, G.; Dean, J. Efficient estimation of word representations in vector space. *arXiv* **2013**, arXiv:1301.3781.
3. Lampert, C.H.; Nickisch, H.; Harmeling, S. Learning to detect unseen object classes by between-class attribute transfer. In Proceedings of the 2009 IEEE Conference on Computer Vision and Pattern Recognition, Miami, FL, USA, 20–25 June 2009; pp. 951–958.

4. Palatucci, M.; Pomerleau, D.; Hinton, G.E.; Mitchell, T.M. Zero-shot learning with semantic output codes. *Adv. Neural Inf. Process. Syst.* **2009**, *22*, 1410–1418.

5. Chao, W.L.; Changpinyo, S.; Gong, B.; Sha, F. An empirical study and analysis of generalized zero-shot learning for object recognition in the wild. In *Proceedings of the Computer Vision—ECCV 2016: 14th European Conference, Amsterdam, The Netherlands, 11–14 October 2016, Proceedings, Part II 14*; Springer: Cham, Switzerland, 2016; pp. 52–68.

6. Saad, E.; Paprzycki, M.; Ganzha, M.; Bădică, A.; Bădică, C.; Fidanova, S.; Lirkov, I.; Ivanović, M. Generalized Zero-Shot Learning for Image Classification—Comparing Performance of Popular Approaches. *Information* **2022**, *13*, 561. [CrossRef]

7. Li, X.; Xu, Z.; Wei, K.; Deng, C. Generalized zero-shot learning via disentangled representation. In Proceedings of the AAAI Conference on Artificial Intelligence, Virtual, 2–9 February 2021; Volume 35, pp. 1966–1974.

8. Chen, L.; Zhang, H.; Xiao, J.; Liu, W.; Chang, S.F. Zero-shot visual recognition using semantics-preserving adversarial embedding networks. In Proceedings of the IEEE Conference on Computer Vision and Pattern Recognition, Salt Lake City, UT, USA, 18–23 June 2018; pp. 1043–1052.

9. Schonfeld, E.; Ebrahimi, S.; Sinha, S.; Darrell, T.; Akata, Z. Generalized zero-and few-shot learning via aligned variational autoencoders. In Proceedings of the IEEE/CVF Conference on Computer Vision and Pattern Recognition, Long Beach, CA, USA, 16–17 June 2019; pp. 8247–8255.

10. Chen, Z.; Luo, Y.; Qiu, R.; Wang, S.; Huang, Z.; Li, J.; Zhang, Z. Semantics disentangling for generalized zero-shot learning. In Proceedings of the IEEE/CVF International Conference on Computer Vision, Montreal, BC, Canada, 11–17 October 2021; pp. 8712–8720.

11. Tong, B.; Wang, C.; Klinkigt, M.; Kobayashi, Y.; Nonaka, Y. Hierarchical disentanglement of discriminative latent features for zero-shot learning. In Proceedings of the IEEE/CVF Conference on Computer Vision and Pattern Recognition, Long Beach, CA, USA, 16–17 June 2019; pp. 11467–11476.

12. Chou, Y.Y.; Lin, H.T.; Liu, T.L. Adaptive and generative zero-shot learning. In Proceedings of the International Conference on Learning Representations, Virtual Event, 3–7 May 2021.

13. Han, Z.; Fu, Z.; Chen, S.; Yang, J. Contrastive embedding for generalized zero-shot learning. In Proceedings of the IEEE/CVF Conference on Computer Vision and Pattern Recognition, Virtual Conference, 19–25 June 2021; pp. 2371–2381.

14. Chen, S.; Wang, W.; Xia, B.; Peng, Q.; You, X.; Zheng, F.; Shao, L. Free: Feature refinement for generalized zero-shot learning. In Proceedings of the IEEE/CVF International Conference on Computer Vision, Montreal, BC, Canada, 11–17 October 2021; pp. 122–131.

15. Bui, M.H.; Tran, T.; Tran, A.; Phung, D. Exploiting domain-specific features to enhance domain generalization. *Adv. Neural Inf. Process. Syst.* **2021**, *34*, 21189–21201.

16. Milbich, T.; Roth, K.; Bharadhwaj, H.; Sinha, S.; Bengio, Y.; Ommer, B.; Cohen, J.P. Diva: Diverse visual feature aggregation for deep metric learning. In *Proceedings of the Computer Vision—ECCV 2020: 16th European Conference, Glasgow, UK, 23–28 August 2020, Proceedings, Part VIII 16*; Springer: Cham, Switzerland, 2020; pp. 590–607.

17. Ma, J.; Zhao, Z.; Yi, X.; Chen, J.; Hong, L.; Chi, E.H. Modeling task relationships in multi-task learning with multi-gate mixture-of-experts. In Proceedings of the 24th ACM SIGKDD International Conference on Knowledge Discovery & Data Mining, London, UK, 19–23 August 2018; pp. 1930–1939.

18. Jacobs, R.A.; Jordan, M.I.; Nowlan, S.J.; Hinton, G.E. Adaptive mixtures of local experts. *Neural Comput.* **1991**, *3*, 79–87. [CrossRef] [PubMed]

19. Khosla, P.; Teterwak, P.; Wang, C.; Sarna, A.; Tian, Y.; Isola, P.; Maschinot, A.; Liu, C.; Krishnan, D. Supervised contrastive learning. *Adv. Neural Inf. Process. Syst.* **2020**, *33*, 18661–18673.

20. Sung, F.; Yang, Y.; Zhang, L.; Xiang, T.; Torr, P.H.; Hospedales, T.M. Learning to compare: Relation network for few-shot learning. In Proceedings of the IEEE Conference on Computer Vision and Pattern Recognition, Salt Lake City, UT, USA, 18–23 June 2018; pp. 1199–1208.

21. Xian, Y.; Lorenz, T.; Schiele, B.; Akata, Z. Feature generating networks for zero-shot learning. In Proceedings of the IEEE Conference on Computer Vision and Pattern Recognition, Salt Lake City, UT, USA, 18–22 June 2018; pp. 5542–5551.

22. Frome, A.; Corrado, G.S.; Shlens, J.; Bengio, S.; Dean, J.; Ranzato, M.; Mikolov, T. Devise: A deep visual-semantic embedding model. *Adv. Neural Inf. Process. Syst.* **2013**, *26*, 2121–2129.

23. Akata, Z.; Reed, S.; Walter, D.; Lee, H.; Schiele, B. Evaluation of output embeddings for fine-grained image classification. In Proceedings of the IEEE Conference on Computer Vision and Pattern Recognition, Boston, MA, USA, 7–12 June 2015; pp. 2927–2936.

24. Akata, Z.; Perronnin, F.; Harchaoui, Z.; Schmid, C. Label-embedding for attribute-based classification. In Proceedings of the IEEE Conference on Computer Vision and Pattern Recognition, Portland, OR, USA, 23–28 June 2013; pp. 819–826.

25. Romera-Paredes, B.; Torr, P. An embarrassingly simple approach to zero-shot learning. In Proceedings of the International Conference on Machine Learning, Lille, France, 6–11 July 2015; pp. 2152–2161.

26. Liu, S.; Long, M.; Wang, J.; Jordan, M.I. Generalized zero-shot learning with deep calibration network. *Adv. Neural Inf. Process. Syst.* **2018**, *31*, 2009–2019.

27. Yang, G.; Han, A.; Liu, X.; Liu, Y.; Wei, T.; Zhang, Z. Enhancing Semantic-Consistent Features and Transforming Discriminative Features for Generalized Zero-Shot Classifications. *Appl. Sci.* **2022**, *12*, 12642. [CrossRef]

28. Li, J.; Jing, M.; Lu, K.; Ding, Z.; Zhu, L.; Huang, Z. Leveraging the invariant side of generative zero-shot learning. In Proceedings of the IEEE/CVF Conference on Computer Vision and Pattern Recognition, Long Beach, CA, USA, 15–20 June 2019; pp. 7402–7411.
29. Xian, Y.; Sharma, S.; Schiele, B.; Akata, Z. f-vaegan-d2: A feature generating framework for any-shot learning. In Proceedings of the IEEE/CVF Conference on Computer Vision and Pattern Recognition, Long Beach, CA, USA, 15–20 June 2019; pp. 10275–10284.
30. Felix, R.; Reid, I.; Carneiro, G. Multi-modal cycle-consistent generalized zero-shot learning. In Proceedings of the European Conference on Computer Vision (ECCV), Munich, Germany, 8–14 September 2018; pp. 21–37.
31. Vyas, M.R.; Venkateswara, H.; Panchanathan, S. Leveraging seen and unseen semantic relationships for generative zero-shot learning. In *Proceedings of the Computer Vision—ECCV 2020: 16th European Conference, Glasgow, UK, 23–28 August 2020, Proceedings, Part XXX 16*; Springer: Cham, Switzerland, 2020; pp. 70–86.
32. Li, Z.; Zhang, D.; Wang, Y.; Lin, D.; Zhang, J. Generative Adversarial Networks for Zero-Shot Remote Sensing Scene Classification. *Appl. Sci.* **2022**, *12*, 3760. [CrossRef]
33. Sohn, K.; Lee, H.; Yan, X. Learning structured output representation using deep conditional generative models. *Adv. Neural Inf. Process. Syst.* **2015**, *28*, 3483–3491.
34. Verma, V.K.; Arora, G.; Mishra, A.; Rai, P. Generalized zero-shot learning via synthesized examples. In Proceedings of the IEEE Conference on Computer Vision and Pattern Recognition, Salt Lake City, UT, USA, 18–23 June 2018; pp. 4281–4289.
35. Kim, J.; Shim, K.; Shim, B. Semantic feature extraction for generalized zero-shot learning. In Proceedings of the AAAI Conference on Artificial Intelligence, Virtual, 22 February–1 March 2022; Volume 36, pp. 1166–1173.
36. Tang, H.; Liu, J.; Zhao, M.; Gong, X. Progressive layered extraction (ple): A novel multi-task learning (mtl) model for personalized recommendations. In Proceedings of the 14th ACM Conference on Recommender Systems, Virtual Event, 22–26 September 2020; pp. 269–278.
37. Park, H.; Yeo, J.; Wang, G.; Hwang, S.W. Soft representation learning for sparse transfer. In Proceedings of the 57th Annual Meeting of the Association for Computational Linguistics, Florence, Italy, 28 July–2 August 2019; pp. 1560–1568.
38. Xin, S.; Jiao, Y.; Long, C.; Wang, Y.; Wang, X.; Yang, S.; Liu, J.; Zhang, J. Prototype Feature Extraction for Multi-task Learning. In Proceedings of the ACM Web Conference 2022, Lyon France, 25–29 April 2022; pp. 2472–2481.
39. Narayan, S.; Gupta, A.; Khan, F.S.; Snoek, C.G.; Shao, L. Latent embedding feedback and discriminative features for zero-shot classification. In *Proceedings of the Computer Vision—ECCV 2020: 16th European Conference, Glasgow, UK, 23–28 August 2020, Proceedings, Part XXII 16*; Springer: Cham, Switzerland, 2020; pp. 479–495.
40. Wah, C.; Branson, S.; Welinder, P.; Perona, P.; Belongie, S. The Caltech-Ucsd Birds-200-2011 Dataset. 2011. Available online: https://authors.library.caltech.edu/27452/ (accessed on 29 March 2023).
41. Nilsback, M.E.; Zisserman, A. Automated flower classification over a large number of classes. In Proceedings of the 2008 Sixth Indian Conference on Computer Vision, Graphics & Image Processing, Bhubaneswar, India, 16–19 December 2008; pp. 722–729.
42. Patterson, G.; Hays, J. Sun attribute database: Discovering, annotating, and recognizing scene attributes. In Proceedings of the 2012 IEEE Conference on Computer Vision and Pattern Recognition, Providence, RI, USA, 16–21 June 2012; pp. 2751–2758.
43. He, K.; Zhang, X.; Ren, S.; Sun, J. Deep residual learning for image recognition. In Proceedings of the IEEE Conference on Computer Vision and Pattern Recognition, Las Vegas, NV, USA, 27–30 June 2016; pp. 770–778.
44. Krizhevsky, A.; Sutskever, I.; Hinton, G.E. Imagenet classification with deep convolutional neural networks. *Commun. ACM* **2017**, *60*, 84–90. [CrossRef]
45. Reed, S.; Akata, Z.; Lee, H.; Schiele, B. Learning deep representations of fine-grained visual descriptions. In Proceedings of the IEEE Conference on Computer Vision and Pattern Recognition, Las Vegas, NV, USA, 27–30 June 2016; pp. 49–58.
46. Jiang, H.; Wang, R.; Shan, S.; Chen, X. Transferable contrastive network for generalized zero-shot learning. In Proceedings of the IEEE/CVF International Conference on Computer Vision, Seoul, Republic of Korea, 27 October–2 November 2019; pp. 9765–9774.
47. Min, S.; Yao, H.; Xie, H.; Wang, C.; Zha, Z.J.; Zhang, Y. Domain-aware visual bias eliminating for generalized zero-shot learning. In Proceedings of the IEEE/CVF Conference on Computer Vision and Pattern Recognition, Seattle, WA, USA, 14–19 June 2020; pp. 12664–12673.
48. Li, K.; Min, M.R.; Fu, Y. Rethinking zero-shot learning: A conditional visual classification perspective. In Proceedings of the IEEE/CVF International Conference on Computer Vision, Seoul, Republic of Korea, 27 October–2 November 2019; pp. 3583–3592.

Article

Transformer-Based Global PointPillars 3D Object Detection Method

Lin Zhang, Hua Meng, Yunbing Yan * and Xiaowei Xu

School of Automobile and Traffic Engineering, Wuhan University of Science and Technology, Wuhan 430065, China; zhanglin4025@wust.edu.cn (L.Z.)
* Correspondence: yyb@wust.edu.cn

Abstract: The PointPillars algorithm can detect vehicles, pedestrians, and cyclists on the road, and is widely used in the field of environmental awareness in autonomous driving. However, its feature encoding network only uses a minimalist PointNet network for feature extraction of point cloud information, which does not consider the global context information of the point cloud, and the local structure features are not sufficiently extracted, and these feature losses can seriously affect the performance of the object detection network. To address this problem, this paper proposes an improved PointPillars algorithm named TGPP: Transformer-based Global PointPillars. After the point cloud is divided into several pillars, global context features and local structure features are extracted through a multi-head attention mechanism, so that the point cloud after feature coding has global context features and local structure features; the two-dimensional pseudo-image generated by this feature is used for feature learning using a two-dimensional convolutional neural network. Finally, the SSD detection head is used to achieve 3D object detection. It is demonstrated that the TGPP achieves an average accuracy improvement of 2.64% in the KITTI test set.

Keywords: deep learning; 3D object detection; lidar point cloud; transformer

1. Introduction

The 3D object detection technology is an important part of the environment perception module in the automatic driving system. Accurately identifying objects such as vehicles, pedestrians, and cyclists on the road is the basis for vehicle planning and control. To better accomplish this goal, self-driving cars need to rely on a variety of sensors, among which lidar is one of the most important sensors. Lidar can measure the distance to the surrounding environment through a scanner, and directly generate sparse 3D point cloud information, which has inherent advantages in the task of 3D object detection. Traditional methods usually down-sample the point cloud information first, then remove the ground, and then use European, DBSCAN and other clustering methods combined with 3D bounding boxes to detect objects [1–5]. The traditional method requires cumbersome parameter adjustment work during the deployment process, making it difficult to apply in practice. With the rapid development of deep learning technology and parallel computing units, the end-to-end 3D object detection method based on deep learning has become the current key research content.

With the rapid development of computer vision and deep learning, 2D object detection technology has made great progress, but there are essential differences between the two data forms of the point cloud and image. In order, direct convolution of the point cloud will lead to severe distortion of the features [6], so the excellent 2D object detection algorithm cannot be directly applied to the 3D object detection task. In 2017, Qi et al. proposed PointNet [7] and PointNet++ [8] deep convolutional neural networks, which take the original point cloud as input and can be applied to point cloud point-by-point feature extraction, point cloud recognition, and point cloud semantics segmentation, and other fields also provide feature

Citation: Zhang, L.; Meng, H.; Yan, Y.; Xu, X. Transformer-Based Global PointPillars 3D Object Detection Method. *Electronics* **2023**, *12*, 3092. https://doi.org/10.3390/electronics12143092

Academic Editor: Donghyeon Cho

Received: 2 July 2023
Revised: 13 July 2023
Accepted: 14 July 2023
Published: 16 July 2023

extraction tools for 3D object detection tasks based on point cloud data. Subsequently, a point-based 3D object detection method was proposed. PointRCNN [9] is a more classic point-based method. The main idea is to extract point-by-point features from the PointNet network and predict 3D proposals to achieve 3D object detection. This type of method takes a lot of time to retrieve points, so the calculation is very large, and the detection efficiency is very low. In response to this problem, Zhou Y and others proposed the VoxelNet [10] algorithm, which is the earliest voxel-based method. This algorithm represents the point cloud as voxels, and the follow-up work in voxels reduces the amount of calculation and is based on the voxel method, which is more convenient for the extraction of target features, but due to the slow inference speed of the 3D convolutional neural network, its detection efficiency is still not ideal. As an upgraded version of VoxelNet, the SECOND [11] algorithm replaces the ordinary 3D convolution with sparse 3D convolution to speed up the reasoning time, but it still cannot eliminate the disadvantage of the slow calculation speed of 3D convolution. To this end, the PointPillars [12] algorithm proposes a novel encoder, which realizes end-to-end learning on 3D object detection tasks using only 2D convolutional neural networks. Its unique pillars-based encoding method greatly speeds up the detection speed. In addition, its simple algorithm framework can be easily deployed to a variety of laser radars. At present, it is one of the most widely used methods in engineering practice, and the research and improvement of the algorithm have practical application value and engineering significance.

At present, the detection rate of the PointPillars algorithm still has a large advantage, but its detection accuracy is inferior to the later excellent works. For example, Li Y et al. proposed the UVTR [13], which explicitly expresses and interacts with image and point cloud features in voxel space; Lai X et al. proposed the SphereFormer [14] method, which solved the problem of discontinuous information and limited receptive field. Therefore, in the past two years, some scholars have proposed some methods to improve PointPillars. For example, in 2021, Xinwei He et al. [15] proposed an intra-pillar multi-scale feature extraction module to enhance the overall learning ability of the PointPillars algorithm, thereby improving detection accuracy. This work improved the local structural feature extraction method of the point cloud, but still does not consider the global context feature information of the point cloud; in 2022, Dejiang Chen et al. [16] improved the 2D convolutional down-sampling module of the PointPillars algorithm based on Swin Transformer [17], optimized the original 2D convolutional neural network, and improved the Average Orientation Similarity (AOS) accuracy to a certain level. The improvement of this work optimizes the 2D convolutional neural network to improve the learning ability of point cloud features.

However, the above improvement scheme still does not make full use of point cloud features: its feature encoding process is to divide all point clouds into uniform pillars, where each pillar can be understood as a combination of voxels at the corresponding position on the z-axis, and then pass a minimalist PointNet network, which performs local feature extraction and uses Max Pooling to obtain points representing the features of each pillar, and finally generates a sparse 2D pseudo-image through position mapping. In this encoding process, the local feature extraction is insufficient and does not consider the global features causing a loss to the features of the point cloud.

To solve the appealing problem, this paper proposes an improved PointPillars algorithm named TGPP (Transformer-based Global PointPillars), based on Transformer [18], to improve the feature encoding network: after the point cloud is divided into pillars, the global position feature calculation and local structure feature calculation are performed based on the improved Transformer module, and each the rich global context features and local features of the pillars enable the local features of the point cloud and accurate global position information to be preserved in the feature encoding process to improve the accuracy of the algorithm object detection.

2. TGPP Algorithm Network

TGPP is an improvement on PintPillars. The reasoning speed of the PointPillars algorithm is very fast, exceeding the scanning frequency of the radar, so its real-time detection is very good. The algorithm uses 3D point clouds as input, which can realize end-to-end learning and can detect road vehicles, pedestrians, and cyclists. These three common objects are identified.

The TGPP algorithm structure is shown in Figure 1 below. The algorithm can be divided into three main parts: (1) Pillar Feature Net: divide the 3D point cloud into pillars, and generate the 2D pseudo image. (2) Two-dimensional convolutional neural network: use multiple down-sampling of 2D pseudo-images to obtain feature maps of different resolutions, and then up-sample multiple feature maps after down-sampling to the same size for splicing to generate the final feature map. (3) Object detection head: generate a 3D detection frame and object classification for the feature map, and obtain the position and type of the object. The main difference between this method and the original method lies in the feature encoding network, and the structure of this algorithm will be introduced in detail below.

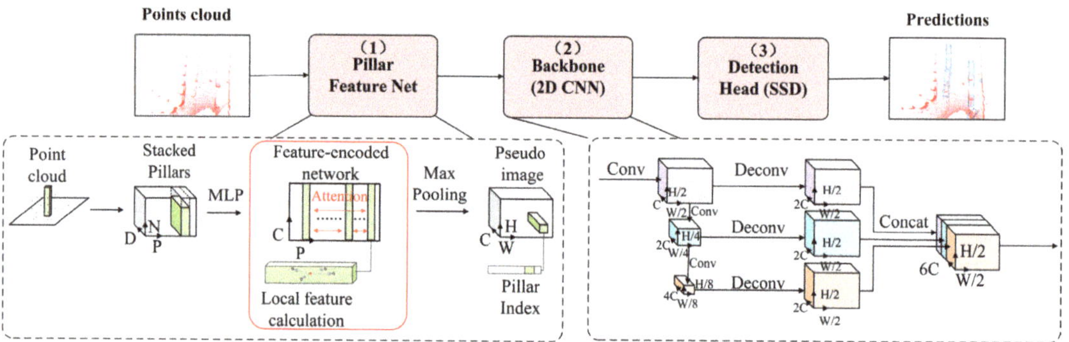

Figure 1. TGPP algorithm network structure.

2.1. Overall Algorithm Process

This algorithm takes the original point cloud data information as input, and first expresses the point cloud as a uniformly distributed pillar: the three-dimensional point cloud information is directly obtained from the top view, all points are discretized into a uniform square network on the x–y plane in the grid, and each pillar is cubic with infinite extension in the z-axis direction of each grid.

Due to the sparsity of the point cloud, most of the pillars are empty, and there are usually only a small number of points in the non-empty pillars. This sparsity is used to create a size density tensor (D, P, N), where D represents the feature dimension of each pillar, P represents the number of pillars, and N represents the maximum number of points in each pillar. When the points in the pillars exceed N, random sampling selects N points; when the points are less than N, it will be filled with 0 samples.

After obtaining a (D, P, N) tensor, the input is based on the improved Transformer feature an encoding network for feature extraction. First, we use the MLP (Multi-Layer Perceptron) operation for position encoding and dimension-up processing, changing from (D, P, N) tensors to (C, P, N) tensors, and C represents the feature dimension after dimension enhancement (256); then, based on the multi-head attention mechanism, we calculate the global context features for each pillars and calculate the local structural features for the points in each pillar, so that the point cloud information in each pillars has global context features and local structural features, in particular, in order to fully extract the local structural features of the point cloud, a combination of local and global position encoding is used; then, the maximum pooling is used to extract the feature points that best represent the features of the pillars; finally, according to the pillars index, we remap the point cloud to the corresponding position of the original grid, and generate a 2D pseudo-image of size (C, H, W), where H and W represent the height and width of the image.

The generated 2D pseudo-image will be input into the 2D convolutional neural network for feature learning, and finally, the detection head based on the design of SSD (Single Shot Multibox Detector) [19] is used to realize the classification and regression of 3D object detection and generate a 3D object detection frame.

2.2. Feature Encoding Network Based on Transformer

The Transformer model is a deep learning model based on the attention mechanism, which has been widely used in natural language processing (NLP), image processing, and other fields. Its core idea is to split the input sequence into a set of vector representations, and then use the attention mechanism to learn the dependencies between positions. Through the multi-head attention mechanism, Transformer can perform more comprehensive and accurate feature extraction on point cloud data, and its application in 3D object detection tasks based on point cloud data has gradually become a trend. PCT [20], Point Transformer [21], SOE-Net [22], VoxSeT [23], FlatFormer [24] and other works have achieved good results. Therefore, it is feasible to improve the feature encoding network based on Transformer.

The feature encoding network structure of this algorithm is shown in Figure 2 below, which is a network structure based on the encoder–decoder. The input of the feature encoding network is the point cloud information represented by the pillar distribution, and a vector sequence is generated through position encoding, and input to the multi-head attention module for calculation; each element of the input sequence is compared with other elements in the sequence Elements interact and give different weights according to their relevance. This interaction is realized by calculating the attention weight matrix; then it is input into the feedforward neural network module, and the output of the attention layer is further nonlinearly transformed; additionally, in order to prevent degradation problems during the training process, we add ResNet residual neural network [25] and LN layer [26] (Figure 2: Add&Norm module); the difference between the decoder and the encoder is that there is add a masked multi-head attention module, whose input is the predicted output of the entire feature calculation process; and the final output layer converts the output of the decoder into the final probability distribution through a linear transformation and Softmax function for generating prediction results. Nx represents the number of encoders and decoders, that is, the number of layers of the Transformer. Each layer independently processes the input and passes its output to the next layer.

The core content of the feature encoding network is to use the multi-head attention module to calculate the global context feature of the pillars and the local feature calculation of the internal structure of the pillars:

(1)   Global Feature Calculation:

The calculation formula of the global attention [18] is

$$\text{Attention}(Q, K, V) = \text{Softmax}(\frac{QK^T}{\sqrt{d_k}}) \tag{1}$$

In Formula (1), Q, K, and V are feature codes of point cloud columns. First, calculate the dot product of the Q matrix and K matrix, and divide it by the scale $\sqrt{d_k}$ to prevent overflow of the dot product result, where d_k is the vector in the Q and K dimension. After calculating the dot product, use the Softmax function to normalize the dot product to a probability distribution, and finally multiply it by the matrix V to obtain the attention score matrix between different pillars.

The multi-head attention allows the model to simultaneously focus on information from different pillars and different locations. The representation of the multi-head attention [18] is as follows:

$$\text{MultiHead}(Q, K, V) = \text{Concat}(\text{head}_1, \ldots, \text{head}_h)W^o$$
$$\text{head}_i = \text{Attention}\left(QW_i^Q, KW_i^K, VW_i^V\right) \tag{2}$$

Among them, $W_i^K \in R^{d_{model} \times d_k}$,$W_i^V \in R^{d_{model} \times d_v}$,$d_v = d_k$,$d_{model} = d_k \times h$; h is the number of attention heads.

All non-empty pillars perform global context feature calculation through the multi-head attention mechanism, which can add global attention to each pillar.

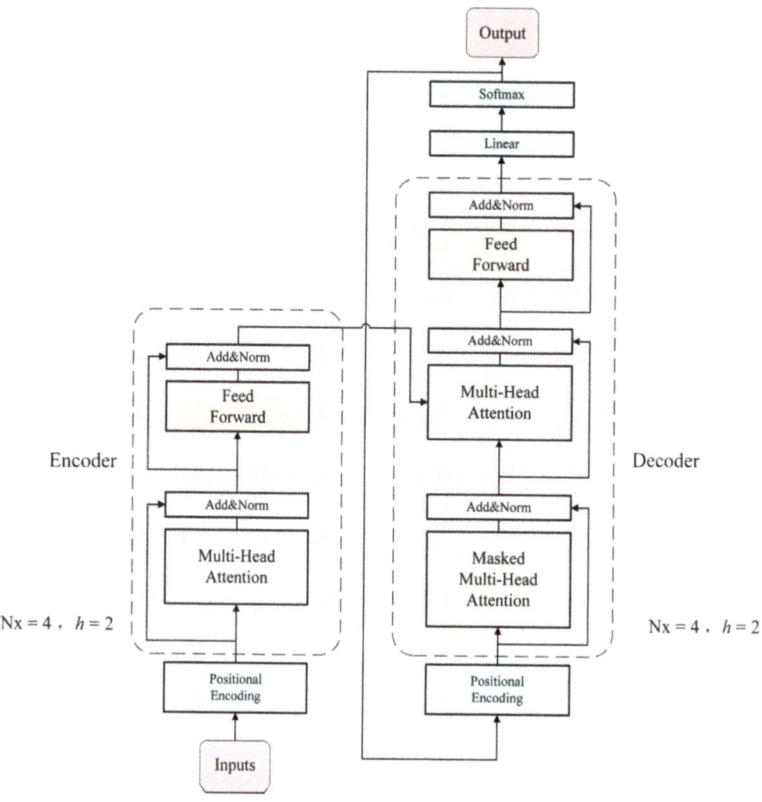

Figure 2. TGPP algorithm feature-encoded network structure.

(2) Local Feature Calculation:

In the process of calculating the global feature of the pillars, the local feature calculation is added, and the local geometric relationship between the center point and the adjacent point is used to effectively aggregate the local features by learning the attention weight. The specific method is to use the subtraction relationship, and at the same time add the local position information δ to the attention vector γ and the feature vector α to aggregate features. The overall calculation expression [27] is

$$g_i = \sum_{\mu_j \in x(i)} \rho(\gamma(\varphi(\mu_i) - \varphi(\mu_j) + \delta))(\alpha(\mu_j) + \delta) \tag{3}$$

Among them, $\mu = \{ f_i | i = 1, 2, \ldots, n \}$ is a set of feature vectors composed of points in the pillars, $\mu(i) \in \mu$; g_i is the feature output after adding local attention; φ, ϕ, and α are point-by-point feature transformation functions, similar to linear projection functions; and γ is the attention-generating mapping function. The calculation formula of local position information δ is

$$\delta = \varepsilon(p_i - p_j) \tag{4}$$

Among them, p_i, p_j is the coordinates of the 3D point cloud; and ε is composed of two ReLU functions [28].

After the above processing, the point cloud features after feature encoding will have global position features and local structure features, which reduce the feature loss caused by feature encoding, and the subsequent generated 2D pseudo-images are more conducive to subsequent feature learning to improve object detection accuracy.

2.3. 2D Convolutional Neural Network and SSD Detection Head

After the original point cloud information passes through the feature encoding network, a 2D pseudo image is generated, and the 2D convolutional neural network can be used very conveniently for feature learning. The structure of the 2D convolutional neural network is shown in Figure 1. The backbone network consists of two sub-networks: a top-down feature extraction network and an up-sampling and feature stitching network. The top-down sub-network uses a gradually decreasing spatial resolution to acquire features, and consists of a series of block structures, where each block structure contains three parameters (S, L, F), each block contains L 3×3 2D convolutional layers, F output channels, and the step size of the convolutional layer is S. The network that performs up-sampling and feature splicing is responsible for up-sampling the features from the first sub-network and applying the BN Layer [29] and the ReLU function to form the final output features. The use of a 2D convolutional neural network avoids the disadvantages of slow inference speed of algorithms such as VoxelNet using a 3D convolutional neural network, simplifies the structure of the model, reduces the amount of calculation, has good detection accuracy, and greatly improves detection speed.

The SSD detection head is used to predict the position, category, and orientation of 3D objects. We use the 2D intersection-over-union ratio (IOU) to match the prior frame with the real label frame, regardless of the height information, but use it as an additional regression object, because in the real road object, all objects can be considered to be in the same plane of the three-dimensional space, the height difference between all categories of objects is not very large, and better results can be obtained by directly using the SmoothL1 function [30] for regression. At the same time, the FPN (feature pyramid network) [31] operation is also introduced in the detection head to handle objects of different sizes. By extracting features at different scales, objects of different sizes can be located more accurately.

3. Algorithm Implementation Details

3.1. Details of Feature Encoding Network Structure Parameters

The cross-section of each pillar is a square with a side length of 0.16 m. In the actual feature encoding process, only the front view part is intercepted to generate a pseudo-image, because the real label information of the KITTI dataset is only in the front view captured by the camera. It is marked in the image, so the points of the original point cloud information in the negative direction of the x-axis should be discarded, and the points that are too far away should be removed. Refer to the original algorithm to take the maximum and minimum values of (x, y, z) in the point cloud space. It is min: (0, -39.68, -3), max: (69.12, 39.68, 1), in meters; the maximum value P of the number of pillars is 12,000, and the maximum value of point sampling in each pillar is N, which is set to 32. The number of Transformer layers is 4, the number of heads is 2, and a 2-layer learnable MLP is used for position encoding.

3.2. Loss Calculation

This article uses the same loss calculation method as the original algorithm. Each real label box contains $(x, y, z, w, l, h, \theta)$ 7 parameters, where (x, y, z) represents the three-dimensional coordinates of the object center; (z, w, l) represents the label of the length, width, and height of the frame, and θ represents the rotation angle. The regression residual of the positioning task between the prior box and the ground truth box is defined as

$$\Delta x = \frac{x^{gt} - x^a}{d^a}, \Delta y = \frac{y^{gt} - y^a}{d^a}, \Delta z = \frac{z^{gt} - z^a}{d^a}$$
$$\Delta w = \log \frac{w^{gt}}{w^a}, \Delta l = \log \frac{l^{gt}}{l^a}, \Delta h = \log \frac{h^{gt}}{h^a},$$
$$\Delta \theta = \sin(\theta^{gt} - \theta^a) \tag{5}$$

Among them, x^{gt} represents the x value of the label box; x^a represents the x value of the prior frame; y, z, w, l, h, θ are the same; and d^a represents the diagonal distance between the length and width of the prior frame, defined as $d^a = \sqrt{(w^a)^2 + (l^a)^2}$. The total localization loss is

$$L_{loc} = \sum_{b \in (x,y,z,w,l,h,\theta)} \text{SmoothL1}(\Delta b) \tag{6}$$

Since it is not possible to completely distinguish between two a priori boxes with completely opposite directions during angle regression, it is necessary to add direction classification to the a priori box. The direction classification loss function uses the Softmax function, denoted as L_{dir} [11]. The object classification loss function uses Focal Loss [32]:

$$L_{cls} = -\lambda_a (1 - p^a)^r \log p^a \tag{7}$$

Among them, p^a represents the probability that the predicted prior box belongs to the positive class, $\lambda = 0.25$, $r = 2$. Finally, the total loss function is obtained as

$$L = \frac{1}{N_{pos}} (\beta_{loc} L_{loc} + \beta_{cls} L_{cls} + \beta_{dir} L_{dir}) \tag{8}$$

Among them, N_{pos} is the number of correct prior frames. The values of β_{loc}, β_{cls} and β_{dir} we refer to SECOND algorithm, so $\beta_{loc} = 2$, $\beta_{cls} = 1$, $\beta_{dir} = 0.2$.

4. Testing Results

4.1. KITTI Dataset Division

The training and testing of the model use KITTI's 3D object detection dataset [33], which consists of lidar point clouds and image samples. It is only trained on the lidar point cloud, but the lidar point cloud and image fusion are used. The method to realize the comparison between the prior frame and the true value. The sample data have 7481 training samples. For the convenience of comparison, the same data set division method as the PointPillars algorithm is used: the training samples are divided into 3712 training samples and 3769 testing samples.

4.2. Experiment Analysis

4.2.1. Model Training

The computer environment used for the training and testing of this algorithm is Ubuntu 20.04 system, the processor is Intel® Core™ i9-9900 CPU @ 3.10 GHz × 16, the graphics card is Nvidia A40, and the video memory is 48G. TGPP is improved based on the PointPillars algorithm model in the OpenPCDet framework and written in Python3.8.

OpenPCDet is an open-source point cloud object detection algorithm library based on Pytorch. The PointPillars algorithm in this framework adopts more advanced data enhancement methods, optimizers, learning strategies, and other methods to optimize the model. The trained model has better detection accuracy.

The optimizer selected during training is Adam_onecycle, and the maximum learning rate LR is 0.002. This algorithm and PointPillars are trained under the same conditions. During the training process, the loss change curve before and after the model improvement is shown in Figure 3. It can be seen from the figure that the TGPP has a stronger feature learning ability.

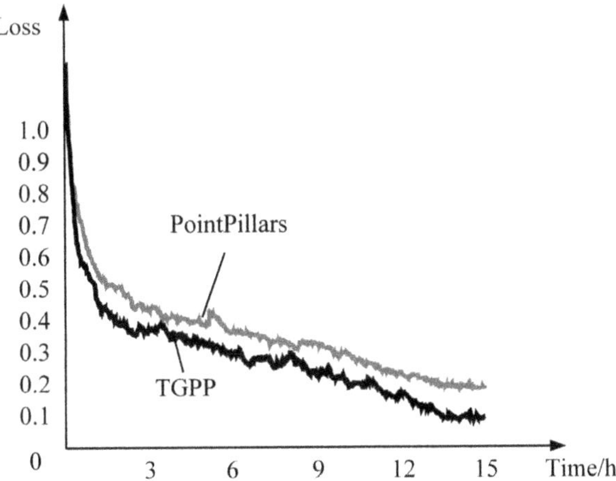

Figure 3. Loss change curve.

4.2.2. Model Training

The test is performed on the trained model based on KITTI's 3D object detection testing set. The test scenarios are divided into three types: simple, medium, and difficult. The test mainly uses the average precision (AP) of 3D object detection as the evaluation index. During the test, the detection of vehicles adopts the standard of IoU = 0.7, and the detection of pedestrians and cyclists adopts the standard of IoU = 0.5.

(1) Compared with PointPillars:

The test results of this algorithm and the PointPillars algorithm are shown in Table 1. It can be seen from Table 1 that compared with PointPillars, this method has improved the 3D object detection performance of vehicles, pedestrians, and cyclists. The vehicle detection AP in the three difficulty scenarios increased by 2.68%, 1.84%, and 2.62%, respectively; the pedestrian detection AP increased by 4.84%, 3.97%, and 3.42%, respectively; the cyclist detection AP increased by 1.41%, 2.12%, and 2.24%, respectively. To better evaluate the overall detection performance, the mAP of vehicles, pedestrians, and cyclists detected under medium difficulty are calculated. The TGPP mAP is 63.56%, and the PointPillars mAP is 60.92%. TGPP has improved by 2.64% mAP in the testing set, which is equivalent to a performance improvement of about 4.3% for PointPillars.

Table 1. Comparison of average precision of 3D object detection (%).

Method	Runtime (ms)	Car (IoU = 0.7)			Pedestrian (IoU = 0.5)			Cyclist (IoU = 0.5)			mAP Mod.
		Easy	Mod.	Hard	Easy	Mod.	Hard	Easy	Mod.	Hard	
PointPillars	16	85.06	76.05	72.03	52.08	45.88	41.67	78.64	60.83	57.43	60.92
TGPP	21	87.74	77.89	74.65	56.92	49.85	45.09	80.05	62.95	59.67	63.56

In terms of detection speed, the average time for PointPillars to process a frame of point cloud data is only 16 ms. Compared with PointPillars, the detection speed has decreased, and the average time is 21 ms, which is 47 Hz when converted into Hz. Considering the vehicle-mounted laser, the scanning frequency of the lidar is usually 10–20 Hz, so this method can still meet the real-time detection requirements.

(2)   Comparison with Other 3D Object Detection Methods:

Comparing this method with the more excellent methods in recent years, as shown in Table 2, the 3D detection performance of other methods is derived from their own papers, some of which did not give the detection speed, taken from KITTI's 3D object detection method performance leaderboard. It can be seen from the table that this method is compared with commonly used methods based on the fusion of image and point cloud data such as MV3D [34], RoarNet [35], AVoD-FPN [36], and F-PointNet [37]. There are no small advantages in the speed or detection of AP. Among lidar-based methods, this method also has certain advantages compared with voxel-based methods. For example, compared with VoxelNet, SECOND, TANe [38], and PSA-Det3D [39] the mAP is 14.51%, 7.17%, 2.93%, and 2.43% higher; the detection accuracy of the point-based method is usually higher, but this method also has advantages compared with it, such as PointRCNN and STD [40], where the mAP is 4.51% and 2.85% higher, respectively. At the same time, this method is superior to all the methods mentioned above in terms of detection speed.

Table 2. Comparison of 3D object detection accuracy with other methods (%).

Method	Runtime (ms)	Car (IoU = 0.7)			Pedestrian (IoU = 0.5)			Cyclist (IoU = 0.5)			mAP Mod.
		Easy	Mod.	Hard	Easy	Mod.	Hard	Easy	Mod.	Hard	
Lidar& Img.											
MV3D	360	71.29	62.68	56.56	-	-	-	-	-	-	-
RoarNet	100	83.71	73.04	59.16	-	-	-	-	-	-	-
AVOD-FPN	100	81.94	71.88	66.38	50.80	42.81	40.88	64.00	52.18	46.61	55.62
F-PointNet	169	81.20	70.39	62.19	51.21	44.89	40.23	71.96	56.77	50.39	57.35
Only Lidar Voxel-base											
VoxelNet	220	77.47	65.11	57.73	39.48	33.69	31.50	61.22	48.36	44.37	49.05
SECOND	50	83.13	73.66	66.20	51.07	42.56	37.29	70.51	53.85	46.90	56.39
TANet	35	83.81	75.38	67.66	54.92	46.67	42.42	73.84	59.86	53.46	60.63
PSA-Det3D	80	87.46	78.80	74.47	49.72	42.81	39.58	75.82	61.79	55.12	61.13
Point-base											
PointRCNN	100	85.94	75.76	68.32	49.43	41.78	38.63	73.93	59.60	53.59	59.05
STD	80	86.61	77.63	76.06	53.08	44.24	41.97	78.89	62.53	55.77	60.71
TGPP	21	87.74	77.89	74.65	56.92	49.85	45.09	80.05	62.95	59.67	63.56

In summary, this method maintains the advantages of the PointPillars algorithm in detection speed and is also superior to the current mainstream methods in detection accuracy. Therefore, it is proved that the feature encoding network improvement scheme proposed in this paper is feasible and practical.

4.3. Comparison of Actual Road Environment Test Results

We use this method and the original method to test the effect of target detection in the same road environment, as shown in Figure 4. In Figure 4 (scenario a), it can be seen that the false detection rate of the original method is higher, and many non-object point clouds are recognized as vehicles and cyclists; in Figure 4 (scenario b), it can be seen that the original method has a higher impact on pedestrians. The false detection rate is high, and the point cloud of non-pedestrians is recognized as pedestrians. From this, it can be seen that the detection accuracy of this method is better than that of the original method in the actual road environment test.

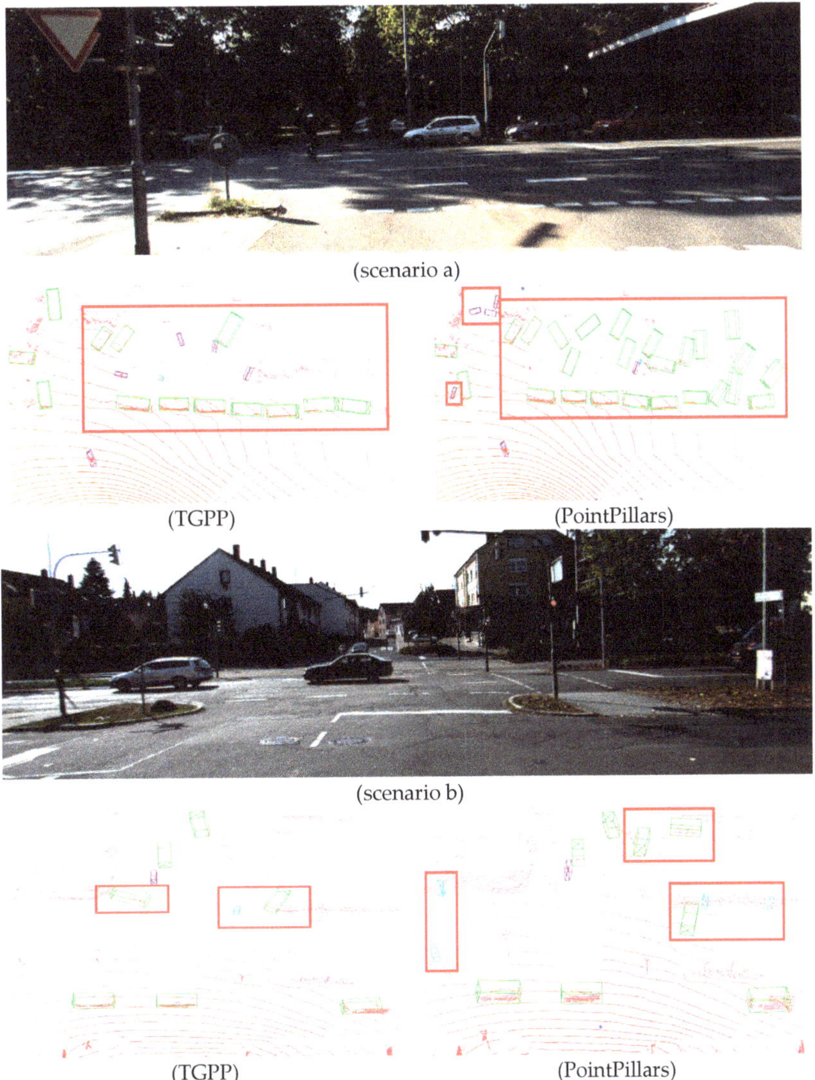

Figure 4. Object detection results in the actual road environment. Car: green bounding boxes; pedestrian: blue bounding boxes; cyclist: purple bounding boxes. The red box displays the difference in detection performance between TGPP and PointPillars. By comparing the object in the red box, it can be found that TGPP has better detection performance.

5. Ablation Experiments

In order to verify the effectiveness of the improved Transformer-based feature encoding network, an ablation experiment is performed. The hyperparameters of the feature encoding network include the number of Transformer layers and the number of heads. Change these two parameters to observe the impact on detection performance. In order to avoid the influence of random number seeds, each set of parameters was trained five times. For the convenience of the experiment, Epoch was set to 120, and the average mAP under medium difficulty was used as the evaluation index. The results are shown in Table 3. It can be seen from the table that when the number of layers and heads is small, the detection performance is not as good as PointPillars. Increasing the number of layers and heads within a certain range can improve the detection performance. When the number of layers is 4 and the number of heads is 2, the effect is the best. Continuing to increase the number of layers and the number of heads will degrade the detection performance.

Table 3. Results of ablation experiments.

Group	Transformer Layer	Transformer Head	Epoch	Training Time	mAP Mod.
1 (PointPillars)	--	--	120	5	60.42
2 (TGPP)	1	1	120	5	59.76
3 (TGPP)	2	1	120	5	59.95
4 (TGPP)	2	2	120	5	61.34
5 (TGPP)	4	1	120	5	62.26
6 (TGPP)	4	2	120	5	63.32
7 (TGPP)	6	1	120	5	62.81
8 (TGPP)	6	2	120	5	62.94

6. Conclusions

In this paper, for the 3D object detection algorithm, an improved PointPillars feature encoding network based on Transformer is proposed. This improved PointPillars algorithm is named TGPP. The improved feature encoding network uses a multi-head attention mechanism to extract global context features and local structure features from pillars. The feature extraction ability of the original algorithm in the feature encoding process is improved, and the feature loss is reduced. Experimental results prove that this algorithm has better object detection performance than PointPillars, and the average object detection accuracy on the KITTI testing set has increased by 2.64%, which is also competitive with other methods in recent years.

Author Contributions: Conceptualization, L.Z. and H.M.; methodology, H.M.; software, H.M.; validation, L.Z., H.M., Y.Y. and X.X.; formal analysis, L.Z. and H.M.; resources, Y.Y. and X.X.; writing—original draft preparation, L.Z. and H.M.; writing—review and editing, Y.Y. and X.X.; project administration, L.Z. and Y.Y.; funding acquisition, L.Z. and Y.Y. All authors have read and agreed to the published version of the manuscript.

Funding: This research was funded by National Natural Science Foundation of China [51975428], "Chunhui Plan" Cooperative Scientific Research Project of the Education Department of China [HZKY20220330] and Guidance Project of Scientific Research Plan of the Education Department of Hubei Province of China [B2022027].

Data Availability Statement: The data presented in this study are available on request from the corresponding author.

Conflicts of Interest: The authors declare no conflict of interest.

References

1. Himmelsbach, M.; Mueller, A.; Lüttel, T.; Wünsche, H.-J. LIDAR-based 3D object perception. In Proceedings of the 1st International Workshop on Cognition for Technical Systems, Munich, Germany, 6–7 October 2008.
2. Xie, D.; Xu, Y.; Wang, R.; Su, Z. Obstacle Detection and Tracking for Unmanned Vehicles Based on 3D Laser Radar. *Automot. Eng.* **2018**, *40*, 952–959. [CrossRef]
3. Xia, X.; Zhu, S.; Zhou, Y.; Ye, M.; Zhao, Y. LiDAR K-means Clustering Algorithm Based on Threshold. *J. Beijing Univ. Aeronaut. Astronaut.* **2020**, *46*, 115–121. [CrossRef]
4. Zong, C.; Wen, L.; He, L. Object Detection Based on Euclidean Clustering Algorithm with 3D Laser Scanner. *J. Jilin Univ. Eng. Technol. Ed.* **2020**, *50*, 107–113. [CrossRef]
5. Ning, X.; Gong, L.; Zhang, J. Detection Method of Passable Road Areas Based on Laser Point Clouds. *Comput. Eng.* **2022**, *48*, 22–29. [CrossRef]

6. Qian, R.; Lai, X.; Li, X. 3D Object Detection for Autonomous Driving: A Survey. *Pattern Recognit.* **2021**, *130*, 108796. [CrossRef]
7. Qi, C.; Su, H.; Mo, K.; Guibas, L. PointNet: Deep Learning on Point Sets for 3D Classification and Segmentation. In Proceedings of the IEEE Conference on Computer Vision and Pattern Recognition, Las Vegas, NV, USA, 27–30 June 2016; pp. 652–660.
8. Qi, C.; Yi, L.; Su, H.; Guibas, L. PointNet++: Deep Hierarchical Feature Learning on Point Sets in a Metric Space. In Proceedings of the NIPS, Long Beach, CA, USA, 4–9 December 2017.
9. Shi, S.; Wang, X.; Li, H. PointRCNN: 3D Object Proposal Generation and Detection From Point Cloud. In Proceedings of the IEEE/CVF Conference on Computer Vision and Pattern Recognition, Salt Lake City, UT, USA, 18–23 June 2018; pp. 770–779.
10. Zhou, Y.; Tuzel, O. VoxelNet: End-to-End Learning for Point Cloud Based 3D Object Detection. In Proceedings of the IEEE/CVF Conference on Computer Vision and Pattern Recognition, Salt Lake City, UT, USA, 18–23 June 2018; pp. 4490–4499.
11. Yan, Y.; Mao, Y.; Li, B. SECOND: Sparsely Embedded Convolutional Detection. In Proceedings of the Italian National Conference on Sensors, Catania, Italy, 21–23 February 2018; Volume 18, p. 3337.
12. Lang, A.H.; Vora, S.; Caesar, H.; Zhou, L.; Yang, J.; Beijbom, O. PointPillars: Fast Encoders for Object Detection From Point Clouds. In Proceedings of the IEEE/CVF Conference on Computer Vision and Pattern Recognition, Long Beach, CA, USA, 15–20 June 2019; pp. 12697–12705.
13. Li, Y.; Chen, Y.; Qi, X.; Li, Z.; Sun, J.; Jia, J. Unifying voxel-based representation with transformer for 3d object detection. *Adv. Neural Inf. Process. Syst.* **2022**, *35*, 18442–18455.
14. Lai, X.; Chen, Y.; Lu, F.; Liu, J.; Jia, J. Spherical transformer for lidar-based 3d recognition. In Proceedings of the IEEE/CVF Conference on Computer Vision and Pattern Recognition, Vancouver, BC, Canada, 18–22 June 2023; pp. 17545–17555.
15. He, X.; Yan, A.; Chen, L.; Hou, P.; Dong, D.; Ma, Y. An Improved PointPillars for Fast and Accurate 3D Object Detection. In Proceedings of the 2021 Unmanned Systems Summit Forum (USS 2021), Changsha, China, 23–24 September 2021; pp. 115–120.
16. Chen, D.; Yu, W.; Gao, Y. Lidar 3D Object Detection Based on Improved Point Pillars. *Laser Optoelectron. Prog.* **2023**, *60*, 447–453.
17. Liu, Z.; Lin, Y.; Cao, Y.; Hu, H.; Wei, Y.; Zhang, Z.; Lin, S.; Guo, B. Swin Transformer: Hierarchical Vision Transformer using Shifted Windows. In Proceedings of the IEEE International Conference on Computer Vision, Montreal, BC, Canada, 11–17 October 2021; pp. 10012–10022.
18. Vaswani, A.; Shazeer, N.M.; Parmar, N.; Uszkoreit, J.; Jones, L.; Gomez, A.N.; Kaiser, L.; Polosukhin, I. Attention is All you Need. In Proceedings of the NIPS, Long Beach, CA, USA, 4–9 December 2017; Volume 30.
19. Liu, W.; Anguelov, D.; Erhan, D.; Szegedy, C.; Reed, S.E.; Fu, C.-Y.; Berg, A. SSD: Single Shot MultiBox Detector. In Proceedings of the Computer Vision—ECCV 2016: 14th European Conference, Amsterdam, The Netherlands, 11–14 October 2016; pp. 21–37.
20. Guo, M.-H.; Cai, J.; Liu, Z.-N.; Mu, T.-J.; Martin, R.R.; Hu, S. PCT: Point cloud transformer. *Comput. Vis. Media* **2020**, *7*, 187–199. [CrossRef]
21. Engel, N.; Belagiannis, V.; Dietmayer, K. Point Transformer. *IEEE Access* **2020**, *9*, 16259–16268. [CrossRef]
22. Xia, Y.; Xu, Y.; Li, S.; Wang, R.; Du, J.; Cremers, D.; Stilla, U. SOE-Net: A self-attention and orientation encoding network for point cloud based place recognition. In Proceedings of the IEEE/CVF Conference on Computer Vision and Pattern Recognition, Nashville, TN, USA, 20–25 June 2021; pp. 11348–11357.
23. He, C.; Li, R.; Li, S.; Zhang, L. Voxel Set Transformer: A Set-to-Set Approach to 3D Object Detection from Point Clouds. In Proceedings of the IEEE/CVF Conference on Computer Vision and Pattern Recognition, New Orleans, LA, USA, 18–24 June 2022; pp. 8417–8427.
24. Liu, Z.; Yang, X.; Tang, H.; Yang, S.; Han, S. FlatFormer: Flattened Window Attention for Efficient Point Cloud Transformer. In Proceedings of the IEEE/CVF Conference on Computer Vision and Pattern Recognition, Vancouver, BC, Canada, 18–22 June 2023; pp. 1200–1211.
25. He, K.; Zhang, X.; Ren, S.; Sun, J. Deep Residual Learning for Image Recognition. In Proceedings of the IEEE Conference on Computer Vision and Pattern Recognition, Las Vegas, NV, USA, 26 June–1 July 2016; pp. 770–778.
26. Ba, J.L.; Kiros, J.R.; Hinton, G.E. Layer normalization. *arXiv* **2016**, arXiv:1607.06450.
27. Qiu, S.; Wu, Y.; Anwar, S.; Li, C. Investigating Attention Mechanism in 3D Point Cloud Object Detection. In Proceedings of the 2021 International Conference on 3D Vision (3DV), Prague, Czech Republic, 12–15 September 2022; pp. 403–412.
28. Glorot, X.; Bordes, A.; Bengio, Y. Deep Sparse Rectifier Neural Networks. In Proceedings of the Fourteenth International Conference on Artificial Intelligence and Statistics, Fort Lauderdale, FL, USA, 11–13 April 2011; pp. 315–323.
29. Ioffe, S.; Szegedy, C. Batch Normalization: Accelerating Deep Network Training by Reducing Internal Covariate Shift. In Proceedings of the International Conference on Machine Learning, Lille, France, 6–11 July 2015; pp. 448–456.
30. Girshick, R.B. Fast R-CNN. In Proceedings of the IEEE International Conference on Computer Vision, Santiago, Chile, 11–18 December 2015; pp. 1440–1448.
31. Lin, T.-Y.; Dollár, P.; Girshick, R.B.; He, K.; Hariharan, B.; Belongie, S.J. Feature Pyramid Networks for Object Detection. In Proceedings of the Computer Vision and Pattern Recognition, Las Vegas, NV, USA, 27–30 June 2016; pp. 2117–2125.
32. Lin, T.-Y.; Goyal, P.; Girshick, R.B.; He, K.; Dollár, P. Focal Loss for Dense Object Detection. In Proceedings of the IEEE Transactions on Pattern Analysis and Machine Intelligence, Venice, Italy, 22–29 October 2017; pp. 2980–2988.
33. Geiger, A.; Lenz, P.; Urtasun, R. Are we ready for autonomous driving? The KITTI vision benchmark suite. In Proceedings of the 2012 IEEE Conference on Computer Vision and Pattern Recognition, Providence, RI, USA, 16–21 June 2012; pp. 3354–3361.
34. Chen, X.; Ma, H.; Wan, J.; Li, B.; Xia, T. Multi-view 3D Object Detection Network for Autonomous Driving. In Proceedings of the Computer Vision and Pattern Recognition, Las Vegas, NV, USA, 27–30 June 2016; pp. 1907–1915.

35. Shin, K.; Kwon, Y.; Tomizuka, M. RoarNet: A Robust 3D Object Detection based on RegiOn Approximation Refinement. In Proceedings of the 2019 IEEE Intelligent Vehicles Symposium (IV), Paris, France, 9–12 June 2019; pp. 2510–2515.
36. Ku, J.; Mozifian, M.; Lee, J.; Harakeh, A.; Waslander, S.L. Joint 3D Proposal Generation and Object Detection from View Aggregation. In Proceedings of the IEEE/RJS International Conference on Intelligent Robots and Systems, Vancouver, BC, Canada, 24–28 September 2017; pp. 1–8.
37. Qi, C.; Liu, W.; Wu, C.; Su, H.; Guibas, L. Frustum PointNets for 3D Object Detection from RGB-D Data. In Proceedings of the 2018 IEEE/CVF Conference on Computer Vision and Pattern Recognition, Salt Lake City, UT, USA, 18–22 June 2018; pp. 918–927.
38. Liu, Z.; Zhao, X.; Huang, T.; Hu, R.; Zhou, Y.; Bai, X. TANet: Robust 3D Object Detection from Point Clouds with Triple Attention. In Proceedings of the AAAI Conference on Artificial Intelligence, Honolulu, HI, USA, 27 January–1 February 2019; pp. 11677–11684.
39. Huang, Z.; Zhao, J.; Zheng, Z.; Chena, D.; Hu, H. PSA-Det3D: Pillar Set Abstraction for 3D object Detection. *Pattern Recognit. Lett.* **2022**, *168*, 138–145. [CrossRef]
40. Yang, Z.; Sun, Y.; Liu, S.; Shen, X.; Jia, J. STD: Sparse-to-Dense 3D Object Detector for Point Cloud. In Proceedings of the IEEE International Conference on Computer Vision, Seoul, Republic of Korea, 27 October–2 November 2019; pp. 1951–1960.

electronics

Article

A Railway Track Extraction Method Based on Improved DeepLabV3+

Yanbin Weng [1,*], Zuochuang Li [1,*], Xiahu Chen [1,2], Jing He [1], Fengnian Liu [1], Xiaobin Huang [1] and Hua Yang [1]

[1] Hunan University of Technology, Tianyuan District, Zhuzhou 412007, China; chenxh@hut.edu.cn (X.C.); hejing@hut.edu.cn (J.H.); bingxuewa@jgsys.wecom.work (F.L.); m21085400010@stu.hut.edu.cn (X.H.); yanghua@hut.edu.cn (H.Y.)

[2] Zhuzhou Taichang Electronic Information Technology Co., Ltd., Zhuzhou 412007, China

[*] Correspondence: wengyb@hut.edu.cn (Y.W.); m21085400017@stu.hut.edu.cn (Z.L.)

Abstract: Extracting railway tracks is crucial for creating electronic railway maps. Traditional methods require significant manual labor and resources while existing neural networks have limitations in efficiency and precision. To address these challenges, a railway track extraction method using an improved DeepLabV3+ model is proposed, which incorporates several key enhancements. Firstly, the encoder part of the method utilizes the lightweight network MobileNetV3 as the backbone extraction network for DeepLabV3+. Secondly, the decoder part adopts the lightweight, universal upsampling operator CARAFE for upsampling. Lastly, to address any potential extraction errors, morphological algorithms are applied to optimize the extraction results. A dedicated railway track segmentation dataset is also created to train and evaluate the proposed method. The experimental results demonstrate that the model achieves impressive performance on the railway track segmentation dataset and DeepGlobe dataset. The MIoU scores are 88.93% and 84.72%, with Recall values of 89.02% and 86.96%. Moreover, the overall accuracy stands at 97.69% and 94.84%. The algorithm's operation time is about 5% lower than the original network. Furthermore, the morphological algorithm effectively eliminates errors like holes and spots. These findings indicate the model's accuracy, efficiency, and enhancement brought by the morphological algorithm in error elimination.

Keywords: deep learning; MobileNetV3; morphological algorithm; railway extraction; aerial imagery

Citation: Weng, Y.; Li, Z.; Chen, X.; He, J.; Liu, F.; Huang, X.; Yang, H. A Railway Track Extraction Method Based on Improved DeepLabV3+. *Electronics* **2023**, *12*, 3500. https://doi.org/10.3390/electronics12163500

Academic Editor: George A. Tsihrintzis

Received: 10 July 2023
Revised: 8 August 2023
Accepted: 16 August 2023
Published: 18 August 2023

1. Introduction

Railway transportation plays a vital role in China's economic development and is a fundamental component of its transportation system. Extracting railway tracks is essential for creating railway electronic maps, ensuring smooth railway operations, and safeguarding people's lives and property. Traditionally, drawing railway track maps involved processing satellite positioning data, which required extensive expertise and involved a substantial workload [1–3]. However, with the availability of remote sensing and UAV aerial images [4], deep learning methods can now be applied to extract railway tracks and generate railway electronic maps, offering convenience and efficiency.

The process of extracting railway tracks and roads shares several similarities. Traditional methods for obtaining railway or road information involve setting specific conditions based on texture, spectral, and geometric features. These conditions are then used to extract deeper features and acquire the desired information. For road extraction, Xiao Chi et al. [5] proposed using road color features to obtain initial road segments, which were refined using a region merging algorithm to achieve complete road information. Shi W. et al. [6] introduced a general adaptive neighborhood approach to perform spectrum-space classification, distinguishing road and non-road regions. Xiaoyu Liu et al. [7] utilized grid approximation and adaptive filtering parameter calculation, combined with the spatial distribution characteristics of roads, to extract roads through clustering fitting and other techniques. Lingran Kong et al. [8] constructed feature points based on the spectral characteristics of road areas, connecting them to form an initial road network. Various constraints were added, and the road centerline was extracted by maximizing the Posterior probability criterion. These studies demonstrate different approaches to extracting road information using color features, spatial analysis, adaptive methods, and probabilistic criteria. Similarly, railway track extraction methods employ comparable principles to analyze relevant features and obtain accurate railway track information.

Traditional extraction methods for railway tracks often rely on researchers possessing significant prior knowledge. However, these methods face challenges in distinguishing features that share similar characteristics. For instance, rivers and railways may exhibit similar geometric features, and buildings and railways may have comparable spectral features. Consequently, traditional extraction methods are prone to inaccuracies and lack robustness when confronted with such complexities. As a result, these methods are not well-suited for the intricate environments found in modern cities.

In recent years, deep learning has experienced rapid advancements and found widespread applications in various fields, including facial expression recognition [9], lane detection [10], railway foreign body detection, track defect detection, catenary detection, and road extraction [11]. Convolutional Neural Networks (CNNs), as a classical deep learning architecture, have significantly contributed to road segmentation research. Zuoming She et al. [12] proposed a CNN model for road extraction, optimizing extracted data to obtain comprehensive road features. Jiguang Dai et al. [13] introduced a method based on multi-scale CNNs for road extraction in remote sensing images. They employed a sub-image training model and incorporated residual connections to address resolution reduction and gradient disappearance issues during extraction. Zhang X. et al. [14] developed an FCN network utilizing a spatially consistent integration algorithm to determine loss function weights for extracting road regions. Xiangwen Kong et al. [15] introduced an SM-Unet semantic segmentation network with a stripe pooling module to enhance road extraction performance. Hao Qi et al. [16] proposed the MBv2-DPPM model, considering segmentation accuracy and speed. However, this model still exhibits some errors, such as convex points and spots. Overall, the rapid development of deep learning technology has significantly improved road segmentation and extraction tasks, but some challenges persist, such as handling complex road structures and enhancing accuracy in challenging scenarios.

Compared to traditional extraction methods, deep learning-based extraction methods offer several advantages. They require less prior knowledge and workload for researchers, and the overall process is relatively straightforward. This makes deep learning methods more suitable for handling the complexities of modern urban environments. While the accuracy of deep learning-based extraction methods has been improved through extensive research by scholars, there are still certain challenges. Deep networks often have many layers and parameters, leading to inefficient network performance. Additionally, the pixel-level nature of deep learning extraction may result in spots, holes, or breakpoints in the final extraction results.

The errors in extracting tracks caused by holes and breakpoints using deep learning methods are attributed to multiple factors. Firstly, it could be due to data-related issues. Both unmanned aerial imagery and remote sensing images inevitably suffer from occlusions, such as trees and buildings, in the scene. Moreover, during extraction, the pixels representing tracks often constitute a small proportion of the entire image, leading to a class imbalance where non-track pixels dominate. This imbalance might cause the model to be biased towards predicting non-track categories, thus affecting the accuracy of track extraction. Secondly, limitations in the principles of the methods used could be a contributing factor. Most deep-learning techniques aim to identify underlying patterns between images and labels. However, in the case of pixel-level semantic segmentation, variations in lighting conditions around the tracks can result in misclassification. Lastly, issues with the annotated data could play a role. Errors in the labeled data may cause the deep learning model to learn incorrect features, leading to recognition inaccuracies. Therefore, the simplest way to address recognition errors is through post-processing. After obtaining initial results, applying post-processing techniques like removing isolated pixels and eliminating noise can help refine the track extraction outcome.

The paper presents several contributions to address the challenges in deep learning-based extraction methods:

- Segmentation Dataset: A railway track segmentation dataset is established, consisting of 7892 original images and their corresponding label images. These images are collected from aerial shots of railway UAVs in various stations in China, capturing different environmental conditions and railway track information.
- Improved DeepLabV3+ Model: The paper proposes an enhanced DeepLabV3+ network model. It replaces the original backbone network with a lightweight MobileNetV3 network module, which helps mitigate the efficiency issues caused by the deep network hierarchy and large parameter quantity. The bilinear upsampling module is also replaced with CARAFE, improving both extraction process accuracy.
- Morphological Algorithm Optimization: The paper introduces an optimization method using morphological algorithms. After obtaining initial extraction results from the improved model, morphological operations, such as erosion and expansion, are applied to eliminate potential errors like spots and holes. This optimization process enhances the accuracy of railway track extraction.

The remaining sections of the manuscript are organized as follows: The methodology flow and network structure employed in this study are presented in Section 2. The experimental data, experimental environment, and evaluation metrics are introduced in Section 3. The experimental process is outlined, and the obtained results are presented in Section 4. A comprehensive discussion of the results obtained in this study is provided in Section 5. Finally, the overall conclusions are presented in Section 6.

2. Materials and Methods

This section presents an exposition of the methodology flow and network architecture employed in this study, providing a comprehensive account of the key components involved. The subsequent elucidation aims to facilitate a deeper understanding of the underlying processes and techniques utilized in this research endeavor.

2.1. Algorithm Flow

This study is divided into five main stages: data acquisition, data preprocessing, trajectory extraction, morphological optimization, and calculation evaluation index. In the data acquisition stage, aerial images captured by orbital UAVs in various domestic stations are selected. These images encompass different weather conditions and terrains, providing a diverse dataset for analysis. During the data preprocessing stage, corresponding labels are created for the original images. The dataset is then divided into training, test, and validation sets using a specific proportion. The training set is further augmented to increase its size and enhance the model's learning capability. In the Railway track segmentation and extraction stage, the prepared dataset is fed into the improved network proposed in this paper for training. The network is trained to optimize its weights, which are utilized to extract binary images of the railway tracks from the original images. The morphological optimization stage involves applying morphological binary operations to refine the extracted railway track images obtained in the previous stage. This process helps eliminate defects such as holes and spots in the extraction results, improving the overall quality of the extracted tracks. Finally, in calculating the evaluation index stage, common semantic segmentation evaluation metrics are employed to assess the quality of the results. Metrics such as MIoU, recall, and accuracy are calculated to evaluate the performance of the railway track extraction and segmentation method. Overall, the study follows a systematic approach, starting from data acquisition and preprocessing, progressing to railway track segmentation and extraction using the improved network, morphological optimization, and finally, evaluating the results using standard evaluation metrics. The method flow of this article is shown in Figure 1.

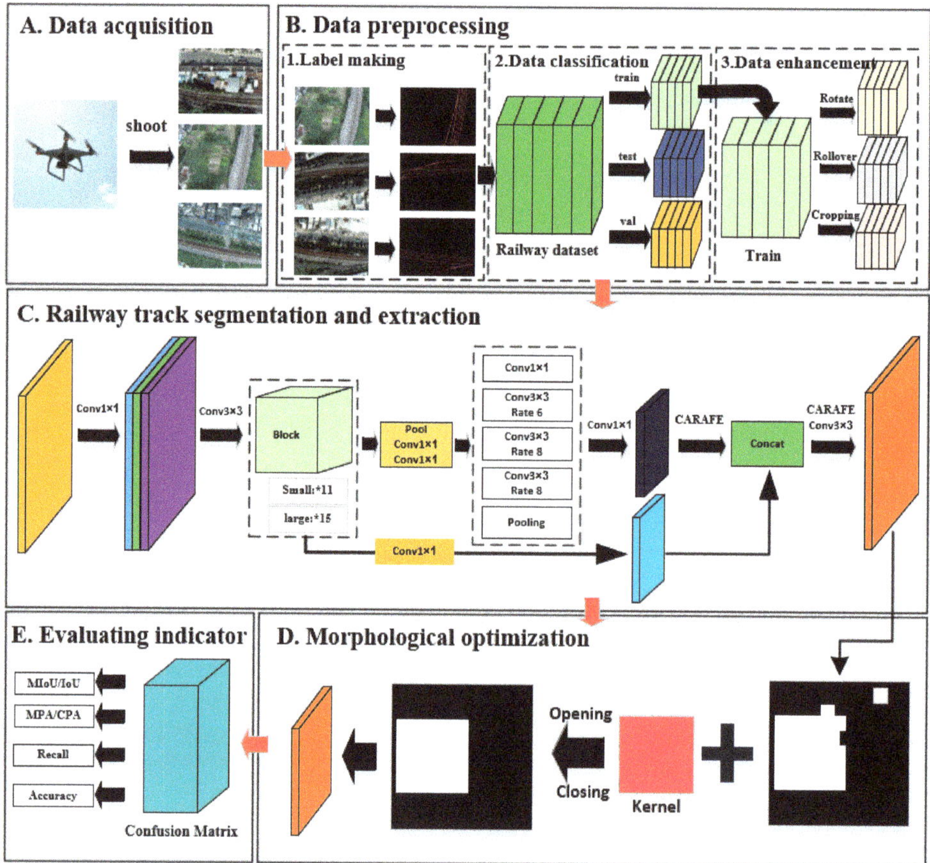

Figure 1. Flowchart of the algorithm proposed in this article.

2.2. Improved DeepLabv3+ Network Structure

The network architecture used in this paper is an encoder-decoder structure. This structure mainly consists of an encoder and a decoder. The encoder is the most important component in the entire model, responsible for the feature extraction and compression process. Through multiple layers of convolution and pooling operations, the encoder can capture various low-level and high-level features in the input data. These features provide meaningful information for subsequent tasks and help the model learn the crucial characteristics of the input data. The design and performance of the encoder directly impact the overall effectiveness of the model. A well-designed encoder can assist the model in better understanding and processing the data, thereby improving the model's performance and generalization ability.

On the other hand, the decoder is responsible for feature restoration. By utilizing operations such as upsampling and deconvolution, the decoder can reverse the encoder process and restore the original data's dimensions and resolution. The close collaboration between the encoder and decoder drives the model to perform excellently in various tasks.

In the original DeepLabv3+ model, the encoder structure utilizes DeepLabv3 [17], while the decoder replaces the direct 16x bilinear upsampling used in DeepLabv3 with a specialized upsampling module. The decoder is responsible for processing, fusing, and upsampling the input features from the encoder, ultimately generating the extraction result [18]. While this approach addresses the drawback of missing details in direct 16x bilinear upsampling, the upsampling module in the decoder also introduces additional computational complexity, resulting in decreased efficiency during the extraction process. Despite these modifications, the enhancement in extraction accuracy is not substantial.

To address these challenges, this paper introduces an improved DeepLabv3+ model. The encoder incorporates the lightweight MobileNetV3 [19] network, proposed by Google, as the backbone network for initial feature extraction. Subsequently, the ASPP (Atrous Spatial Pyramid Pooling) module is employed to consider context information and fuse features from different receptive fields. In the decoder, both lower-level and higher-level semantic features generated by the encoder are utilized. The lower-level semantic features undergo a 1×1 convolution to increase their dimension, while the higher-level semantic features are upsampled using a $4\times$ CARAFE (Content-Aware ReAssembly of FEatures) module. The features are fused, followed by two consecutive 3×3 convolutions. A final 4-fold CARAFE upsampling is performed to obtain the initial extraction results. However, the initial extraction results may still contain imperfections such as spots, voids, bumps, or pits. To improve the completeness of the extraction results, the morphological processing technique is applied to optimize the initial extraction results. This processing helps eliminate these imperfections and enhance the overall quality of the extraction. Figure 2 illustrates the proposed structure of the improved DeepLabv3+ network, showcasing the flow and components of the model.

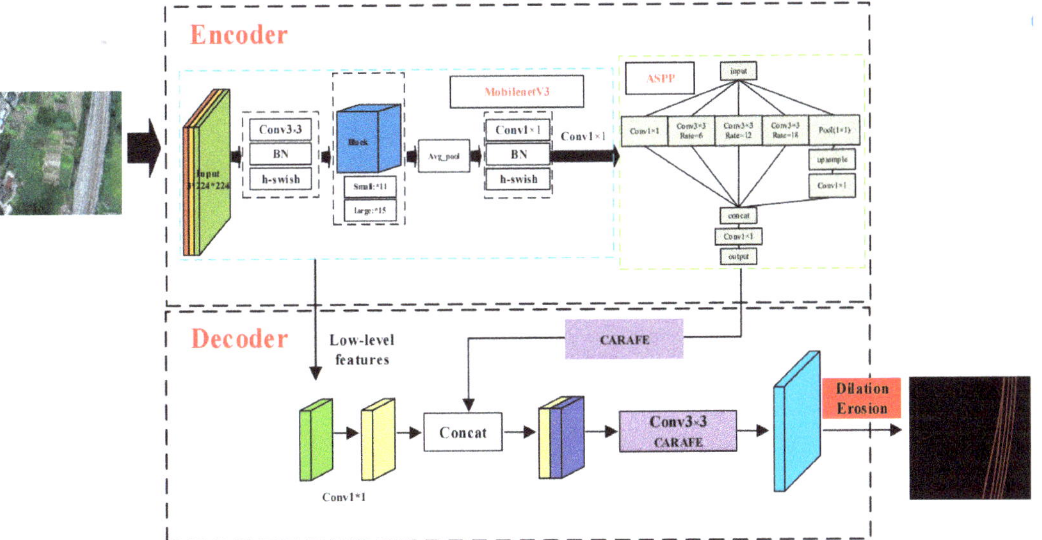

Figure 2. Improved DeepLabv3+ network structure diagram.

The process of railway track extraction in this paper consists of three stages: the training stage, the extraction stage, and the optimization stage. In the training phase, pre-trained weights are utilized to expedite the convergence of the model. The loss function is then applied to calculate the error between the predicted output and the ground truth labels. This loss value serves as a feedback signal used to adjust the weight values of each layer in the network through back-propagation facilitated by the optimizer. Multiple iterations of training are performed until the loss value reaches its minimum and stops decreasing. At this point, the predicted values closely resemble the real values, and the model weights are considered optimal. The model training results thus achieve the best performance. Moving on to the extraction stage, the input image is processed sequentially with each layer's features based on the trained weights. The various semantic features extracted from each layer are then fused and upsampled in the decoder section of the model. This process ultimately yields the preliminary extraction results, which represent the initial segmentation of the railway track. In the optimization stage, the preliminary extraction results undergo morphological operations such as erosion, dilation, opening, and closing. These operations are employed to rectify possible errors and enhance the completeness of the results. By applying these morphological operations, more accurate and comprehensive extraction results are obtained.

2.3. Mobilenetv3 Network

MobileNetV3 is the latest lightweight network proposed by Google. It offers several advantages, including fewer parameters, lower computation requirements, and reduced time consumption. In MobileNetV3, the convolution kernel of the first layer has been modified from 32 to 16, further reducing time consumption without compromising accuracy. In MobileNetV2 [20], the Swish function replaced the ReLU function, resulting in a significant improvement in accuracy. However, the computation and derivation of the Swish activation function and other non-linear functions were more complex, leading to an increased time burden. To address these limitations, MobileNetV3 introduces the $h - swish$ function as the activation function. The $h - swish$ function is similar to ReLU6 but offers easier calculations. The expression for the $h - swish$ activation function is as follows:

$$h - swish(x) = x \frac{ReLU6(x + 3)}{6} \tag{1}$$

The performance of a nonlinear activation function can vary based on the depth of the network layer. Generally, the $h - swish$ function performs better as the number of network layers increases. Therefore, in the MobileNetV3 structure, the $h - swish$ activation function is used exclusively in the first and subsequent layers of the network. This approach leverages the strengths of the $h - swish$ function, such as its computational efficiency and ability to maintain high accuracy. However, as the depth of the network increases, other factors, such as the complexity of the task and the characteristics of the data, may come into play. To optimize the network's overall performance, it is common to employ different activation functions in different layers based on specific requirements and performance characteristics. By selectively applying the $h - swish$ activation function to the initial layers of MobileNetV3, the trade-off between accuracy and computational efficiency is effectively managed.

In MobileNetV3, the core of the network is the Block structure, which incorporates the channel attention mechanism [21] and updates the activation function. This module enables explicit modeling of interdependencies between channels and adaptive recalibration of channel-level characteristic responses. When the Block structure in MobileNetV3 is activated, the input feature matrix undergoes processing through 1×1 convolution and 3×3 convolution. It is then passed to the attention module for further processing. The channel attention module pools each channel of the input feature matrix, transforming it into a vector using two fully connected layers. In the first fully connected layer, the number of channels is reduced to $1/4$ of the input feature, while the number of channels remains unchanged in the second fully connected layer. The output vector from the attention module represents the weight relationships among different channel features in the preceding input feature matrix. This weight signifies the importance of each channel feature, with higher weights assigned to more significant features. The Block structure of MobileNetV3 is illustrated in Figure 3.

Utilizing the channel attention module to calculate feature channel weights can introduce additional time consumption to the overall model. The conventional channel attention module employs the sigmoid activation function with an exponent, which demands significant computational resources. Moreover, during backpropagation, there is a risk of vanishing gradients. To address these concerns, MobileNetV3 employs the h-sigmoid function as the activation function in the attention module. This choice helps to mitigate the additional computational burden. Using the h-sigmoid function, MobileNetV3 reduces the computational overhead while effectively modeling the channel weights within the attention module. The expression of the h-sigmoid activation function is as follows:

$$f(x) = \begin{cases} 1, (x > 3) \\ \frac{x}{6} + 0.5, (3 \geq x \geq -3) \\ 0, (x < -3) \end{cases} \tag{2}$$

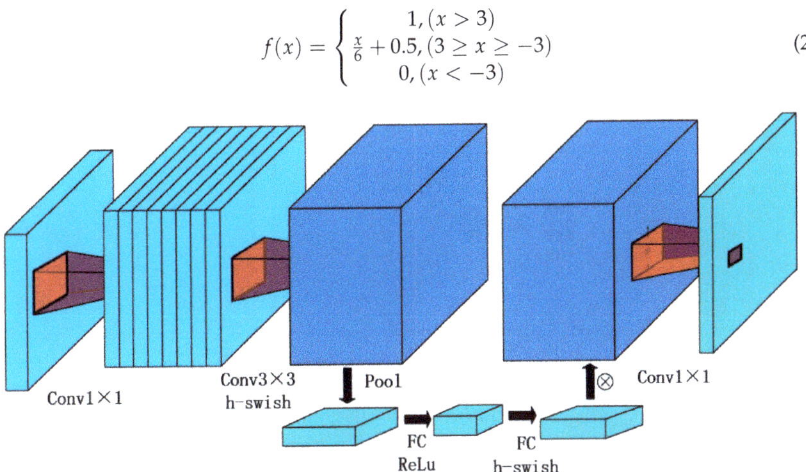

Figure 3. The block structure of MobileNetV3.

In summary, in MobileNetV3, reducing the convolutional size decreased the number of parameters, updating the non-linear activation function improved model computational efficiency, and utilizing network architecture search (NAS) and NetAdapt further enhanced computational efficiency by finding optimal network structures. The experiments conducted by Howard et al. [19] demonstrated that MobileNetV3-Large and MobileNetV3-Small versions achieved varying degrees of improvement in tasks such as ImageNet classification, COCO detection, and semantic segmentation compared to other networks. These improvements mainly manifested as increased accuracy and reduced processing time.

2.4. Morphological Algorithm

The morphology algorithm is an image-processing technique that relies on lattice theory and topology. It consists of four fundamental operations: erosion, dilation, opening, and closing. Opening and closing operations are composite operations that combine erosion and dilation [22].

At the core of the morphology algorithm is a convolution kernel-like structure, which can be designed in a square or circular shape around a reference point, depending on the requirements. During the execution of the algorithm, this "kernel" moves systematically across the input binary image. By analyzing the pixel values, the algorithm determines the relationships between different parts of the image, allowing for an understanding of its structural characteristics. Subsequently, appropriate processing of the binary image can be performed based on this analysis.

The morphological erosion operation in the morphology algorithm involves finding the minimum value among the pixels in a specific area of a binary image. In the case of a binary input image consisting of values [0, 1], the morphology algorithm's "kernel" traverses the image. If only pixel 0 or pixel 1 is present within the range of the kernel, no changes are made to that region. However, if both pixel 0 and pixel 1 are present within the kernel's range, the corresponding region in the binary image, centered around the reference point of the kernel, is assigned a value of 0. The operation's effect is demonstrated in Figure 4a.

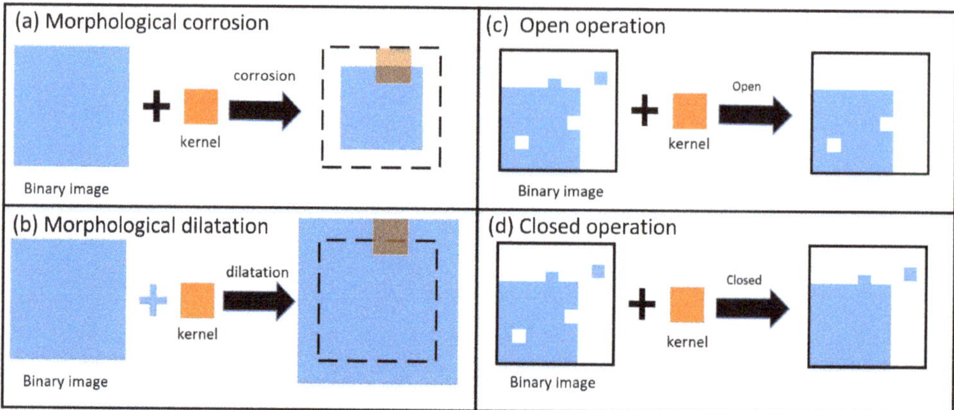

Figure 4. Schematic diagram of morphological algorithm effect.

On the other hand, the morphological dilation algorithm performs a local maximum operation. It operates similarly to the erosion algorithm mentioned above, where regions with either pixel value 0 or pixel value 1 undergo no processing. However, if both pixel values 0 and 1 are present simultaneously, the binary image region centered around the reference point defined in the "kernel" is copied as pixel 1. The operation's effect is depicted in Figure 4b.

The open and close operations in the morphology algorithm are composite operations that combine erosion and dilation. The open operation involves applying erosion followed by dilation. This operation effectively eliminates small spots and convex areas within a specified region, as depicted in Figure 4c. On the other hand, the close operation applies dilation first and then erosion. It is useful for filling holes and depressions in the image, as illustrated in Figure 4d. In the context of the paper, the model first performs the open operation to eliminate spots and convex areas. This step helps remove small artifacts and irregularities. Then, the close operation is applied to connect fragmented structures and fill in any remaining holes or gaps, thereby achieving a more complete and refined image representation.

2.5. Lightweight up Sampling Structure CARAFE

Upsampling allows for the enlargement of extracted features, making it an essential process in feature extraction. However, many upsampling algorithms suffer from small receptive fields, high computational complexity, and a lack of consideration for the contextual information within the feature maps. To address these limitations, Wang J. et al. [23] proposed a lightweight upsampling operator called CARAFE. This operator automatically generates different upsampling kernels to handle pixel information within the input feature map. CARAFE consists of the upsampling kernel prediction module and the feature recombination module. The upsampling kernel prediction module predicts the appropriate upsampling kernels based on the input feature map, considering its content information. The feature recombination module then utilizes the predicted kernels to recombine the feature maps, effectively capturing and preserving more detailed information during upsampling. By incorporating the CARAFE upsampling operator, the network can better adapt to the pixel information and content characteristics of the feature map, leading to improved performance in tasks such as semantic segmentation.

Taking the upsampling process of the advanced semantic features of the model in this article as an example. When the railway track segmentation data set is used for extraction, the input original image is $512 \times 512 \times 3$. That is, the size is 512×512, and the number of channels is 3. The image is processed by the backbone network and ASPP module, and the output advanced feature map is $32 \times 32 \times 256$. In the upsampling kernel prediction module, the advanced feature graph output by the decoder is first compressed by 1×1 convolution channel to obtain a feature graph with a size of $32 \times 32 \times 64$. Then, according to the multiple of 4 times the upsampling, the compressed feature graph is re-encoded by 3×3 convolution to obtain a feature graph with several channels of $4^2 \times 5^2$. Then dimension expansion is carried out to obtain $4 \times 32 \times 4 \times 32 \times 5^2$ upsampled kernel. In the feature recombination module, a point with a size of 5×5 is selected from the input feature map and the corresponding region of the point in the upper sampling kernel at the prediction point is dot product operation, and finally a feature map with a size of $128 \times 128 \times 256$ is obtained. The working principle of the up-sampling module in this paper is shown in Figure 5.

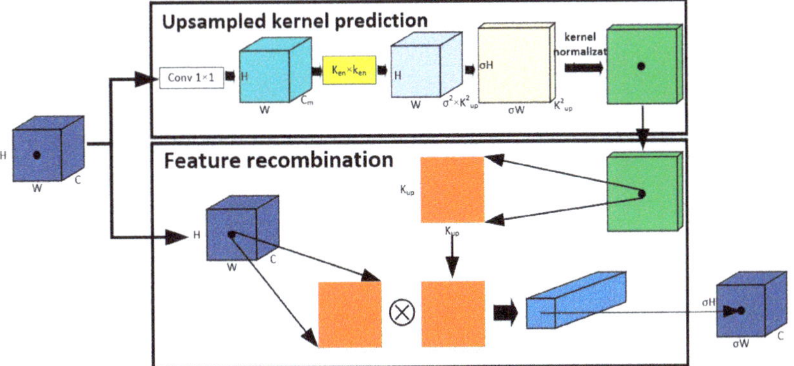

Figure 5. Working principle of upsampling module.

Currently, there are various upsampling methods used in deep learning, including "Nearest", "Bilinear", "Deconvolution", "Pixel Shuffle", "Gumbel" and "Spatial Attention". The "Nearest" and "Bilinear" methods are similar, as they only determine the upsampling kernel based on the spatial position of the pixels and do not utilize semantic information from the feature maps. Additionally, their receptive fields are usually small. On the other hand, "Deconvolution", "Pixel Shuffle" and "Gumbel" are learning-based upsampling methods, which involve a higher number of parameters and require significant computational resources. "Spatial Attention" employs a learned attention weight for each upsampled pixel to guide the sampling and interpolation of pixels from the original image, resulting in more accurate and clear reconstructions. Nevertheless, this method is task-dependent and may not be suitable for certain tasks, and it could also potentially suffer from the issue of vanishing gradients.

To compare the performance of the various upsampling algorithms mentioned above, extensive experiments were conducted on Faster R-CNN in reference [23]. Different operators were used to perform the upsampling operation in Feature Pyramid Network (FPN). The reference [23] experimental results are shown in Table 1.

Table 1. Experimental Comparison of Different Upsampling Algorithms in Reference [23].

Algorithms	AP	AP_S	AP_M	AP_L	$FLOP_S$	Params
Nearest	36.50%	21.30%	40.30%	47.20%	0	0
Bilinear	36.70%	21.00%	40.50%	47.50%	8 K	0
Deconvolution	36.40%	21.30%	39.90%	46.50%	1.2 M	590 K
Pixel Shuffle	36.50%	20.90%	40.40%	46.70%	4.7 M	2.4 M
Gumbel	36.90%	21.50%	40.60%	48.10%	1.1 M	132 K
Spatial Attention	36.90%	21.70%	40.80%	47.00%	28 K	2.3 K
CARAFE	**37.80%**	**23.10%**	**41.70%**	**48.50%**	**199 K**	**74 K**

The experimental results demonstrate that CARAFE upsampling outperforms other upsampling methods, showing significant improvements in Average Precision (AP), APS, APM, and APL. This indicates that CARAFE exhibits superior performance and is effective across different object sizes.

3. Experimental Data and Evaluation Indexes

In this section, we present the experimental data, experimental setup, and evaluation metrics employed in this study, aiming to provide a comprehensive overview of the empirical aspects of our research.

3.1. Experimental Data

In this paper, the railway track segmentation dataset is created using UAV aerial images provided by a company. The dataset primarily consists of railway UAV aerial images captured in various railway stations in different cities in China, encompassing diverse railway environments. The original aerial images do not have corresponding label images, so it was necessary to manually annotate the label files using LabelMe software. The railway track segmentation dataset needs to be divided into three subsets: the training, test, and validation sets. The training set accounts for 90% of the total data, while the test and validation sets use the remaining 10%. Specifically, the test set and the validation set share the same subset; they are not separately partitioned. To prevent overfitting, improve model robustness, enhance model generalization, and address sample imbalance, several data augmentation techniques were employed on the training set. These techniques included random cropping, rotation, horizontal flip, vertical flip, and center flip. Through these augmentation methods, the divided training set was enriched, resulting in a final training set consisting of 7892 railway images along with their corresponding label images. In the railway label dataset used in this paper, the background pixel value is set as 0, while the target pixel value representing the railway track is set as 1. The original and labeled images are displayed in Figure 6.

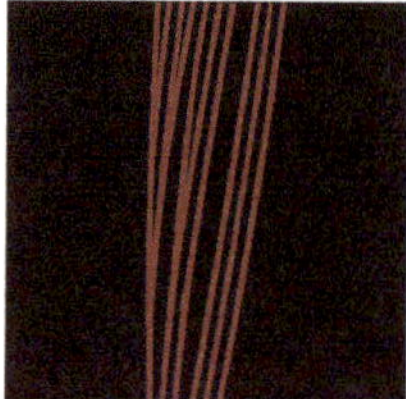

Figure 6. Original image and label image (Railway dataset).

This paper used the DeepGlobe dataset [24] as a public dataset consisting of original and label images. Specifically, 3984 images were selected from this dataset. In the DeepGlobe dataset, the pixel values in the label images range from 0 to 255. However, for the model used in this paper, the label dataset requires pixel values to be 0 to 1. Therefore, preprocessing was performed on the DeepGlobe dataset. The preprocessing involved converting the regions in the label images with a pixel value of 255 to a pixel value of 1 using a binary image processing algorithm. This ensured consistency with the label dataset used in the model. Subsequently, the data images were divided into training and test sets, following a 9:1 ratio. The validation set was not separately divided. The images from the DeepGlobe dataset and their corresponding labels are displayed in Figure 7.

 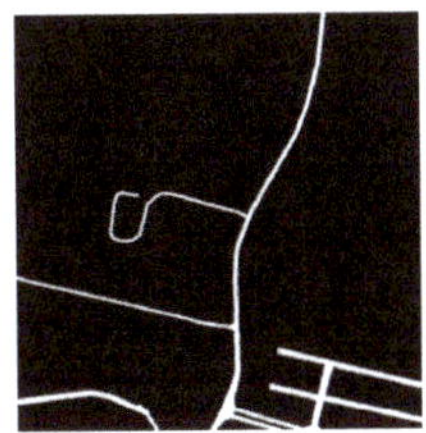  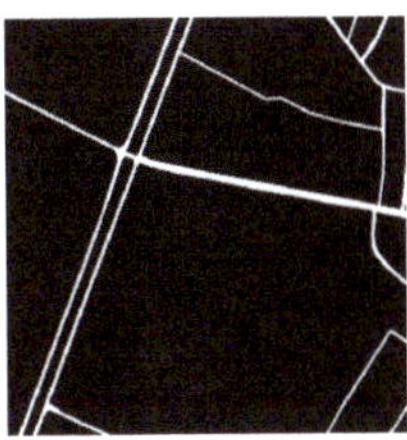

Figure 7. Raw image and label image (Deepglobe dataset).

3.2. Experimental Environment and Parameter Setting

The experimental environment for this paper was a Windows 10 system with a 64-bit operating system, an Intel i5 CPU, and an NVIDIA Tesla T4 graphics card. The model was implemented using Torch 1.2.0 and Python 3.7. During the training process, the initial learning rate was set to 7×10^{-3}. To achieve high accuracy, the Stochastic Gradient Descent (SGD) optimizer was used. A weight decay of 1×10^{-4} was applied to prevent overfitting. The model was trained for 100 epochs, and the weights were saved every five epochs for further analysis and evaluation.

In semantic segmentation tasks, the role of the loss function is crucial. Semantic segmentation refers to the process of assigning each pixel in an image to a specific semantic class, thus requiring corresponding predicted results for every pixel. The loss function measures the difference between the model's predictions and the ground-truth labels. By minimizing the loss function, the optimization algorithm can adjust the model parameters to improve the performance of the semantic segmentation model. The loss function used in this article is the cross-entropy loss function, and its expression is as follows:

$$CE\_loss = -\frac{1}{n}\sum_{x}[ylna + (1-y)ln(1-a)] \tag{3}$$

In semantic segmentation tasks, each pixel's prediction is represented as a class probability vector. The cross-entropy loss function is well-suited for pixel-level classification as it naturally measures the dissimilarity between predicted and ground-truth class distributions. It also possesses desirable gradient properties, enabling common optimization algorithms like stochastic gradient descent to effectively adjust model parameters, accelerate model convergence, and minimize the loss function.

3.3. Evaluation Index

The model presented in this paper is a semantic segmentation model. To assess its performance, several evaluation metrics were employed, including Intersection over Union (IoU) and Mean Intersection over Union (MIoU), Class Pixel Accuracy (CPA) and Mean Pixel Accuracy (MPA), Recall, and Accuracy. IoU measures the overlap between the predicted segmentation mask and the ground truth mask for each class, while MIoU computes the average IoU across all classes. CPA calculates the percentage of correctly classified pixels for each class, and MPA computes the average accuracy across all classes. Recall assesses the ability of the model to identify positive instances correctly, and Accuracy measures the overall accuracy of the model's predictions. By evaluating the model's performance using these metrics, a comprehensive understanding of its merits and limitations can be obtained. The evaluation index formula is as follows:

$$IoU = \frac{TP}{TP + FN + FP} \tag{4}$$

$$MIoU = \frac{1}{N}\sum_{k=1}^{N} IoU_k \tag{5}$$

$$CPA = \frac{TP}{TP + FP} \tag{6}$$

$$MPA = \frac{1}{N}\sum_{k=1}^{N} CPA_k \tag{7}$$

$$Recall = \frac{TP}{TP + FN} \tag{8}$$

$$Accuracy = \frac{TP + TN}{TP + TN + FP + FN} \tag{9}$$

In evaluating the model, the following definitions are used: *TP* (True Positives) represents the cases where the model correctly predicts a positive instance, and the label result is also positive. *FP* (False Positives) represents the cases where the model incorrectly predicts a positive instance, but the label result is negative. *TN* (True Negatives) represents the cases where the model correctly predicts a negative instance, and the label result is also negative. *FN* (False Negatives) represents the cases where the model incorrectly predicts a negative instance, but the label result is positive.

4. Results

This section outlines the experimental procedure employed in this study and presents a comparative analysis of the experimental results.

4.1. Visual Analysis of Loss Function

To further evaluate the performance of the proposed railway track extraction method, the convergence process of the training loss functions for different models was visually analyzed on the DeepGlobe public dataset and the railway track segmentation dataset. The convergence rate and the minimum value reached by the loss function represent the performance of different semantic segmentation models. To eliminate any other interfering factors, the models used in this paper's loss function visualization experiment all employed the same loss function, and the number of training iterations was set to 100. The networks compared in this experiment include U-Net and different backbone networks of the DeepLabv3+ model.

In the experiment using the railway track segmentation dataset, the U-Net network exhibits significant fluctuations in its loss iteration curve. The training loss continuously decreases, while the validation loss slightly increases after the initial decrease, eventually converging to the minimum loss value at around the 60th iteration. On the other hand, the DeepLabv3+ models with MobileNetV2 and Xception backbones show smaller fluctuations in their loss curves, converging to the minimum loss within 50 to 60 iterations. In this paper, the convergence curve of the model's loss function is relatively smooth, with both the training loss and validation loss converging rapidly. The lowest loss value is reached at around the 40th iteration. The loss function curve of the railway track segmentation data set is shown in Figure 8.

The visualization experiments conducted on the DeepGlobe public dataset reveal several observations. In the U-Net model, the training loss decreases rapidly, but the validation loss converges slower. The MobileNetV2 backbone network model exhibits faster convergence speed, although the loss value remains relatively high. On the other hand, the Xception backbone network model shows the slowest convergence speed. Compared to these models, the network proposed in this paper demonstrates improved convergence speed and achieves a lower final loss value. The public data set loss function curve is shown in Figure 9.

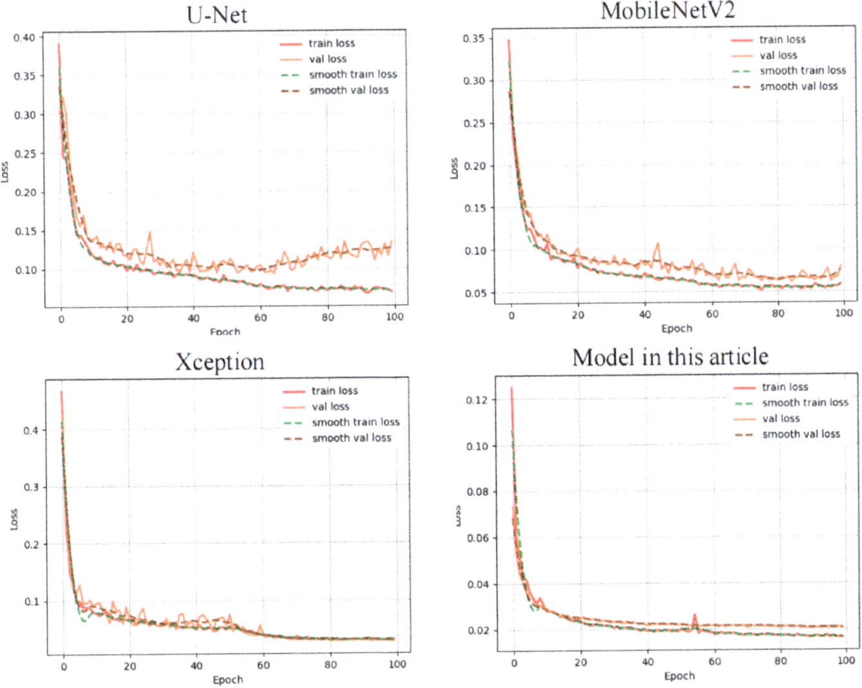

Figure 8. Convergence Comparison of Loss function of Railway Dataset.

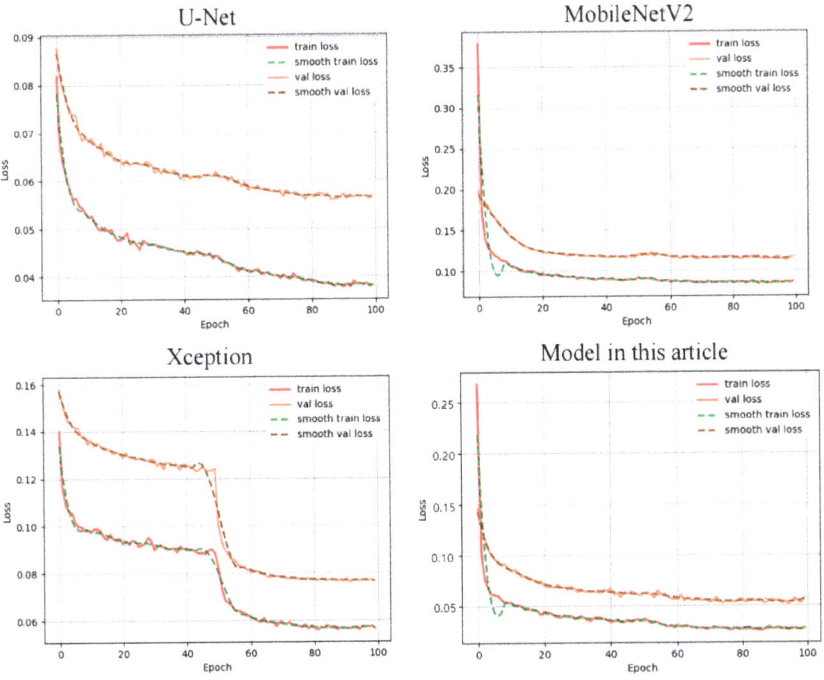

Figure 9. Convergence Comparison of Loss function of Common Dataset.

The visual curve of the loss function demonstrates that the proposed model exhibits faster convergence, a smoother loss curve, and a smaller convergence value. As a result, the railway track extraction model proposed in this paper showcases superior performance.

4.2. Comparative Experimental Analysis

4.2.1. Comparison of Data Augmentation Experiment Results

Data augmentation is a technique that involves transforming and expanding the original training data to increase the diversity and quantity of data samples. This article presents the following reasons for employing data augmentation: Firstly, augmenting data helps the model better capture and learn the invariances and patterns within the data, enhancing its generalization capabilities on unseen data. Secondly, deep learning models are prone to overfitting in the presence of limited samples, and data augmentation can mitigate the risk of overfitting by enlarging the training dataset. The model becomes more adept at handling perturbations and noise in the input data through data augmentation, leading to improved robustness and increased accuracy.

Using the railway track dataset experiment as an example, we conducted experiments with three different training set sizes: 700 images, 3500 images, and 7000 images, to demonstrate the necessity of data augmentation through the extracted results. In the experiments, all other conditions were kept the same, with only the number of training samples varying. The experimental results are shown in Figure 10.

  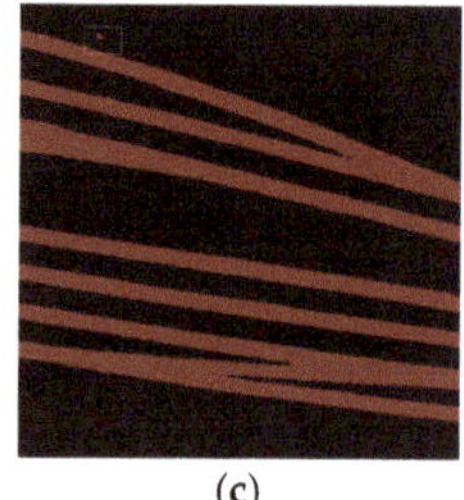 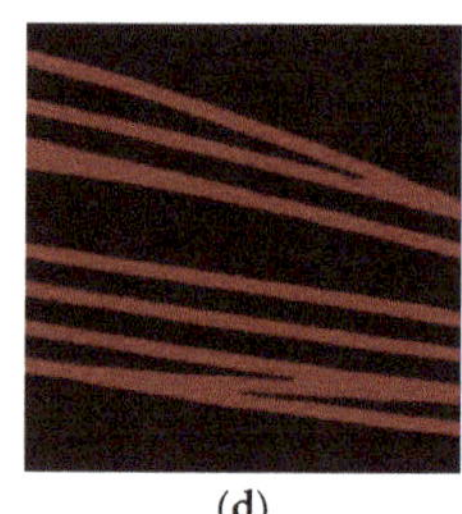

(a)   (b)   (c)   (d)

Figure 10. Different data quantity extraction results. (**a**) Label, (**b**) 700 images, (**c**) 3500 images, (**d**) 7000 images. The red area represents the railway track region. The white box outlines the differences in the image.

From this, it can be observed that with a limited number of Epochs, larger data quantities lead to better model performance. However, it does not imply that more data is always better, as excessively large datasets can consume more time. Therefore, multiple experiments are necessary to find the optimal combination between data quantity and model performance.

4.2.2. Morphological Algorithm Experimental Comparison

The composite operations of morphological algorithms can eliminate specific extraction errors. OpenCV provides an efficient and user-friendly interface for image morphological transformations, where all morphological operations are based on erosion and dilation. During the execution of the *erode()* function, the image undergoes an erosion operation, setting the value at point p to the minimum value within the area covered by the kernel when aligned with point p. On the other hand, the *dilate()* function performs a dilation operation, replacing the minimum value with the maximum value. The two function expressions are as follows:

$$erode(x,y) = \min_{(i,j)\,\in\,kernel} src(x+i, y+j) \tag{10}$$

$$dilate(x,y) = \max_{(i,j)\,\in\,kernel} src(x+i,y+j) \tag{11}$$

When performing morphological processing on the initially extracted results of railway tracks, it is necessary to set the kernel size. This study uses a matrix of size 5×5, and the required operations are specified using the *morphologyEx()* function. First, an opening operation is applied to remove speckles and protuberances, with the *morphologyEx()* function parameter set to MOP_OPEN. Subsequently, a closing operation is performed to connect small fractures and fill in holes, with the *morphologyEx()* function parameter set to MOP_CLOSE. Through optimization with morphological algorithms, a more complete extraction result can be achieved. Taking the Railway track segmentation dataset as an example, the comparison of results before and after morphological processing is shown in Figure 11.

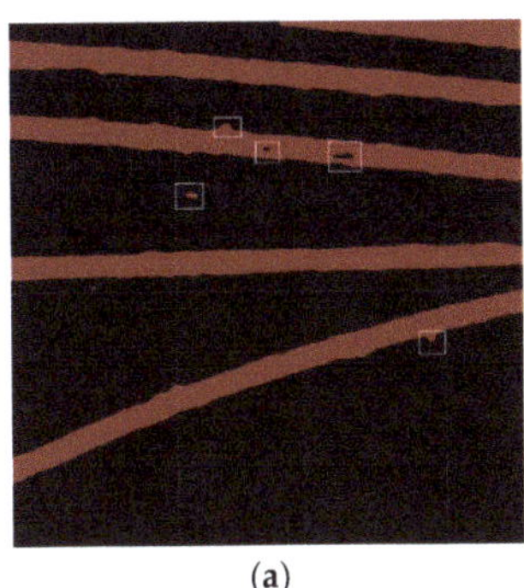
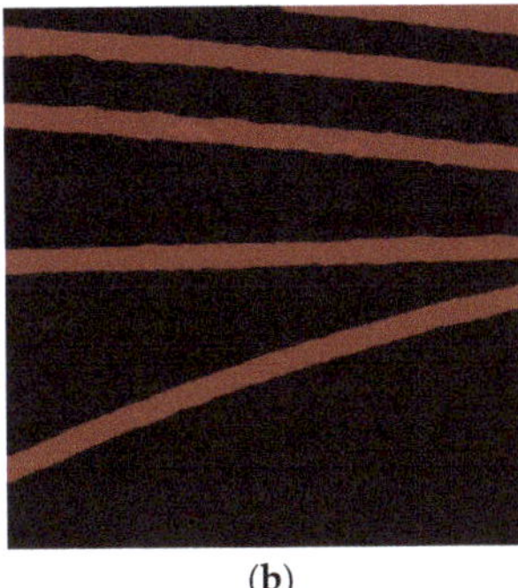

(a) (b)

Figure 11. Comparison of Results Before and After Morphological Processing (Railway track segmentation dataset). (**a**) Initial extraction results, (**b**) Processed results. The red area represents the railway track region. The white box outlines the differences in the image.

4.2.3. Railway Dataset Experiment

To compare the overall accuracy of this algorithm with other algorithms, an evaluation was conducted using the railway track segmentation dataset. The evaluation metrics and runtime of the proposed model were compared with the U-Net semantic segmentation network [25], L-UNet network [26], DeepLabv3 model with ResNet backbone [27], and DeepLabv3+ network [28] using the improved MobileNetV2 backbone network with rectangular contrast. For ease of comparison, the prediction time was selected to measure the model's time difference, and 288 original images from the railway track segmentation dataset were used for prediction. It can be observed that the DeepLabv3+ model with the MobileNetV2 backbone network achieves improved accuracy and lower prediction time compared to the U-Net network and DeepLabv3 network. This indicates that utilizing a lightweight backbone network, such as MobileNetV2, in the DeepLabv3+ model can significantly reduce the network's scale without compromising accuracy and overall runtime.

Furthermore, the MobileNetV3 backbone network was employed in this paper, which further enhances accuracy and reduces runtime by approximately 5% compared to MobileNetV2 before the morphological processing. Based on the above analysis, it can be concluded that the proposed method improves the accuracy of railway track extraction and significantly reduces the overall running time. The comparison results are presented in Table 2.

Despite the improvements made in the DeepLabv3+ model, there are still some imperfections in the extracted railway track area, such as holes, bumps, and spots. This is mainly due to the nature of semantic segmentation, which operates at a pixel level. In non-railway areas of the original image, there may be pixels that are similar or identical to those in the railway area, leading to segmentation errors. To further enhance the semantic segmentation results, morphological erosion and dilation operations can be employed. The open operation (erosion followed by dilation) helps to remove spots and bumps, while the close operation (dilation followed by erosion) can eliminate voids within the railway area without altering the overall shape. The effectiveness of these morphological operations depends on adjusting the size of the kernel in the algorithm to achieve optimal results. By applying these operations to the railway area, a complete and refined railway track can be obtained. Table 2 compares the results obtained by the model proposed in this paper and other extraction models, highlighting the improvements achieved. The comparison between the model in this paper and other model extraction results is shown in Table 3.

Table 2. Comparison of evaluation indicators between this method and other models.

Model	IoU	MIoU	CPA	MPA	Recall	Accuracy	Times
Ning Y. et al. [25]	63.41%	86.16%	74.94%	93.85%	86.52%	94.72%	96 s
Miao X. et al. [26]	65.22%	87.60%	76.71%	95.41%	87.01%	96.11%	80 s
Ling H. et al. [27]	64.13%	86.06%	75.53%	94.07%	86.62%	95.34%	72 s
Yuejuan R. et al. [28]	64.97%	86.87%	76.21%	94.90%	86.99%	95.96%	64 s
Proposed method	**66.21%**	**88.93%**	**76.33%**	**95.51%**	**89.02%**	**97.69%**	**61 s**

Table 3. Comparison of extraction results between this article and other models. The red area represents the railway track region. The white box outlines the differences in the image.

Original	Ning Y. et al. [25]	Miao X. et al. [26]	Ling H. et al. [27]	Yuejuan R. et al. [28]	Proposed Method	Label

4.2.4. DeepGlobe Public Data Set Model Comparative Experimental Analysis

To assess the generalization ability of the proposed model, the same experiment was conducted on the DeepGlobe public dataset, which was preprocessed to enable semantic segmentation. Different models were trained on this dataset, and predictions were made on 398 raw images. The experimental results demonstrate that similar to the results obtained on the railway track segmentation dataset, the proposed model exhibits improved extraction accuracy compared to other models and reduced computational time. The comparison results of evaluation metrics and runtime are presented in Table 4.

Table 4. Comparison of evaluation indicators between this method and other models.

Model	IoU	MIoU	CPA	MPA	Recall	Accuracy	Times
Ning Y. et al. [25]	61.66%	79.91%	72.57%	82.58%	84.35%	90.32%	132 s
Miao X. et al. [26]	63.98%	81.32%	74.92%	84.06%	85.77%	92.66%	110 s
Ling H. et al. [27]	61.96%	80.06%	73.24%	82.97%	84.90%	91.92%	98 s
Yuejuan R. et al. [28]	62.15%	82.35%	74.17%	83.95%	85.03%	93.81%	88 s
Proposed method	**65.21%**	**84.72%**	**75.80%**	**86.60%**	**86.96%**	**94.84%**	**84 s**

The public dataset contains road environments that are more complex, and as a result, the evaluation metrics have decreased to varying degrees compared to the railway track segmentation dataset. However, under the same conditions, the proposed model still exhibits improved extraction performance compared to other models. The road images extracted by the proposed model undergo further processing using a morphological algorithm, resulting in more complete road features. A comparison between the results of the proposed model after morphological processing and other models is presented in Table 5.

Table 5. Comparison of extraction results between this article and other models. The red area represents the railway track region. The white box outlines the differences in the image.

Original	Ning Y. et al. [25]	Miao X. et al. [26]	Ling H. et al. [27]	Yuejuan R. et al. [28]	Proposed Method	Label

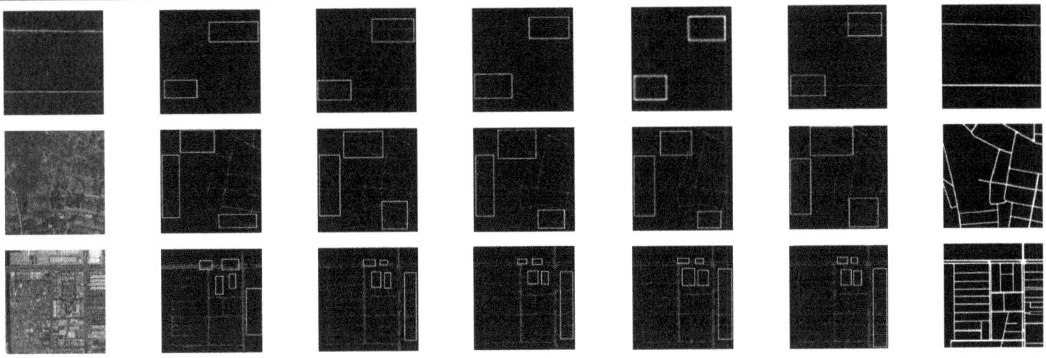

4.3. Analysis of Ablation Experiment

To verify whether the enhanced DeepLabv3+ model proposed in this paper achieves improved accuracy and reduced runtime compared to the original model, ablation experiments were conducted on both the DeepGlobe public and railway track segmentation datasets.

The ablation experiments conducted in this paper primarily focused on examining the impacts of the MobileNetV3 backbone extraction network, MobileNetV2 backbone extraction network, CARAFE, and morphological algorithm on the accuracy and runtime of railway track extraction, while keeping other conditions constant. Initially, experiments were conducted using the MobileNetV2 backbone extraction network to evaluate the effects of CARAFE and the morphological algorithm on extraction accuracy and runtime. Subsequently, the same experimental operations were performed using the MobileNetV3 backbone extraction network.

The experimental results on the railway track segmentation dataset demonstrate that replacing the upsampling module with CARAFE leads to a slight improvement in accuracy and a reduction in runtime. Furthermore, the addition of the morphological algorithm further enhances accuracy but significantly increases the runtime. However, under identical conditions, the MobileNetV3 backbone extraction network is more efficient and consumes less time than the MobileNetV2 extraction network. The comparison results of the ablation experiments on the railway track segmentation dataset are presented in Table 6.

The experimental results on the DeepGlobe public dataset exhibit similar trends to those observed on the railway track segmentation dataset. The incorporation of CARAFE and the morphology algorithms leads to an improvement in extraction accuracy to some extent, while the performance of the MobileNetV3 backbone surpasses that of MobileNetV2. The comparison results of the ablation experiments on the DeepGlobe public dataset are presented in Table 7.

Table 6. Railway Dataset Ablation Experiment.

Mobilenetv2	Mobilenetv3	CARAFE	Morphological	Accuracy	Times
√	×	×	×	93.57%	63 s
√	×	√	×	93.91%	62 s
√	×	√	√	96.15%	65 s
×	√	×	×	94.71%	60 s
×	√	√	×	95.07%	58 s
×	√	√	√	**97.59%**	**61 s**

Table 7. Deepglobal public dataset ablation experiment.

Mobilenetv2	Mobilenetv3	CARAFE	Morphological	Accuracy	Times
√	×	×	×	91.45%	88 s
√	×	√	×	91.93%	86 s
√	×	√	√	93.34%	90 s
×	√	×	×	92.88%	83 s
×	√	√	×	93.21%	82 s
×	√	√	√	**94.84%**	**84 s**

Firstly, the results of the ablation experiments on two datasets show that MobileNetV3 achieves slightly higher accuracy than MobileNetV2 while also reducing processing time by approximately 5%. This experiment provides evidence that MobileNetV3 inherits the strengths of MobileNetV2 while incorporating "lightweight" enhancements.

Furthermore, compared to the default 16x Bilinear upsampling, employing CARAFE upsampling leads to a slight improvement in accuracy without significant differences in time consumption. This experiment demonstrates that CARAFE upsampling, which considers image semantic features and adaptive kernel selection, can effectively enhance model performance.

Finally, the composite operation of morphological algorithms can also significantly improve the extraction results.

5. Discussion

Based on the experimental results, it is evident that the proposed method has significantly enhanced the extraction accuracy while reducing time consumption. This improvement can be attributed to the utilization of NAS (Neural Architecture Search) and NetAdapt algorithms in the MobileNetV3 network. Furthermore, the inclusion of the CARAFE module and morphology algorithm has a noticeable effect on improving the extraction accuracy, albeit at the cost of increased overall running time. The CARAFE module, a lightweight upsampling operator, considers the content information of the feature map while maintaining a low parameter count. This leads to shorter upsampling time and a slight improvement in accuracy. The morphology algorithm, on the other hand, assesses the validity of a region based on the pixel distribution within a binary image. Suppose the number of pixel values surrounding a particular region significantly exceeds the number of pixel values within the region. In that case, the pixel value of the region is modified, effectively eliminating spots and holes.

To enhance the generalization capability of this paper's model, both data and model aspects were addressed. Firstly, in terms of data, a railway segmentation dataset was created using unmanned aerial vehicle (UAV) images taken from various regions and cities in China. The dataset encompasses diverse weather conditions, terrains, and environments, including railway stations in different regions of China. After partitioning the railway segmentation dataset into training, validation, and test sets, data augmentation techniques were applied to the training set. Secondly, for the model aspect, SGD (Stochastic Gradient Descent) was chosen as the optimizer to update and compute the network parameters influencing the model training and outputs, driving them towards or achieving optimal values.

Furthermore, an L2 regularization module was incorporated into the model to prevent overfitting. During the model training process, the best weights and the final training weights were automatically saved for optional selection. To validate the model's generalization ability in different traffic scenarios, experiments were also conducted on the publicly available DeepGlobe dataset. The experimental results demonstrate that this model is suitable for diverse environments and exhibits a certain generalization capability.

6. Conclusions

This paper introduces an improved DeepLabv3+ model for railway track extraction. The study begins by creating a railway track segmentation dataset with diverse scenes using UAV aerial images from several domestic stations. Addressing the low work efficiency caused by excessive layers in existing semantic segmentation networks, the proposed approach utilizes the lightweight MobileNetV3 network as the backbone network for the DeepLabv3+ model. The CARAFE lightweight upsampling operator is also employed for the decoder's upsampling component. This design achieves segmentation accuracy while reducing the scale of the semantic segmentation network and improving model efficiency. To address potential issues such as holes and spots in the initial extraction results, this paper incorporates a morphological algorithm to optimize the outcomes. By employing a combination of morphological erosion and dilation operations, the algorithm effectively eliminates these undesired artifacts. The proposed algorithm is evaluated using the railway track segmentation dataset and the DeepGlobe dataset, comparing it with common semantic segmentation networks. Experimental results demonstrate that the proposed model exhibits significantly improved accuracy while consuming less time.

Although this method has reduced the model's computational time to some extent, it necessitates morphological optimization of the extracted results, which adds a small amount of additional processing. Further optimization of the model is required to minimize time overhead, ensuring faster and more comprehensive extraction of railway tracks.

Author Contributions: Conceptualization, Y.W. and Z.L.; railway image acquisition, Y.W. and X.C.; data preprocessing, Z.L. and X.H.; formal analysis, X.C., J.H. and F.L.; funding acquisition, Y.W., J.H. and F.L.; writing—original draft preparation, Y.W. and Z.L.; experiment, Z.L.; writing—review and editing, Y.W., Z.L., J.H. and H.Y.; paper revision, Y.W., Z.L. and J.H. All authors have read and agreed to the published version of the manuscript.

Funding: This research was funded by the National Key R&D Program of China (2021YFF0501101), The National Natural Science Foundation of China (U1934219), and Hunan Provincial Natural Science Foundation (2021JJ50049, 2022JJ50067).

Data Availability Statement: Public data sets can be found at http://deepglobe.org/index.html (accessed on 20 October 2022). Other data in this study can be contacted by the corresponding author.

Acknowledgments: Firstly, the authors would like to express their gratitude for the financial support provided by the National Key Research and Development Program of China, the National Natural Science Foundation of China, and the Hunan Provincial Natural Science Foundation. Secondly, the authors would like to thank Zhuzhou Taichang Electronic Information Technology Co., Ltd. for their invaluable data and technical assistance.

Conflicts of Interest: The authors declare no conflict of interest.

References

1. Liu, J.; Cai, B.; Wang, J.; Tang, T. Research on Algorithm of Electronic Track Map Data Reduction for Train Locating. *J. Railw. Sci.* **2011**, *33*, 73–79. [CrossRef]
2. Zuo, Z.; Wang, K.; Xu, C.; Liu, C. Data Processing Method for Generating High-Precision Electronic Track Map. *China Railw. Sci.* **2016**, *37*, 134–138. [CrossRef]
3. Lu, D.; Wang, Z.; Wang, J.; Cai, B.; Ding, X. Research on Electronic Track Map Data Reduction Method for Novel Train Control System. *J. Railw. Sci.* **2023**, *45*, 51–61. [CrossRef]

4.   Shang, J.; Wang, J.; Liu, S.; Wang, C.; Zheng, B. Small Target Detection Algorithm for UAV Aerial Photography Based on Improved YOLOv5s. *Electronics* **2023**, *12*, 2434. [CrossRef]
5.   Xiao, C.; Tian, X. Research on Road Extraction Algorithm Based on Color Feature. *Mod. Comput.* **2022**, *28*, 98–102. [CrossRef]
6.   Shi, W.; Miao, Z.; Debayle, J. An Integrated Method for Urban Main-Road Centerline Extraction from Optical Remotely Sensed Imagery. *IEEE Trans. Geosci. Remote Sens.* **2013**, *52*, 3359–3372. [CrossRef]
7.   Liu, X.; Zhang, J.; Liu, N.; Che, Y.; Zhang, C. Urban Road Extraction Based on Morphological Filtering and Trajectory Detection. *Prog. Laser Optoelectron.* **2022**, *59*, 47–54. [CrossRef]
8.   Kong, L.; Liu, L.; Wu, Y.; Zhang, Q.; We, X. Feature Point Process Based Road Centerline Extraction from Remote Sensing Image. *Radio Eng.* **2023**, *53*, 12. [CrossRef]
9.   Laraib, U.; Shaukat, A.; Khan, R.A.; Mustansar, Z.; Akram, M.U.; Asgher, U. Recognition of Children's Facial Expressions Using Deep Learned Features. *Electronics* **2023**, *12*, 2416. [CrossRef]
10.   Liu, B.; Feng, L.; Zhao, Q.; Li, G.; Chen, Y. Improving the Accuracy of Lane Detection by Enhancing the Long-Range Dependence. *Electronics* **2023**, *12*, 2518. [CrossRef]
11.   Li, M.; Jiang, W.M.; Zhang, N.; Liu, H.L. The numerical simulation analysis of nappe-wind generated by flood discharge. *Acta Aerodyn. Sin.* **2003**, *4*, 004. [CrossRef]
12.   She, Z.; Shen, Y.; Song, J.; Xiang, Y. Using the classical CNN network method to construct the automatic extraction model of remote sensing image of Guiyang road elements. *Bull. Surv. Mapp.* **2023**, *4*, 177–182. [CrossRef]
13.   Dai, J.; Du, Y.; Jin, G.; Tao, D. A Road Extraction Method Based on Multiscale Convolutional Neural Network. *Remote Sens. Inf.* **2019**, *35*, 28–37. [CrossRef]
14.   Zhang, X.; Ma, W.; Li, C.; Wu, J.; Tang, X.; Jiao, L. Fully Convolutional Network-Based Ensemble Method for Road Extraction from Aerial Images. *IEEE Geosci. Remote Sens. Lett.* **2019**, *17*, 1777–1781. [CrossRef]
15.   Kong, X.; Wang, C.; Zhang, S.; Li, J.; Sui, Y. Application of Improved U-Net Network in Road Extraction from Remote Sensing Images. *Remote Sens. Inf.* **2022**, *37*, 97–104. [CrossRef]
16.   Qi, H.; Li, Y.; Qi, Y.; Liu, L.; Dong, Z.; Du, X. Research on Track and Obstacle Detection Based on New Lightweight Semantic Segmentation Network. *J. Railw. Sci.* **2019**, *45*, 58–66. [CrossRef]
17.   Chen, L.C.; Papandreou, G.; Schroff, F.; Adam, H. Rethinking Atrous Convolution for Semantic Image Segmentation. *arXiv* **2017**, arXiv:1706.05587. [CrossRef]
18.   Chen, L.C.; Zhu, Y.; Papandreou, G.; Schroff, F.; Adam, H. *Encoder-Decoder with Atrous Separable Convolution for Semantic Image Segmentation*; Springer: Cham, Switzerland, 2018. [CrossRef]
19.   Howard, A.; Sandler, M.; Chen, B.; Wang, W.; Chen, L.-C.; Tan, M.; Chu, G.; Vasudevan, V.; Zhu, Y.; Pang, R.; et al. Searching for MobileNetV3. In Proceedings of the 2019 IEEE/CVF International Conference on Computer Vision (ICCV), Seoul, Republic of Korea, 27 October–2 November 2019; pp. 1314–1324. [CrossRef]
20.   Sandler, M.; Howard, A.; Zhu, M.; Zhmoginov, A.; Chen, L.C. MobileNetV2: Inverted Residuals and Linear Bottlenecks. In Proceedings of the IEEE Conference on Computer Vision and Pattern Recognition, Salt Lake City, UT, USA, 18–23 June 2018; pp. 4510–4520. [CrossRef]
21.   Hu, J.; Shen, L.; Sun, G. Squeeze-and-excitation Networks. In Proceedings of the IEEE Conference on Computer Vision and Pattern Recognition, Salt Lake City, UT, USA, 18–23 June 2018; pp. 7132–7141. [CrossRef]
22.   Dougherty, E. *Mathematical Morphology in Image Processing*; CRC Press: Boca Raton, FL, USA, 2018.
23.   Wang, J.; Chen, K.; Xu, R.; Liu, Z.; Loy, C.C.; Lin, D. Carafe: Content-aware Reassembly of Features. In Proceedings of the IEEE/CVF International Conference on Computer Vision, Seoul, Republic of Korea, 27 October–2 November 2019; pp. 3007–3016. [CrossRef]
24.   Demir, I.; Koperski, K.; Lindenbaum, D.; Pang, G.; Huang, J.; Basu, S.; Hughes, F.; Tuia, D.; Raskar, R. DeepGlobe 2018: A Challenge to Parse the Earth through Satellite Images. In Proceedings of the 2018 IEEE/CVF Conference on Computer Vision and Pattern Recognition Workshops (CVPRW), Salt Lake City, UT, USA, 18–22 June 2018. [CrossRef]
25.   Ning, Y.; Sheng, J. Research and Discussion on Road Extraction Using Deep Learning Network U-Net. *J. Heilongjiang Hydraul. Eng. Coll.* **2020**, *11*, 1–8. [CrossRef]
26.   Miao, X.; Li, Y.; Zhong, J.; Zuo, Z.; Xiong, W. L-UNet:lightweight network for road extraction in cloud occlusion scene. *J. Image Graph.* **2021**, *26*, 2670–2679. [CrossRef]
27.   Han, L.; Yang, C.; Li, L.; Liu, Z.; Huang, B. Road Extraction of High Resolution Remote Sensing Imagery Based on Deeplab v3. *Remote Sens. Inf.* **2021**, *36*, 22–28. [CrossRef]
28.   Ren, Y.; Ge, X. An road synthesis extraction method of remote sensing image based on improved DeepLabV3+ network. *Bull. Surv. Mapp.* **2022**, *6*, 55–61. [CrossRef]

Article

An Improved CNN for Polarization Direction Measurement

Hao Han [1,2], Jin Liu [1,2], Wei Wang [1,2,*], Chao Gao [1,2] and Jianhua Shi [1,2]

[1] College of Advanced Interdisciplinary Studies, National University of Defense Technology, Changsha 410073, China; hanhao.20@nudt.edu.cn (H.H.); liujin1344@nudt.edu.cn (J.L.); gaochao14@nudt.edu.cn (C.G.); gexin7651@nudt.edu.cn (J.S.)

[2] Nanhu Laser Laboratory, National University of Defense Technology, Changsha 410073, China

* Correspondence: wangwei8610@nudt.edu.cn

Abstract: Spatially polarization modulation has been proven to be an efficient and simple method for polarization measurement. Since the polarization information is encoded in the intensity distribution of the modulated light, the task of polarization measurement can be treated as the image processing problem, while the pattern of the light is captured by a camera. However, classical image processing methods could not meet the increasing demand of practical applications due to their poor computational efficiency. To address this issue, in this paper, an improved Convolutional Neural Network is proposed to extract the Stokes parameters of the light from the irradiance image. In our algorithm, residual blocks are adopted and different layers are connected to ensure that the underlying features include more details of the image. Furthermore, refined residual block and Global Average Pooling are introduced to avoid overfitting issues and gradient vanishing problems. Finally, our algorithm is tested on massive synthetic and real data, while the mean square error (MSE) between the extracted values and the true values of the normalized Stokes parameters is counted. Compared to VGG and FAM, the experimental results demonstrate that our algorithm has outstanding performance.

Keywords: polarization direction measurement; CNN; global average pooling

1. Introduction

Polarization measurement is important in many applications, such as in sky polarized light navigation [1], remote sensing [2–4], the food industry [5–8] and ellipsometry [9–11]. Generally, four methods, including interferometric polarimeter [12,13], temporally modulated polarimeter [14,15], division-of-amplitude polarimeter [16,17] and spatially modulated polarimeter [18–21], are well accepted to solve this problem. In a interferometric polarimeter, the coherent optical paths are constructed, and the polarization information can be calculated from the interference pattern. The interferometric polarimeter is robust and stable, however, the analyzing process of the interference pattern is usually complicated. In a temporally modulated polarimeter, the rotatable or active optical elements are utilized to modulate the incident polarized light in a time sequential, and the polarization state of the incident beam can be obtained by analyzing the time varying intensity signal. The temporally modulated polarimeter is simple and easy to implement, however, its measurement speed is limited, and it is sensitive to the power and wavelength fluctuation of the light sources. In a division-of-amplitude polarimeter, the incident beams are analyzed by several channels with different polarization optics, and the polarization information can be obtained in a single shot. The division-of-amplitude polarimeter is competent for real-time monitoring, but its configuration is usually complicated to adjust. To conquer these issues, some researchers developed a fourth method, i.e., spatially modulated polarimeter. In a spatially modulated polarimeter, the spatially modulated polarization optics, such as micro polarizer arrays, polarization grating, azimuthal or radial polarizers, are utilized to modulate the intensity in the spatial domain, the polarization information of the incident beam can be obtained by processing and analyzing the spatial modulated intensity image. The

Citation: Han, H.; Liu, J.; Wang, W.; Gao, C.; Shi, J. An Improved CNN for Polarization Direction Measurement. *Electronics* **2023**, *12*, 3723. https://doi.org/10.3390/electronics12173723

Academic Editor: Byung Cheol Song

Received: 15 July 2023
Revised: 23 August 2023
Accepted: 30 August 2023
Published: 4 September 2023

spatially modulated polarimeter can achieve the polarization measurement in a compact, rapid and stable way. It is not sensitive to changes in the power and wavelength of the light. However, as the core devices, the spatial modulation devices are difficult to deploy.

Fortunately, a vortex retarder based spatial polarization modulated polarimetry method is proposed [1,11,18]. The vortex retarder is a special wave plate, and it has a constant retardance across the clear aperture, but its fast axis rotates continuously along the azimuth, so it can convert an ordinary polarized light into a vectorial optical beam [18]. Then, the polarization information is included in the light intensity distribution while the vector polarized light field is detected by a polarizer. Compared to other spatial modulation methods, this method has the advantages of stable performance, low wavelength sensitivity, good temperature stability, high modulation quality and low cost. In this method, the polarization information can be extracted using image processing when the pattern of the light is captured by a camera. Consequently, the accuracy of the polarization measurement is determined by the performance of the image processing algorithms.

Recently, image processing algorithms are divided into two categories: traditional methods and machine learning. In traditional methods, the design of feature extractors relies on the designers' professional knowledge. Furthermore, the methods usually need complex parameter tuning processes. To the best of our knowledge, two papers use tradition methods to calculate the polarization state from the irradiance image. In reference [21], an image correlation operation is proposed to extract the polarization direction from the hour-glass-shaped intensity image. However, the measurement accuracy is decided in the step of correlation operation, and numerous calculations need to be performed to ensure a high accuracy, which is very time consuming. What is more, the method can only obtain the polarization direction, and other polarization information, such as ellipticity and polarization handiness are lost. To obtain the Stokes parameters of the polarized light, a Fourier analysis method (FAM) is proposed [18]. In this method, a series of Randon transformations is performed to obtain the modulation curve of the intensity image, and the Stokes parameters of the incident light can be measured by Fouricr analysis of the modulation curve. However, the computational efficiency of the Fourier analysis method is rather poor due to numerous redundant calculations in the Randon transformation. Due to a series of advantages, such as excellent performance, better generalization, end-to-end training and no need for complex parameter tuning, the machine learning method has been widely used in image processing. As the most important branch of machine learning, deep learning performs well and has been widely used in image processing. In reference [22], more than 300 research contributions on deep learning techniques for object detection are introduced. More than 100 deep-learning-based methods have been proposed for image segmentation [23]. Some researchers are devoted to achieving image registration depending on deep learning [24]. Zhao et al. [25] designed SpikeSR-Net to super–resolve a high-resolution image sequence from the low-resolution binary spike streams. In particular, in reference [26], a Convolutional Neural Network (CNN) based on VGGNet architecture was trained to obtain the polarization states of light using a single shot of intensity image. Though it has similar accuracy to FAM, it is much less time consuming.

In this paper, a deep learning technique is also adopted to extract polarization information from the irradiance image due to its outstanding performance. In this paper, an efficient deep-learning-based image processing algorithm, named ResNet-GAP, is proposed to extract the polarization direction from the irradiance image of the modulated input light. To prevent overfitting of the network, global average pooling [27] (GAP) is introduced while ResNet [28] is adopted as the main architecture of our network. Furthermore, the residual block is refined in order to extract image features better and avoid gradient vanishing. In addition, the originally full connection layer is divided into two layers, including a FC layer and a ReLU activation function.

The main work of our paper is as follows: Section 2 introduces the theoretical and experimental investigation, and Section 3 describes our experiments and analyzes the results. Section 4 summarizes the main work of this paper and introduces our future work.

2. Theoretical and Experimental Investigation

The schematic of our experiment includes two stages: irradiance image generation and image processing. In the first stage, a spatially modulated scheme using a vortex retarder is built, and the irradiance image including the polarization state of the light is captured by a camera. Then, in the second stage, an improved CNN is proposed to extract the polarization information from the irradiance image.

2.1. Irradiance Image Generation Stage

In 1852, Stokes proposed that the polarization state of the light wave can be represented by four real number parameters, which are called Stokes parameters. Generally, the Stokes parameters can be written as a column vector of one order, i.e., $S = \begin{bmatrix} S_0 & S_1 & S_2 & S_3 \end{bmatrix}^T$. For a certain optical element or an optical system, the relationship between outgoing light and the incident light can be represented by

$$S_{out} = M \cdot S_{in} \tag{1}$$

S_{out} and S_{in} are the Stokes vector of outgoing and incident light. M is the Muller matrix of the optical system.

As shown in Figure 1, an integrating sphere (IS) and a positive lens are utilized to generate a uniform and collimated natural light field. Then, the incident light with different polarization states are captured while altering the azimuth angles of the transmission axis of the polarizer and the fast axis of the wave plate. Subsequently, the polarized light is modulated by a retarder and another analyzer. For the zero-order vortex half-wave retarder (VHWR), with the initial fast axis oriented along 0°, the Equation (1) can be rewritten as

$$S_{out} = \begin{bmatrix} 1 & 0 & 0 & 0 \\ 0 & \cos 2\varphi & \sin 2\varphi & 0 \\ 0 & \sin 2\varphi & -\cos 2\varphi & 0 \\ 0 & 0 & 0 & -1 \end{bmatrix} S_{in} \tag{2}$$

where φ is the azimuth angle. Consequently, when the transmission axis of the analyzer is oriented at 0°, the light intensity can be denoted by

$$I(\varphi) \propto S_{out0} = S_{in0} + S_{in1} \cos 2\varphi + S_{in2} \sin 2\varphi \tag{3}$$

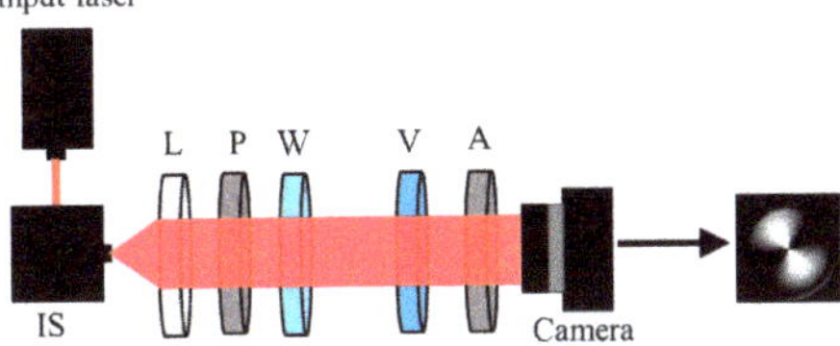

Figure 1. The experiment system. IS: integrating sphere; L: Lens; P: polarizer; W: wave plate; V: vortex retarder; A: analyzer.

Similarly, if the incident light is modulated by a zero-order vortex quarter-wave retarder (VQWR), while the initial fast axis is oriented along 0°, the light intensity is

$$I(\varphi) \propto S_{out0} = S_{in0} + \frac{1}{2}S_{in1} + \frac{1}{2}S_{in1} \cos 2\varphi + \frac{1}{2}S_{in2} \sin 2\varphi - \frac{1}{2}S_{in3} \sin \varphi \tag{4}$$

Equations (2) and (3) point out that, utilizing the optical system shown in Figure 1, the Stokes parameters are encoded in the intensity of the modulated light. In other words, the problem of polarization measurement can be treated as an image processing problem to extract Stokes parameters from the irradiance images.

2.2. ResNet-GAP

However, experiments show that the performance of the classical ResNet50 did not meet our expectation in evaluating the Stokes parameters of the light intensity image. Hence, we modify the ResNet50 architecture to make it suitable for our Stokes parameters evaluation problem. The full specification of our modified network, which we called ResNet-GAP, appears in Table 1. The main architecture of our ResNet-GAP is similar to ResNet50, however, there are some improvements.

Table 1. The architecture of ResNet-GAP.

Layer Name	Output Size	Layers
conv 1	112×112	3×3, stride 2 3×3, stride 1 3×3, stride 1
conv2_x	56×56	3×3 max pool, stride 2 $\begin{bmatrix} 1 \times 1,\ 64 \\ 3 \times 3,\ 64 \\ 1 \times 1,\ 256 \end{bmatrix} \times 3$
conv3_x	28×28	$\begin{bmatrix} 1 \times 1,\ 128 \\ 3 \times 3,\ 128 \\ 1 \times 1,\ 512 \end{bmatrix} \times 4$
conv4_x	14×14	$\begin{bmatrix} 1 \times 1,\ 256 \\ 3 \times 3,\ 256 \\ 1 \times 1,\ 1024 \end{bmatrix} \times 6$
conv5_x	7×7	$\begin{bmatrix} 1 \times 1,\ 512 \\ 3 \times 3,\ 512 \\ 1 \times 1,\ 2048 \end{bmatrix} \times 3$
	1×1	GlobalAveragePooling2D; 1024-d FC, ReLU; dropout; L-d FC, Sigmoid

In conv1 layer, we choose to replace the 7×7 convolution with multiple layers of smaller convolution. Convolutions with larger spatial filters (e.g., 7×7) tend to be time consuming, while convolutions with smaller spatial filters (e.g., 3×3) tend to be much easier in terms of computation. Hence, we replace the 7×7 convolution with a concatenation of three layers of 3×3 convolution (one with stride 2 and two with stride 1). This setup clearly reduces the parameter count and also increases the network depth to maximize the utilization of the network capacity and complexity.

In conv5_x, the first layer is the down sampling module, which, in fact, contains Path A and Path B as in Figure 2. Originally in ResNet50, Path A first completes the channel contraction through 1×1 convolution with stride 2 to realize down sampling, then a 3×3 convolution follows, which keep the number of channels unchanged; the main purpose is to extract features and, at the last step, expand the number of channels through a 1×1 convolution, while Path B is convolved through a 1×1 with stride 2 for down sampling. Here, in Path A, we move the down sampling process into the 3×3 convolution step to avoid information loss in the beginning step as a result of the 1×1 convolution with stride 2, while, in Path B, we use average pooling instead of down sampling.

In the pooling layer, global average pooling (GAP) is used in our modified architecture, which effectively prevents overfitting of the network, strengthens the consistency of feature maps and labels and speeds up network convergence. On the other hand, originally in the ResNet50, there is only one fully connected (FC) layer. Here, we divide it into two FC layers. In the first FC layer, the number of the channels is set to be 1024, followed by a ReLU activation function. Then, dropout operation is employed, which can randomly inactivate some of the nodes to be 0 with probability 0.5 to avoid overfitting. Finally, the channels of the pooling layer are changed from 1000 to L, which is the number of the Stokes parameters. In the second FC layer, the activation function is changed from Softmax to Sigmoid. Though the Softmax activation function is widely used in multi-label image classification, transforming the output of the model into a probability distribution, since the sum of the probabilities of all categories equals 1, if the probability value of the model output is very small or large, problems may occur such as numerical overflow or gradient vanishing when choosing Softmax activation function, which may reduce the effectiveness of the network. Meanwhile, it is unlikely to meet such a problem when choosing the Sigmoid activation function, since the output range of the Sigmoid function is between 0 and 1, which means that the Sigmoid activation function is more stable in the training process compared with the Softmax activation function.

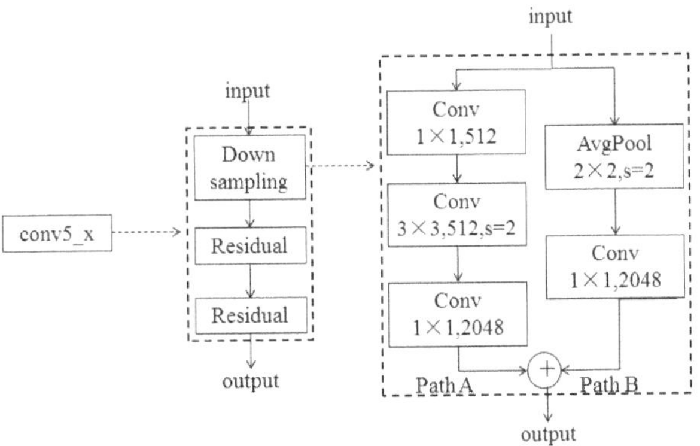

Figure 2. The structure of conv5_x.

3. Experiments and Results

In the subsequent experiments, we employ the Tensorflow framework under Python to construct, configure, train and test the network. And the network is tested in the same hardware configuration. The network is designed following the configuration to train parameters: the Adam optimizer and the MSE function are chosen as the optimizer and the loss function. The custom evaluation function is ownAccuracy. The batch size is 128, i.e., the number of images fed into the network at each time is 128. The accuracy of the network was assessed using the Mean Squared Error (MSE) metric. In order to completely evaluate the performance, in this paper, the algorithms are tested on images when light is modulated by the VHWR and VQWR, respectively. A vast quantity of experimental data are generated by Matlab for network training. Then, the trained network is tested using synthetic and real images, respectively. To evaluate the robustness of the algorithms to noise, Gaussian noise, in which the mean value is 0 and variance values range from 0 to 0.01 in steps o 0.001, is added to the synthetic images. Two state-of-the-art methods (i.e., VGG and FAM) are adopted for comparison.

3.1. Vortex Half-Wave Retarder

3.1.1. Noise-Free Data

Train

We generate noise-free data to train the network. In the training stage, we generate 50,000 perfect, noise-free images. Out of these, 40,000 are randomly chosen for training and 10,000 for validation. As the Stokes parameters are set randomly, 40,000 synthetic images contain enough polarization states of light waves. Theoretically, the trained network can study the characteristics of the training data well.

In the training stage, the size of the batch is set to be 128. MSE of 50 epochs are shown in Figure 3. It can be found that the MSE of the training set is smaller than the validation set. Additionally, the MSE on both sets tend to decrease gradually and converge.

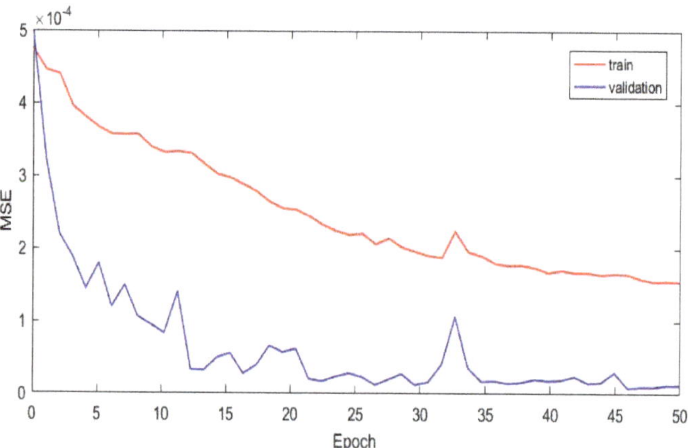

Figure 3. The MSE on the noise-free images corresponding to the training set and the validation set for VHWR.

Test

Based on the results obtained from the above training, the model with the best performance in the training process was selected as the test model. We simulate 11 test sets, each containing 1000 frames of images with zero mean Gaussian noise and variance ranging from 0 to 0.01 in steps of 0.001. We select one image from each group as an example, as shown in Figure 4. The MSE of S1 estimated by ResNet-GAP, VGG and FAM is shown in Figure 5. The MSE of S2 is shown in Figure 6. The training data shows that the MSE of the test set is extremely low, and it can even reach as low as 1×10^{-6}. With an increase in the variance of Gaussian noise, the MSE of S1 and S2 also increases gradually, indicating that the model is sensitive to noise. From Figures 5 and 6, it is evident that the MSE of ResNet-GAP has the slowest growth rate with an increase in noise. That is to say, ResNet-GAP is more robust to noise than VGG and FAM.

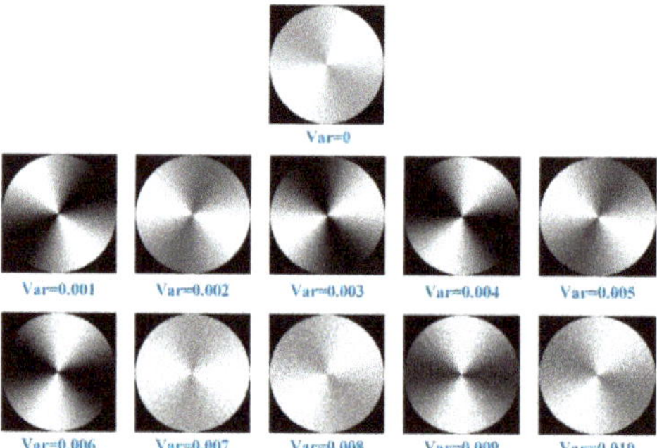

Figure 4. Images disturbed by different levels of noise while the light is modulated by a VHWR.

Furthermore, the performances of different algorithms are tested on real data. A total of 37 real images (shown in Figure 7) are captured by our experimental system, in which the transmission axis of the polarizer was fixed at 20°, and the fast axis of the VHWR was rotated from 0° to 180° with a step of 5°. The results for S1 and S2 are shown in Figure 8. They demonstrate the perfect performance of the algorithms. For ResNet-GAP, the maximum absolute error of S1 is 0.0428, and the average value is 0.0035. For S2, the maximum absolute error is 0.0574, and the average error is 0.0014. The MSE of three algorithms are given in Table 2. This illustrates that the ResNet-GAP network achieved an MSE of 1×10^{-4} when tested on real images, which is marginally larger than the results on the noise-free simulated image. This difference is due to noise in the real image.

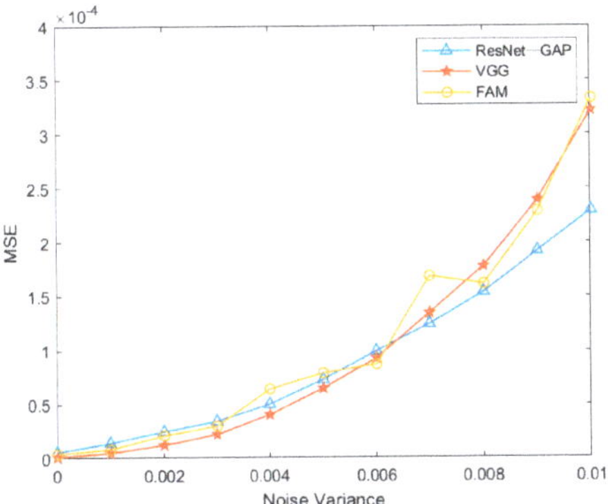

Figure 5. The MSE of S1 with respect to noise while the light is modulated by a VHWR.

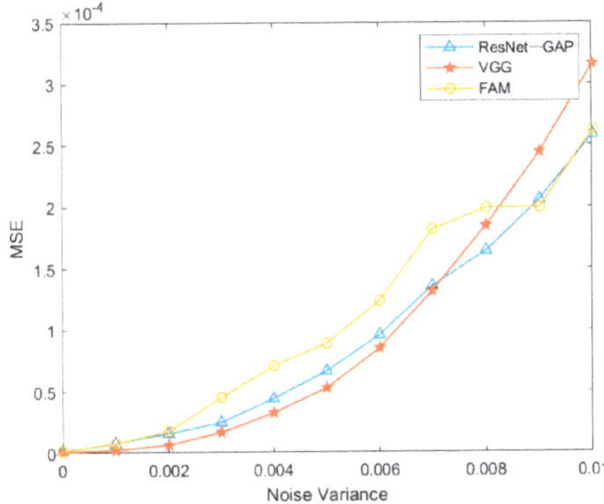

Figure 6. The MSE of S2 to noise while the light is modulated by a VHWR.

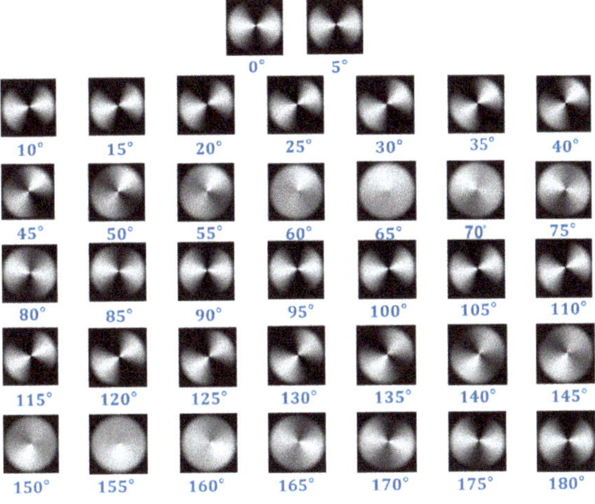

Figure 7. The real images when the fast axis of the VHWR is rotated.

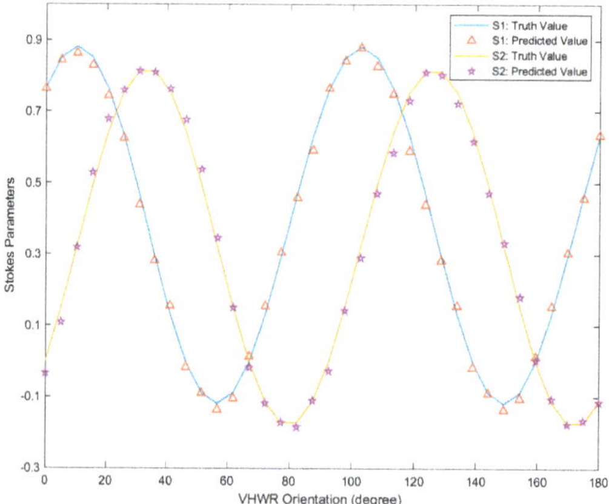

Figure 8. The Stokes parameters real values and their evaluated values by ResNet-GAP.

Table 2. Performance of algorithms on real images for VHWR.

Stokes Parameters	Algorithm	MSE
S1	ResNet-GAP	0.000385
S1	FAM	0.001623
S1	VGG	0.000741
S2	ResNet-GAP	0.000506
S2	FAM	0.000932
S2	VGG	0.000611

The truth values of S1 and S2 are plotted against the predicted values of the three algorithms, as shown in Figures 9 and 10. They indicate that ResNet-GAP has the closest predicted value to the truth value and performs better than VGG and FAM.

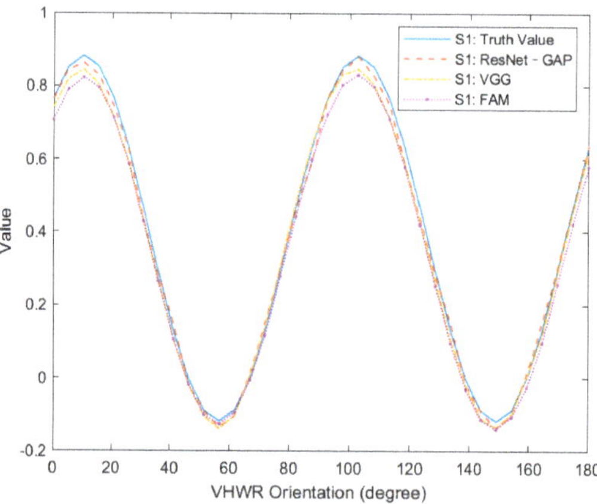

Figure 9. The values of S1 to different orientations of the fast axis for VHWR.

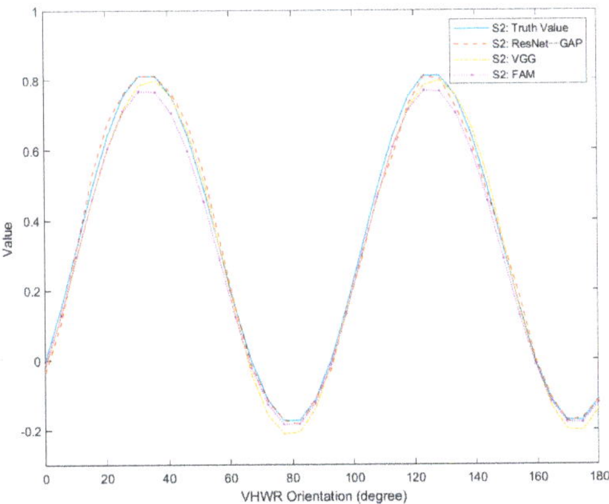

Figure 10. The values of S2 to different orientations of the fast axis for VHWR.

When 37 real images were tested, ResNet-GAP consumed only 0.01443 s, in comparison to VGG and FAM, which took 0.003694 s and 28.763 s, respectively. Obviously, the ResNet-GAP network is much faster than FAM and VGG. It is more suitable for real-time processing.

3.1.2. Noisy Data

Since noise is usually unavoidable in reality, we selected Gaussian noise images with variance 0.01 as the training and validation data sets. The numbers of images for training and validation are 40,000 and 10,000, respectively. Other parameters are kept the same as the noise-free case described above. The MSE of different epochs are shown in Figure 11. It can be seen that the MSE of the training data set is much less than that of the validation data set. The MSE of the training and validation sets converge quickly, approximately in the magnitude of 1×10^{-3}.

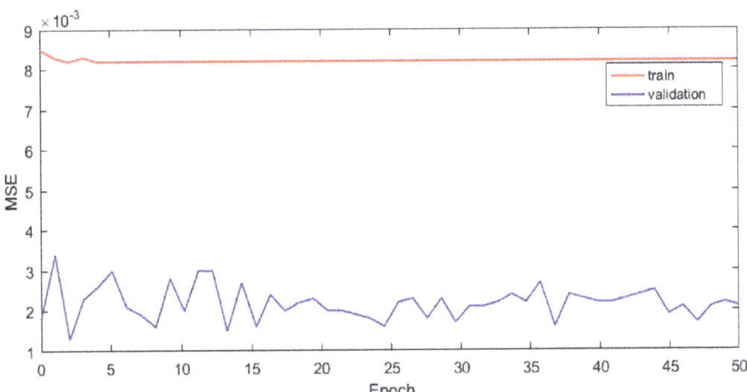

Figure 11. The MSE on the noise images corresponding to the training set and the validation set for VHWR.

However, as can be seen from Figures 5 and 6, the value of MSE obtained by the model with noise-free data is about 2.5×10^{-4} when the noise variance is 0.01. This value is much lower than the MSE obtained by the model with noisy data, as shown in Figure 11. Given the poor performance of the models obtained from training with noisy data, we do not perform further testing experiments.

3.2. Vortex Quarter-Wave Retarder

3.2.1. Noise-Free Data

Train

We examined algorithms on irradiance images when the light is modulated by a VQWR. In this case, the Stokes parameters are S1, S2 and S3. Here, the training and validation data sets are noise-free images. Figure 12 illustrates the MSE corresponding to various epochs, indicating that the MSE of the validation set is lower than that of the training set, which are all in the magnitude of 1×10^{-3}.

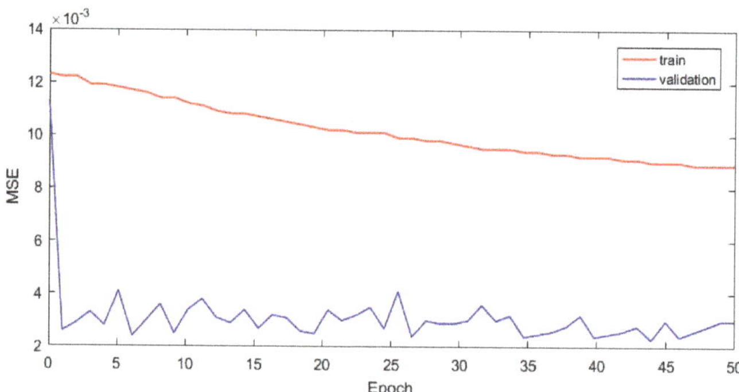

Figure 12. The MSE on the noise-free images corresponding to the training set and the validation set for VQWR.

Test

The training model with the smallest MSE in the training process was chosen as the best model for the test experiment.

To evaluate the robustness of the model to noise, 11 data sets are generated while the images are disturbed by Gaussian noise with zero mean and the variance ranging from 0 to 0.01 in steps of 0.001. Each data set contains 1000 images. We select one image from each data set as an example, and they are shown in Figure 13. The MSE of S1 estimated by ResNet-GAP, VGG and FAM is shown in Figure 14. The MSE of S2 and S3 is shown in Figures 15 and 16, respectively. The results show that, with the increase in the variance of Gaussian noise, the MSE of the Stokes parameters all increase gradually. In other words, all algorithms are sensitive to noise. However, compared to FAM, ResNet-GAP and VGG are more robust.

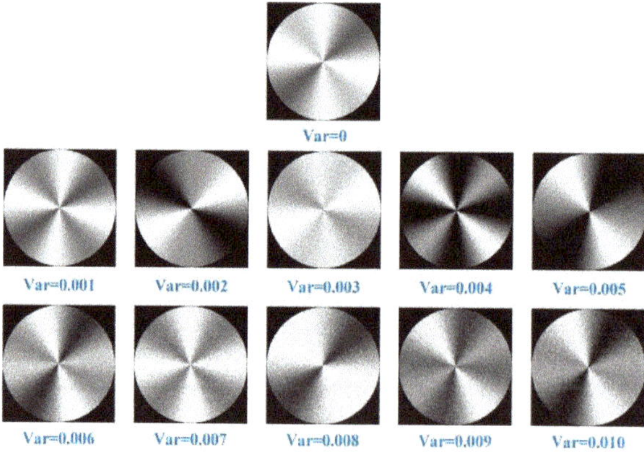

Figure 13. Images disturbed by different level of noise while the light is modulated by a VQWR.

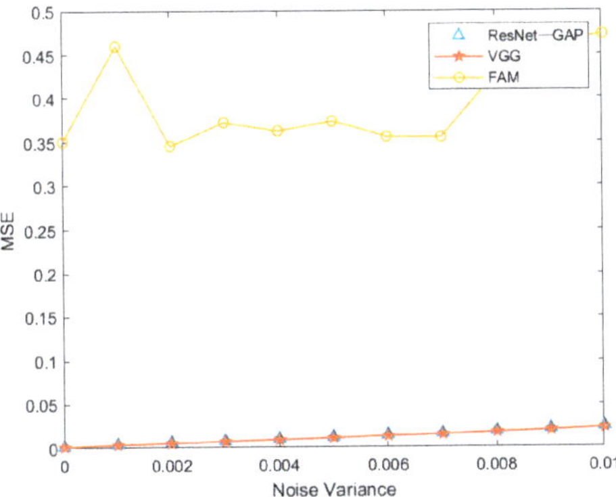

Figure 14. The MSE of S1 with respect to noise while the light is modulated by a VQWR.

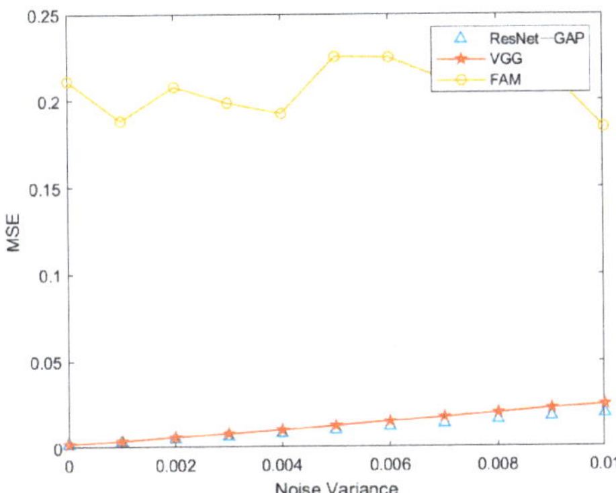

Figure 15. The MSE of S2 with respect to noise while the light is modulated by a VQWR.

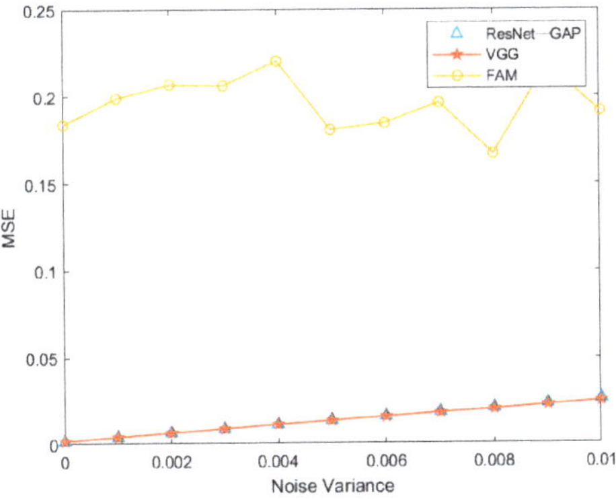

Figure 16. The MSE of S3 with respect to noise while the light is modulated by a VQWR.

Furthermore, the performances of different algorithms are tested on real data. A total of 37 real images are captured by our experimental system. The orientation of the fast axis of VQWR ranges from $0°$ to $180°$ in steps of $5°$. The real images are shown in Figure 17. The results for S1, S2 and S3 shown in Figure 18 demonstrate the perfect performance of the algorithms. For ResNet-GAP, the maximum absolute error of S1 is 0.0794, and the average value is 0.0172. For S2, the maximum absolute error is 0.0855, and the average error is 0.0073. For S3, the maximum absolute error is 0.0735, and the average error is 0.0193.

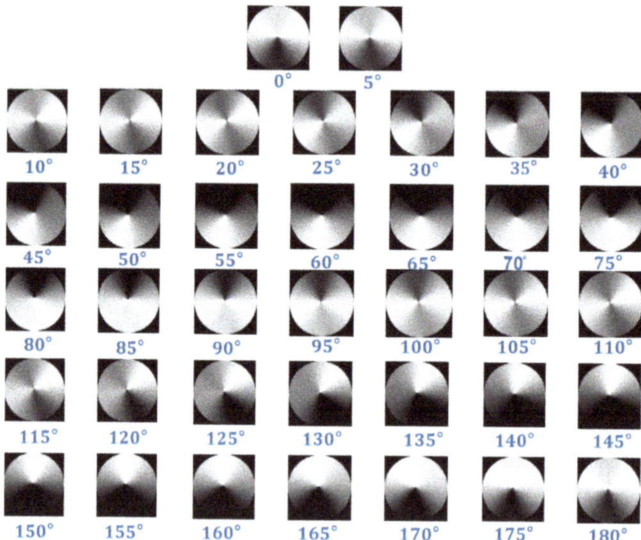

Figure 17. The real images when the fast axis of the VQWR is rotated.

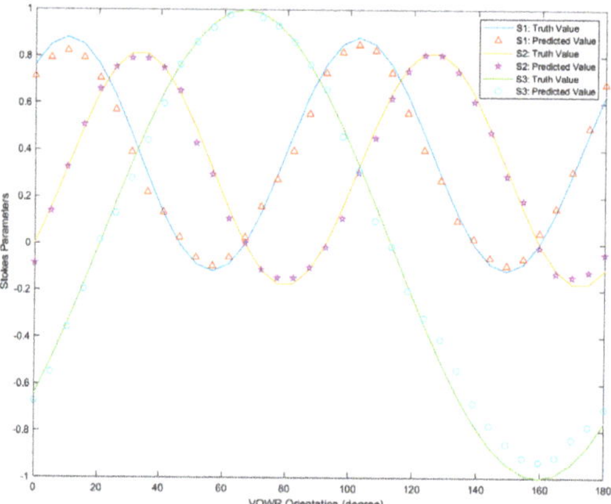

Figure 18. The Stokes parameters' real values and their evaluated values by ResNet-GAP.

The MSE of three algorithms are given in Table 3. As shown in Figures 19–21, the truth values of Stokes parameters are plotted against the predicted values of the three algorithms. They indicate that ResNet-GAP have the closest predicted value to the truth value and performed better than VGG and FAM.

The schematic diagrams of the Poincaré sphere are shown in Figures 22–25, where the blur curves represent the exact continuous distribution expressed in terms of the Stokes parameters, while the red asterisks represent the truth values and the predicted values obtained by ResNet–GAP, VGG and the FAM of the 37 real images respectively. According to the experimental results shown in these figures (from Figures 22–25), we can easily find that the measured results were well consistent with the predicted values in both the Stokes curves and Poincaré sphere, and the fitted curves based on these discrete experimental results were also matched well with the theoretical predicted results.

Table 3. Performance of algorithms on real images for VQWR.

Stokes Parameters	Network	MSE
S1	ResNet-GAP	0.0020
S1	FAM	0.0120
S1	VGG	0.0053
S2	ResNet-GAP	0.0012
S2	FAM	0.0032
S2	VGG	0.0021
S3	ResNet-GAP	0.0031
S3	FAM	0.0046
S3	VGG	0.0027

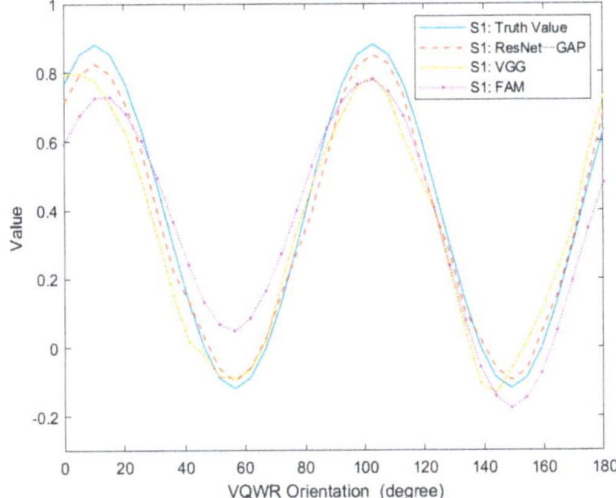

Figure 19. The values of S1 to different orientations of the fast axis for VQWR.

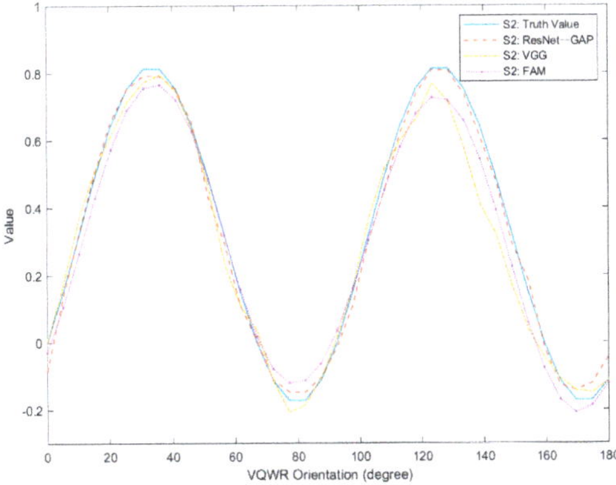

Figure 20. The values of S2 to different orientations of the fast axis for VQWR.

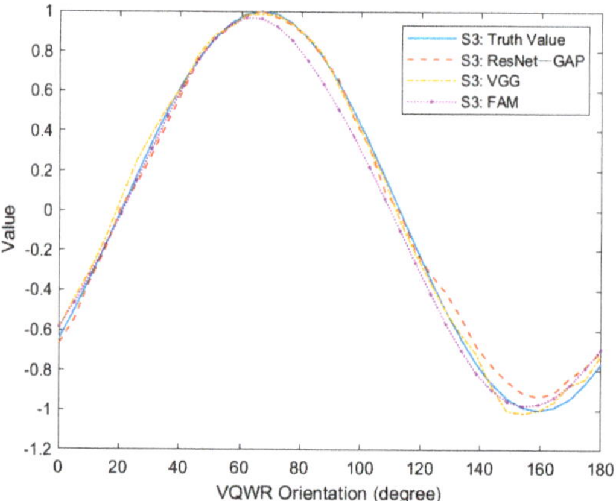

Figure 21. The values of S3 to different orientations of the fast axis for VQWR.

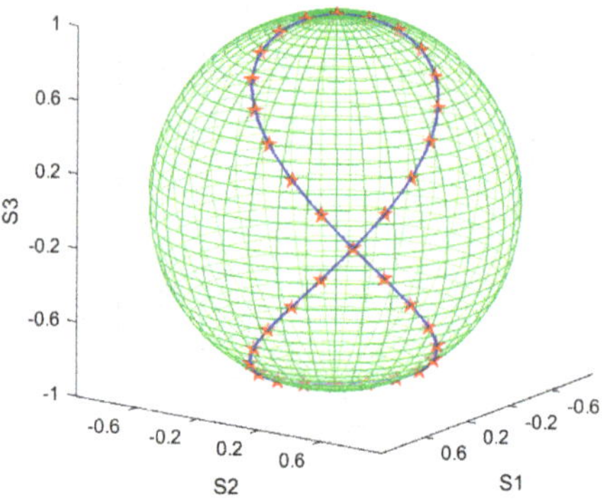

Figure 22. The schematic diagrams of the Poincaré sphere for the truth values.

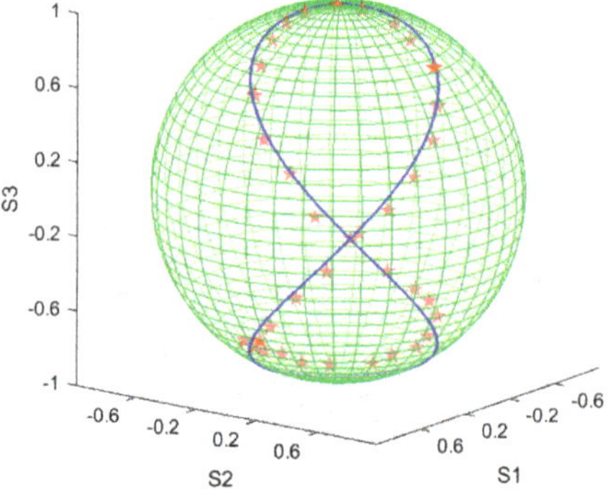

Figure 23. The schematic diagrams of the Poincaré sphere for the predicted values of ResNet-GAP.

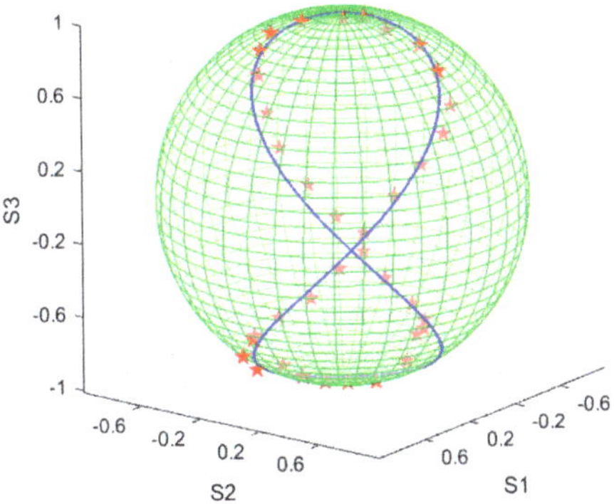

Figure 24. The schematic diagrams of the Poincaré sphere for the predicted values of VGG.

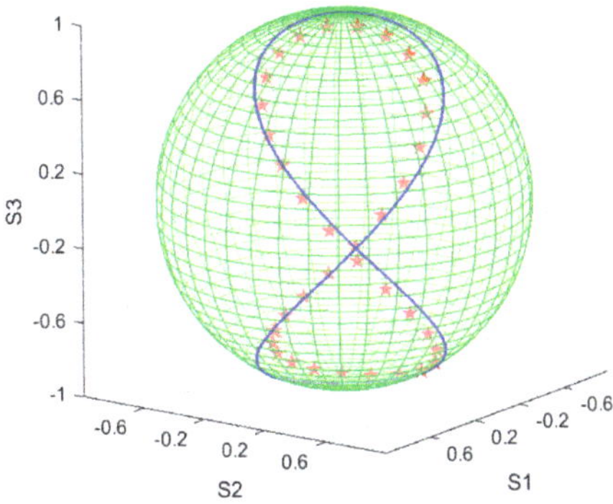

Figure 25. The schematic diagrams of the Poincaré sphere for the predicted values of FAM.

When 37 real images were tested, ResNet-GAP consumed only 0.02035 s, in comparison to VGG and FAM, which took 0.04162 s and 33.57 s, respectively. Obviously, the ResNet-GAP network is much faster than FAM and VGG. It is more suitable for real-time processing.

3.2.2. Noisy Data

Next, we use Gaussian noisy images with a zero mean and 0.01 variance as the training and validation data. All other parameters remain the same as in the VQWR without noise case. Figure 26 shows the MSE for different epochs. We observe that the MSE for the validation set is smaller than that of the training set. Both of them converge quickly, with MSE approximately in the magnitude of 1×10^{-3}.

However, the performance of the model obtained by training with noisy data is worse than that with noiseless data. Therefore, we do not perform further testing experiments.

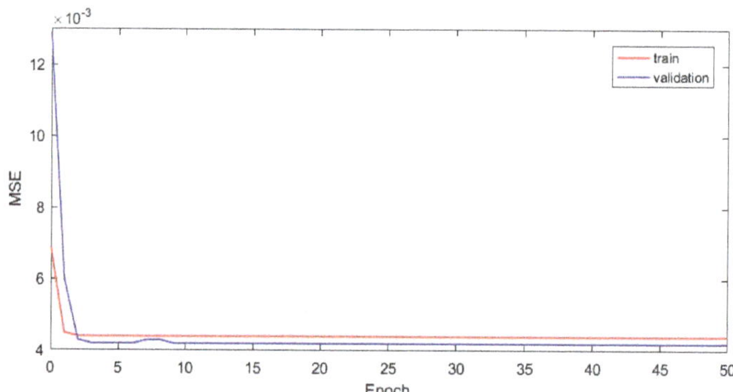

Figure 26. The MSE on the noise-free images corresponding to the training set and the validation set for VQWR.

4. Conclusions

In conclusion, we focused on how to effectively extract Stokes parameters of the light from the irradiance image in this paper. We proposed an improved Convolutional Neural Network which we call ResNet-GAP. The experiment results show that our proposed method can extract Stokes parameters effectively. We have tested on the synthetic and real data obtained from the VHWR and VQWR, respectively. Compared to VGG and FAM, the experiment results demonstrate that our method has outstanding performance with a smaller MSE and with a lower computational cost as well.

Although we have found the effectiveness of our proposed method, during our in-depth research, we also realized that some aspects are worth further research in the future. The first one is that our system uses a He-Ne laser with an operating wavelength of 632.8 nm; in future research, we will discuss the performance of ResNet-GAP under the width-wide wavelength case. Furthermore, we find that our network performs significantly better in testing real half-wavelength slice images than in testing real quarter-wavelength slice images under the criteria of MSE evaluation performance. We will further explore and explain this phenomenon in our subsequent studies. And, based on this, it is hopefully expected that we can obtain a better performance in testing real quarter-wave slice images.

Author Contributions: Conceptualization, W.W.; methodology, J.L. and C.G.; software, H.H.; writing—W.W., J.L., H.H. and C.G.; writing—review and editing, W.W. and J.S. All authors have read and agreed to the published version of the manuscript.

Funding: This research received no external funding.

Data Availability Statement: The data presented in this study are available on request from the corresponding author. The data are not publicly available, as the research group's polarization measurement is still being carried on, the later work will also rely on the current dataset.

Conflicts of Interest: The authors declare no conflict of interest.

References

1. Zhang, W.; Zhang, X.; Cao, Y.; Liu, H.; Liu, Z. Robust sky light polarization detection with an S-wave plate in a light field camera. *Appl. Opt.* **2016**, *55*, 3516–3525. [CrossRef]
2. Sun, Z.; Huang, Y.; Bao, Y.; Wu, D. Polarized remote sensing: A note on the Stokes parameters measurements from natural and man-made targets using a spectrometer. *IEEE Trans. Geosci. Remote Sens.* **2017**, *55*, 4008–4021. [CrossRef]
3. Sun, Z.; Wu, D.; Lv, Y.; Lu, S. Optical properties of reflected light from leaves: A case study from one species. *IEEE Trans. Geosci. Remote Sens.* **2019**, *57*, 4388–4406. [CrossRef]
4. Gu, Q.Y.; Han, Y.; Xu, Y.P.; Yao, H.Y.; Niu, H.F.; Huang, F. Laboratory research on polarized optical properties of saline-alkaline soil based on semi-empirical models and machine learning methods. *Remote Sens.* **2022**, *14*, 226. [CrossRef]
5. Yousaf, M.S.; Khurshid, A.; Ahmad, I.; Mahmood, R.; Alam, M.; Bukhari, S.B.; Khan, J.A.; Rafi, M.; Ikram, M. Label free characterization of soft tissue with optical polarization technique. *Laser Phys.* **2020**, *30*, 075601. [CrossRef]

6.   Yousaf, M.S.; Iftikhar, K.; Ahmat, I.; Khurshid, A. Machine assisted classification of chicken, beef and mutton tissues using optical polarimetry and Bagging model. *Photodiagn. Photodyn. Ther.* **2020**, *31*, 101779. [CrossRef]

7.   Yang, D.; He, H.; He, C.; Ma, H. Monitoring temporal microstructural variations of skeletal muscle tissues by multispectral Mueller matrix polarimetry. *Proc. SPIE* **2017**, *10059*, 21–26.

8.   Muhammad, S.Y.; Ahmat, K.; Rashid, M.; Masroor, I. Polarimetric comparison of fresh and frozen skeletal muscle tissues of goat. *Photodiagn. Photodyn. Ther.* **2020**, *32*, 102071.

9.   Lee, S.W.; Lee, S.Y.; Choi, G.; Pahk, H.J. Co-axial spectroscopic snap-shot ellipsometry for real-time thickness measurements with a small spot size. *Opt. Express* **2020**, *28*, 25879–25893. [CrossRef]

10. Goldstein, D.H. *Polarized Light*, 3rd ed.; CRC Press: Boca Raton, FL, USA, 2017; pp. 327–351.

11. Gao, C.; Lei, B. Spatially polarization-modulated ellipsometry based on the vectorial optical field and image processing. *Appl. Opt.* **2020**, *51*, 5377–5384. [CrossRef]

12. Lemus-Alonso, G.P.; Meneses-Fabian, C.; KantunMontiel, R. One-shot carrier fringe polarimeter in a double aperture common-path interferometer. *Opt. Express* **2018**, *26*, 17624–17634. [CrossRef]

13. Eshaghi, M.; Dogariu, A. Single-shot omnidirectional Stokes polarimetry. *Opt. Lett.* **2020**, *45*, 4340–4343. [CrossRef]

14. Liang, Y.; Qu, Z.; Zhong, Y.; Song, Z.; Li, S. Analysis of errors in polarimetry using a rotating waveplate. *Appl. Opt.* **2019**, *58*, 9883–9895. [CrossRef]

15. Lizana, A.; Campos, J.; van Eeckhout, A.; Márquez, A. Influence of temporal averaging in the performance of a rotating retarder imaging Stokes polarimeter. *Opt. Express* **2020**, *28*, 10981–11000. [CrossRef]

16. Liu, W.; Liao, J.; Yu, Y.; Zhang, X. High-efficient and high-accurate integrated division-of-time polarimeter. *APL Photonics* **2021**, *6*, 071302. [CrossRef]

17. Azzam, R.M.A.; De, A. Optimal beam splitters for the division-of-amplitude photopolarimeter. *J. Opt. Soc. Am. A* **2003**, *20*, 955–958. [CrossRef]

18. Gao, C.; Lei, B. Spatially modulated polarimetry based on a vortex retarder and Fourier analysis. *Chin. Opt. Lett.* **2021**, *19*, 19–24. [CrossRef]

19. Ning, T.; Li, Y.; Zhou, G.; Sun, Y.; Liu, K. Optimized spatially modulated polarimetry with an efficient calibration method and hybrid gradient descent reconstruction. *Appl. Opt.* **2020**, *61*, 2267–2274. [CrossRef]

20. Gao, C.; Wang, F.; Wen, X.; Weng, J.; Lei, B. Error calibration method for a vortex retarder based spatially modulated polarimeter. *Meas. J. Int. Meas. Confed.* **2023**, *212*, 112631. [CrossRef]

21. Lei, B.; Liu, S. Efficient polarization direction measurement by utilizing the polarization axis finder and digital image processing. *Opt. Lett.* **2018**, *43*, 2969–2972. [CrossRef]

22. Liu, L.; Ouyang, W.; Wang, X.; Fieguth, P.; Chen, J.; Liu, X.; Pietikäinen, M. Deep Learning For Generic Object Detection-Survey. *Int. J. Comput. Vis.* **2019**, *128*, 261–318. [CrossRef]

23. Minaee, S.; Boykov, Y.; Porikli, F.; Plaza, A.; Kehtarnavaz, N.; Terzopoulos, D. Image Segmentation Using Deep Learning: A Survey. *IEEE Trans. Pattern Anal. Mach. Intell.* **2022**, *44*, 3523–3542.

24. Hjouj, F.; Jouini, M.S.; Al-Khaleel, M. Advancements in 2D/3D Image Registration Methods. *IEEE Access* **2023**, *11*, 34698–34708. [CrossRef]

25. Zhao, J.; Xiong, R.; Zhang, J.; Zhao, R.; Liu, H.; Huang, T. Learning to Super-resolve Dynamic Scenes for Neuromorphic Spike Camera. In Proceedings of the 37th AAAI Conference on Artificial Intelligence, Washington, DC, USA, 7–14 February 2023.

26. Weng, J.Y.; Gao, C.; Lei, B. Real-time polarization measurement based on spatially modulated polarimeter and deep learning. *Results Phys.* **2023**, *46*, 106280. [CrossRef]

27. Hinton, G.E.; Srivastava, N.; Krizhevsky, A.; Sutskever, I.; Salakhutdinov, R.R. Improving neural networks by preventing co-adaptation of feature detectors. *arXiv* **2012**, arXiv:1207.0580.

28. He, K.; Zhang, X.; Ren, S.; Sun, J. Deep Residual Learning for Image Recognition. In Proceedings of the 2016 IEEE Conference on Computer Vision and Pattern Recognition, CVPR 2016, Las Vegas, NV, USA, 27–30 June 2016.

 *electronics*

MDPI

Article

An Image Unmixing and Stitching Deep Learning Algorithm for In-Screen Fingerprint Recognition Application

Xiaochuan Chen [1,2,*], Xuan Feng [2], Yapeng Li [2], Ran Duan [2], Lei Wang [2], Yangbing Li [2], Minghua Xuan [2], Qiaofeng Tan [1] and Xue Dong [2]

[1] State Key Laboratory of Precision Measurement Technology and Instruments, Department of Precision Instrument, Tsinghua University, Beijing 100084, China
[2] BOE Technology Group Co., Ltd., Beijing 100176, China
* Correspondence: chenxiaochuan@boe.com.cn

Abstract: The market share of organic light-emitting diode (OLED) screens in consumer electronics has grown rapidly in recent years. In order to increase the screen-to-body ratio of OLED phones, under-screen or in-screen fingerprint recognition is a must-have option. Current commercial hardware schemes include adhesive, ultrasonic, and under-screen optical ones. No mature in-screen solution has been proposed. In this work, we designed and manufactured an OLED panel with an in-screen fingerprint recognition system for the first time, by integrating an active sensor array into the OLED panel. The sensor and display module share the same set of fabrication processes when manufactured. Compared with the current widely commercially available under-screen schemes, the proposed in-screen solution can achieve a much larger functional area, better flexibility, and smaller thickness, while significantly reducing module cost. A point light source scheme, implemented by lighting up a single or several adjacent OLED pixels, instead of a conventional area source scheme as in the CMOS image sensor, or a CIS-based solution, has to be adopted since the optical distance is not long enough due to the integration. We designed a pattern for the point light sources and developed an optical unmixing network model to realize the unmixing and stitching of images obtained by each point light source at the same exposure time. After training, data verification of this network model shows that this deep learning algorithm outputs a stitched image of large area and high quality, where FRR = 0.7% given FAR = 1:50 k. In despite of a poorer quality of raw images and a much more complex algorithm compared with current commercial solutions, the proposed algorithm still obtains results comparable to peer studies, proving the effectiveness of our algorithm. Thus, the time required for fingerprint capture in our in-screen scheme is greatly reduced, by which one of the main obstacles for commercial application is overcome.

Keywords: algorithm; deep learning; image stitching; image unmixing; in-screen fingerprint recognition; OLED

Citation: Chen, X.; Feng, X.; Li, Y.; Duan, R.; Wang, L.; Li, Y.; Xuan, M.; Tan, Q.; Dong, X. An Image Unmixing and Stitching Deep Learning Algorithm for In-Screen Fingerprint Recognition Application. *Electronics* **2023**, *12*, 3768. https://doi.org/10.3390/electronics12183768

Academic Editor: Chiman Kwan

Received: 12 July 2023
Revised: 21 August 2023
Accepted: 2 September 2023
Published: 6 September 2023

1. Introduction

In recent years, many categories of consumer electronics, including smart watches, mobile phones, laptops, monitors, and TVs, have adopted OLED screens instead of conventional liquid crystal display (LCD) screens gradually. Compared with LCD, OLED has many obvious advantages [1], such as higher contrast ratio between light and dark, lower power consumption, a wider color gamut, etc. In particular, for mobile phones, OLED screens provide excellent flexibility, opening up the possibility of more module forms for users in mobile applications. For example, foldable and rollable phones with OLED screens have all been implemented [2]. In addition, due to a narrower bezel, the screen-to-body ratio of OLED screens is also higher than its LCD counterpart. In order to further achieve an extremely high screen-to-body ratio for better custom experience, other mobile phone parts related with screens must also be improved in design and implementation [3,4].

For fingerprint recognition, in terms of hardware, a few manufacturers use ultrasonic solutions [5,6]; some other researchers have also proposed a standalone module, which can be adhered to the screen [7]. The mainstream commercial under-screen fingerprint recognition solution is to stack micro lenses and CIS chips under the OLED screen to act like a camera module to capture fingerprint images [8], as shown in Figure 1a. The obtained images are then compared with the pre-captured "standard" fingerprint pattern and conclude the matching degree [9]. Such a scheme has three disadvantages. First, the screen will become thicker due to the stack of CIS and lenses, and will also add weight to the phone. Moreover, under some extreme conditions such as strong ambient light, the module can be seen, as shown in inset of Figure 1b. Second, CIS is fabricated on silicon, which is a rigid module. Inconvenience in design will be caused if some flexible features such as rollability, are needed. Third, due to the consideration on weight, power consumption, and especially cost, the fingerprint recognition module is limited to a small area of the screen. Large-area or even full-screen fingerprint recognition cannot be achieved. Therefore, many researchers and companies are developing and optimizing solutions with sensors integrated under OLED [10–12]. Although the under-screen solution solves the problem of CIS not being able to achieve flexibility, it still causes the screen thickness to increase; in addition, due to a low transmittance of OLED, and obviously due to the sensor performance under the panel process being worse than that of CIS under the standard silicon process, the low light and noise challenges are more serious [13,14]. In contrast, the fingerprint recognition solution integrated in the screen, rather than under the screen, does not increase the module thickness, as can be roughly seen in their cross-sectional schematic views in Figure 1c,d. The in-screen scheme also reduces the transmission requirement, while the cost is lower because the manufacturing process is shared with the backplane of the OLED panel, though crosstalk of display and sensors would be more severe. However, the difficulty of the integration process causes degradation in the sensor performance; thus, the integration time needed is longer. As the optical distance decreases due to the integration, the imaging area becomes smaller, so multi-frame image acquisition is required. Both of these problems require a much longer customer waiting time, which cannot be tolerated. Thus, no mature in-screen scheme has been proposed yet.

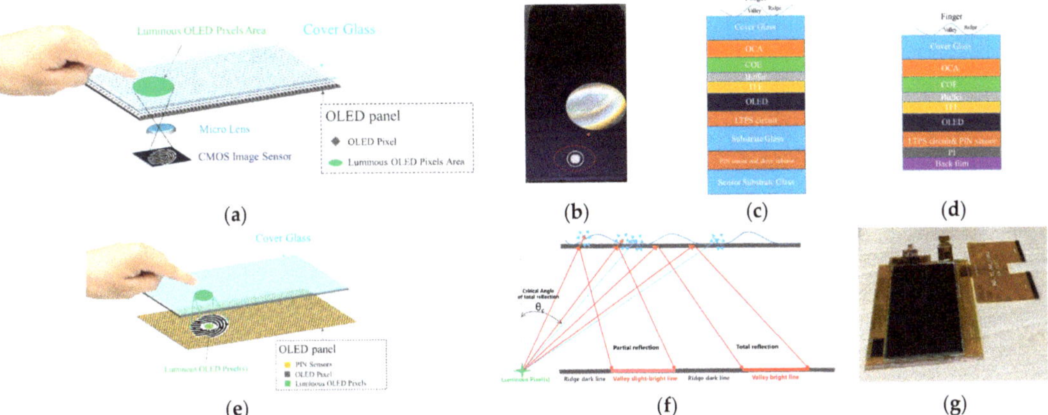

Figure 1. Comparison of commercial CIS-based solution for fingerprint recognition and our in-screen solution demonstrated in this work. (**a**) Schematic of CIS-based solution. (**b**) A photo of mobile phone with OLED screen and under-screen fingerprint recognition system integrated. The inset shows vaguely the CIS module under strong ambient light. (**c**) Common cross-sectional schematic of under-screen scheme. (**d**) Cross-sectional schematic in this work. The thickness is significantly lower.

(**e**) Schematic for our in-screen solution. (**f**) Optical path of incident light reflected to panel in our solution. Clearly, light intensity varies with the distance from the center of light source to the sensor. (**g**) A photo of fabricated OLED panel integrated with in-screen fingerprint recognition scheme in this work.

In terms of algorithms, in recent years, a lot of progress has been made in fingerprint image enhancement based on deep learning. Examples include the use of adversarial learning and edge loss to solve fingerprint sensor interoperability or cross-sensor matching problems [15,16], and the use of cyclic generation adversarial networks to enhance low-quality fingerprint images [17]. When applied to under-screen fingerprint images, due to restrictions in the acquisition environment (sometimes in low temperature or strong ambient light), image quality (sensor density is generally limited by the display size and resolution), algorithm model size (generally about 20 MB or below since consumer electronics require a rapid unlock), and so on, these algorithms have to be optimized to meet higher requirements. Huang et al. proposed an image preprocessing process for fingerprints under OLED screens [18]. To overcome the image quality problems, researchers from Samsung proposed a matching algorithm using multi-scale texture descriptors, A-KAZE, to improve the accuracy of matching [19]. Wu et al. presents a fingerprint alignment algorithm based on the latest under-screen optical fingerprint image sensors in order to avoid very similar but wrong alignment results [20]. To address blurring issues in fingerprint images due to wet fingers, Zhang proposes an algorithm named EMEE (Ellipse Model Extrapolation Equalization) based on an elliptical model [21]. In spite of these works, there are still many problems to be solved, such as handling strategies for different types of lighting conditions, reducing noise, etc. In addition, the above research results are mostly aimed at the under-screen fingerprint scheme, and the algorithm development for the in-screen fingerprint scheme is still missing.

In this study, in terms of hardware, we introduced an in-screen scheme by integrating the optical sensors and its driving circuits into the OLED panel. The sensors and the OLED pixels are fabricated on the same glass substrate and share the same set of masks when manufactured. The test results show that the integrated hardware shows excellent performance. A point light source scheme is adopted since the optical distance is not long enough due to the integration. We found that such a scheme would lead to insufficiency of the effective fingerprint image area, which makes it difficult to achieve key indicators for matching. Therefore, we designed a lighting pattern of multi-point sources for image capture. In terms of algorithms, in order to solve the key problem of the long unlocking time of the in-screen solution, we developed and trained a deep convolutional network model based on a cross grid structure, to extract, enhance, and stitch fingerprint information in multi-point light source fingerprint images. Additionally, we designed a multi-level preprocessing strategy that separately handles regions of ambient light noise, fingermark reflectance, and point sources in order to reduce the impact of environmental light on the quality of fingerprint images. The results of data verification show that the model can achieve the customers' target. Our work overcomes an important problem towards the mass production of in-screen fingerprint recognition schemes for OLED.

2. Device and Optical Methods

We developed an OLED panel with sensors integrated to realize the function of in-screen fingerprint recognition. As can be seen in Figure 1e, the sensor array is nested between OLED pixels. Cover glass above the panel is used to support and protect the panel. The sensor, known as PIN, consists of a p-type and n-type silicon, sandwiching a layer of amorphous silicon as the photosensitive material, converting light into electrons [22]. TFT circuits, used as pixel driving scheme for sensors, were fabricated by the low-temperature polycrystalline silicon (LTPS) process. OLED-related designs such as pixel layout, pixel circuits, and gate on array (GOA) circuits in this study generally follow mature commercial schemes, and also share the LTPS process. A specific designed driver IC for PIN sensors was adopted to realize the accumulation of photo-generated charges and the following processes, such as analog-to-digital conversion (ADC).

For CIS or under-screen fingerprint recognition schemes, the sensors are under the display screen, while the thickness of the display module could serve as optical distance of the image system. However, for in-screen scheme, the optical distance will be significantly reduced; thus, the conventional area source scheme is no longer applicable. We developed a point light source scheme for fingerprint image capture to adapt the proposed in-screen solution. The details are as follows.

When a finger presses down on the surface of the cover glass, one "point light source" (an OLED pixel or several adjacent pixels) right under the finger lights up and starts to luminesce towards the interface of finger and cover glass. At the valley of fingerprint, the interface is glass and air. Some of the incident light from the pixels reflects back into panel and the rest is released into the air through the interface. Since the refractive index of glass is greater than that of air, incident light with an angle greater than the critical angle of total reflection (typical value here is $42°$) will be entirely reflected at the interface, and then this light will be collected by the sensors. Meanwhile, at the ridge of the fingerprint, the interface is glass and human skin. Due to the small difference in refractive index between skin and glass, reflection here will be largely reduced, and a major part of this incident light escapes through the finger side. Therefore, sensors will capture more light reflected by fingerprint valley than that of ridge. A fingerprint image then can be captured by the sensor array, as depicted in Figure 1f. In Figure 1g, we demonstrate a photo of our fabricated OLED panel, with fingerprint recognition module integrated.

There are some incidental problems with this scheme. At the center of the point light source, in the sensor plane, there will be a non-imaging area because the light intensity reflected back is too strong, which exceeds the full well capacity of the sensors (Area I in Figure 2a). Part of the area around the center of the point light source (Area II in Figure 2a), also could not obtain valid data since the reflected ratio of light versus the incident light is too small; thus, the light intensity difference between fingerprint valley and ridge is beyond the resolution limit of the sensor. Moreover, areas far away from the center of the point light source (the outside areas beyond Area III in Figure 2a) could not image well either, due to the long optical propagation distance. The light intensity difference would thus be too small for sensors to distinguish. Therefore, in the sensor plane, for a single-point light source, the valid imaging area of in this scheme is limited to a specific range (Area III in Figure 2a). Experimental data prove that the valid imaging area is too small to provide enough fingerprint features to realize effective fingerprint matching [23]. This is also one of the main challenges faced by researchers when developing an in-screen scheme. Typically, the valid imaging area by a single-point light source scheme is about 53% of that of current commercial CIS solution (which is around 6.5 mm $\times$ 6.5 mm). To solve this problem, naturally, stitching of images from multiple point light sources is considered. There are two ways to achieve this goal, as shown in Figure 2b. One is to light up the multiple point light sources sequentially, obtaining multiple frames of images, and then stitch the images into one. This method could obtain a large image of high quality, but the disadvantage is that it requires multiple exposure times. Based on the characteristic limits of sensor and system, and the principle that the display effect should not be influenced, exposure and processing time of each point light source image needs at least 67 ms. The total image capture time of this sequential scheme would exceed the commercial standard, since a long waiting time (image capture time together with processing time such like fingerprint matching) for fingerprint recognition would not be acceptable by customers. The second method is to light up multiple point light sources at the same time to obtain a large-area image at one-time exposure and capture. However, due to image-to-object optical amplification (roughly $\times 2$ here), the obtained image in the sensor array is enlarged. Therefore, when multiple point light sources are lit up at the same time, the images of each light source will be mixed in the sensor plane, and the correct image cannot be obtained, as depicted in Figure 2c. We cannot avoid this problem by separating the point light sources far apart until they do not mix with each other, because the contact area of the finger and the cover glass is limited.

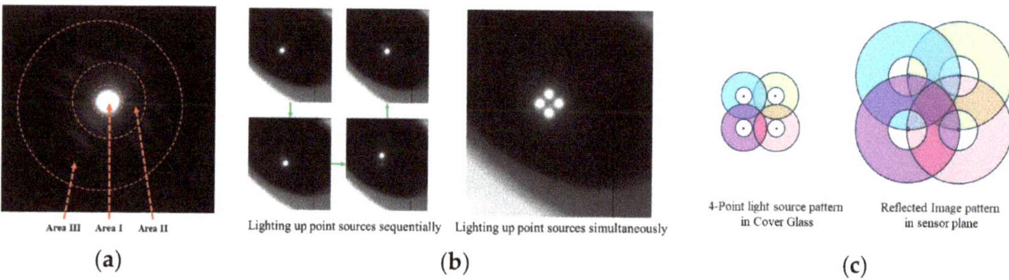

Figure 2. Image of the single-point light source scheme (**a**); lighting timing and pattern for multiple point light source scheme (**b**); and (**c**) schematic of images from different point light sources mixing with each other, due to the image-to-object optical amplification.

In order to solve the mentioned problems, we designed a light-up pattern, and developed an optical unmixing network model to realize unmixing and stitching of images obtained by each point light source at the same exposure time. We adopted a 4-point mixed lighting pattern here, as can be seen in Figure 2c. Image quality difference in spatial distribution caused by optical noise was also taken into consideration in the pattern design. The reasons for choosing this pattern will be discussed in Section 4.

3. Data Acquisition and Processing Methods

Data acquisition and processing methods include the three following parts: Section 3.1. fingerprint image acquisition by the 4-point light source scheme; Section 3.2. image preprocessing; and Section 3.3. unmixing and stitching network model. In addition, in order to train the proposed network model, we first lit up the 4-point light sources sequentially and stitched the 4 corresponding images together, and then set it as ground truth.

3.1. Data Acquisition, Ground Truth, and Dataset Preparation

A low-noise data acquisition system based on a customized analog front end was set up for charge accumulation of the PIN sensors. The system works simultaneously with a commercially used driving scheme for OLED panel. For 4-point light source image acquisition, we first lit up the 4-point light sources at the same time, and testees were required to press a finger onto the screen right upon the light sources. After image acquisition, the testees were required to keep their finger still upon the screen, and the 4-point light sources were then lit up sequentially to obtain the fingerprint images of the corresponding single-point light source. Image acquisition for each finger was conducted for several times repeatedly, and statistically abundant quantity of fingerprint images from fingers of different testees were obtained. The image was generated by the system into a 16-bit PNG format.

Fingerprint images of 18 volunteers (12 male and 6 female) in total were collected. Images of 6 fingerprints for each volunteer were obtained, and the numbering sequence is shown in Figure 3a. Due to the limited area of the sensors, multiple entries are required to fully cover the entire fingerprint feature, which is generally set to 20 entries for each finger. Therefore, it is necessary to collect as many fragments of fingerprints as possible in the first 20 times to achieve full coverage. Then, 40 more tests were conducted as validation test sets. Due to the lack of publicly accessible datasets of in-screen fingerprint images, we use these captured data to train and evaluate our algorithm.

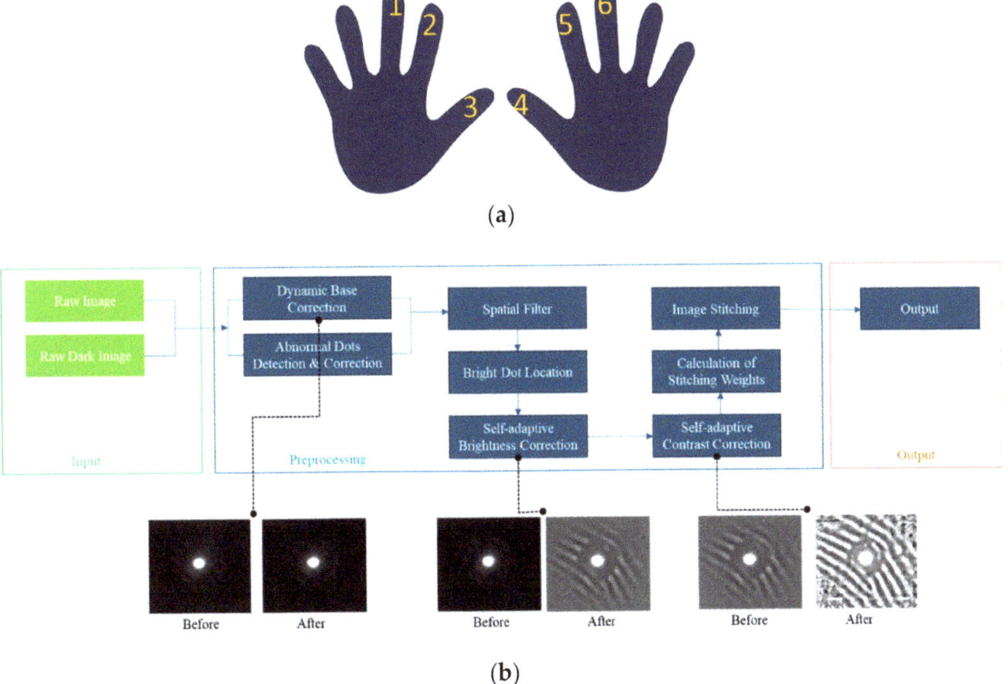

Figure 3. (**a**) Fingerprint data acquired in this work and (**b**) methods applied for preprocessing and stitching of single-point light source images.

Schematic diagram in Figure 3b shows the methods taken in data processing of each single light source image. The problems to be solved in image preprocessing mainly include three aspects: (i) The obtained image by the sensor array itself not being uniform. In the point light source scheme, except for the non-imaging area in the center of light source as depicted in Figure 2c, the luminous intensity of OLED decreases with the increase in the light exit angle, and moreover, the light intensity decreases with the increase in propagation distance when total reflection happens. (ii) Defects such as dead pixels, defective lines, and dynamic patterns. This is due to the process deviations of TFT or metal layers. (iii) Noise in the image. Noise here comes from the sensor and the image acquisition system, including the shot noise of the PIN sensor, electro-magnetic signal noise, reading noise of the system, etc. Among all the processing steps taken, self-adaptive brightness correction is used to solve problem (i). Anomaly detection and correction and image filtering are used to solve the problem (ii). Since valleys and ridges in a fingerprint image are required to be distinguished, self-adaptive contrast correction can significantly enhance the quality of the image. Then, these four images were stitched as one and set as ground truth.

3.2. Image Preprocessing

Similarly, the 4-point mixed light source images obtained by lighting up the light source pattern at the same time not only contain fingerprint information obtained by the sensor through collection of reflected light, but also contain noise information introduced by ambient light and optical signals near the light source points. Before training the proposed convolutional neural networks, it is necessary to preprocess the mixed light image, and strip and segment the primary data and secondary data. As shown in Figure 4a, region A contains major fingerprint information; region B contains a large amount of ambient light noise and a small amount of weak fingerprint data; and region C contains a strong light source signal and a small amount of fingerprint information. Here, we use the threshold segmentation method. The average value range of region A, B, C is calculated through the statistics of a large number of image data. Each pixel and threshold of the mixed light image is compared to segment the image region, as shown in Formula (1):

$$I^a_{(x,y)} = \begin{cases} I_{(x,y)} & \min_a \leq I_{(x,y)} \leq \max_a \\ 0 & I_{(x,y)} \leq \min_a \text{ or } I_{(x,y)} \geq \max_a \end{cases} \tag{1}$$

where $I^a_{(x,y)}$ stands for pixel value at position (x, y) in region A; $I_{(x,y)}$ stands for pixel value in the acquired image at position (x, y); $\min_a$ and $\max_a$ are the minimum and maximum pixel values, respectively, in an artificially selected region that is relatively smooth in region A. They are regarded as the lower limit and upper limit of the range of region A. The Laplacian edge detection algorithm is used to detect the edge in the image. The lengths and positions of the detected edges are compared, and the final extracted edge is considered as the boundary between neighboring regions. The image is then segmented. Since fingerprint data are mainly distributed in the small value range of region A, it is necessary to normalize each segmented image. The results of a processed image are shown in Figure 4b–d. Both the mixed light image and the unmixed light image are gray images in a single channel.

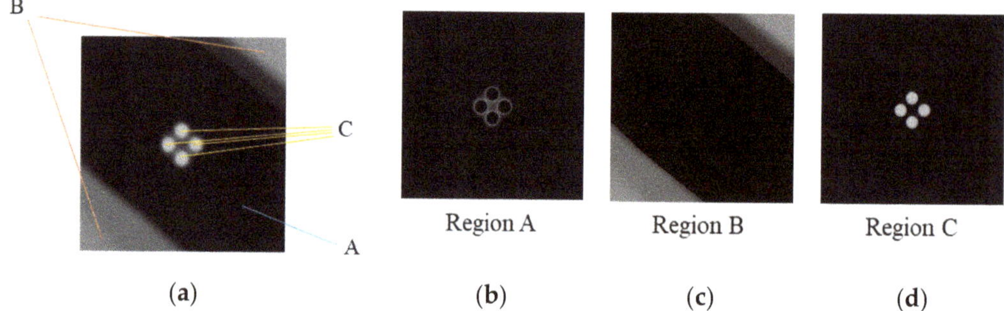

Figure 4. Preprocessing of the mixed light image. (**a**) Partition of the mixed light image. (**b**–**d**) show the images after processing of segmentation method for Region A, B and C respectively.

3.3. Network Structure of Unmixing and Stitching Algorithm

The structure design of the optical unmixing network model is shown in Figure 5. The network structure is like a cross grid, deepening the fusion between deep features and shallow features, making full use of the limited fingerprint information in the mixed optical image. The network uses spatial separable convolution to realize most of the convolution operations, with the convolution kernel size k = 5. The down-sampling layer is composed of spatially separable max pooling layer, min pooling layer, and concat layer, while the filtering kernel size of pooling layer k = 5. The network takes the above mentioned preprocessed three-layer image as input. Concat processing of the input is performed on the channel dimension. That is, the size of the input image matrix is B × 3 × H × W, where B is the number of mixed light images in a training batch, H is the height of the image, W is the width of the image. The output of the network is a B × 1 × H × W unmixing image matrix.

The function of the unmixing optical network model is to correctly recover the fingerprint image containing clear ridge and valley information from the input multi-point unmixing image. L1 loss is used to calculate the pixel-by-pixel gap between the unmixing image generated by the model and the ground truth image, and the gap can be narrowed by optimizing the network parameters, which can guide the fingerprint image generated by the model to have a more accurate ridge and valley direction. However, models that only use L1 loss training focus more on enhancing the thick and obvious ridges and valleys, while small ridges and valleys in local areas are easy to be blurred. Therefore, this paper adds the edge loss function based on the Sobel operator to improve the model's attention to small ridges and valleys. The calculation process is shown in Formulas (2)–(4):

$$L_{L1}(Y,\hat{Y}) = \frac{1}{W \times H \times C} \sum_{x=1}^{W} \sum_{y=1}^{H} \sum_{z=1}^{C} \left| Y_{(x,y,z)} - \hat{Y}_{(x,y,z)} \right| \tag{2}$$

$$L_{sobel}(Y,\hat{Y}) = L_{L1}\big(S(Y), S(\hat{Y})\big) \tag{3}$$

$$\mathrm{L}(Y,\hat{Y}) = L_{L1}(Y,\hat{Y}) + \lambda \times L_{sobel}(Y,\hat{Y}) \tag{4}$$

Y stands for the network output of the mixed light images; $\hat{Y}$ for ground truth; W, H, C, for the width and height of the image and channel number, respectively; $S(x)$ for image edge figure based on Sobel edge detection algorithm; λ value is 1.

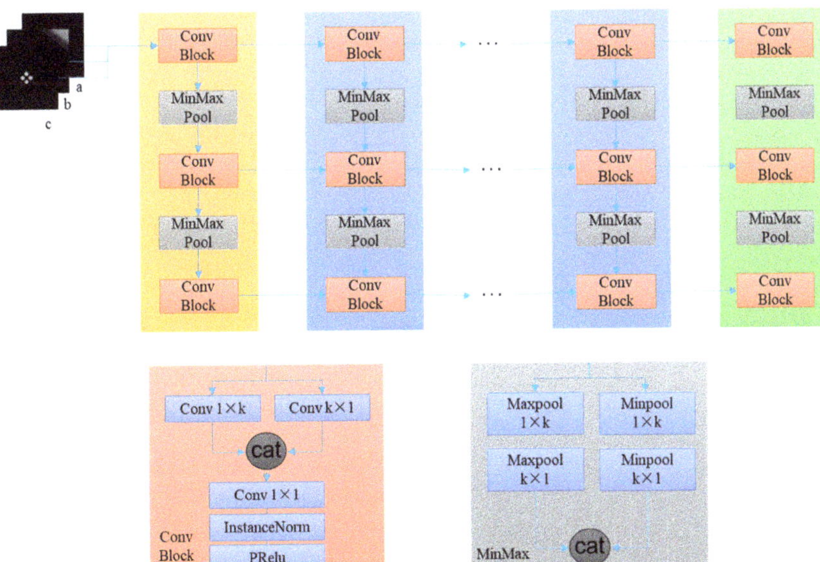

Figure 5. Network structure of the proposed unmixing algorithm.

AdamW algorithm [24] was used to optimize the model parameters of the network model. The initial learning rate was set to 10^{-4} and the batch size of the training data was set to 48.

After the fingerprint image of the point light source is preprocessed and unmixed, we intercept the ring region as the effective fingerprint region, according to the full reflection angle and the requirements of the image signal-to-noise ratio. This is usually realized by using a fixed ring mask. Several independent ring fingerprint regions are obtained. Due to the image amplification oriented by this point light source scheme, the size of the image is larger than the object. The magnification factor is M. In order to stitch, it is necessary to expand the spacing of the center position of each ring region image by M times. Since the position of each point source is set by us, its position on the image can also be calculated by a linear mapping. Therefore, its position on the Mosaic image is also obtained. For the overlapping regions of multiple ring regions, a weighted average can be used to complete the fusion of each ring region, and finally stitched into a complete fingerprint image.

4. Results and Discussion

Figure 6 shows the performance of the in-screen PIN sensor obtained by our test. As can be seen from the figure, although the sensor area of each pixel is small due to the layout limitation of the integration, it still maintains a linear response to light intensity until it reaches full well capacity (about 1800 lx, for the DUT). This shows that the process optimization of the sensor is effective enough to significantly reduce the influence of the edge effect. According to the sensor area, integration time, and IC-related parameters, further calculation can obtain that the optical response degree of our sensor is about 0.0076 fC/(lx·um^2). These data are higher than the previous work on using glass-based PIN in an under-screen scheme [4], and are also consistent with the data that we calculated based on the transmittance of the stack of film layers. This proves that the in-screen scheme can effectively increase the semaphore and reduce power consumption compared with the under-screen scheme. However, our data show that the sensor noise of the in-screen scheme is about twice that of the under-screen scheme due to interference from the display signal because of signal coupling. This issue needs to be addressed through further hardware and design optimization.

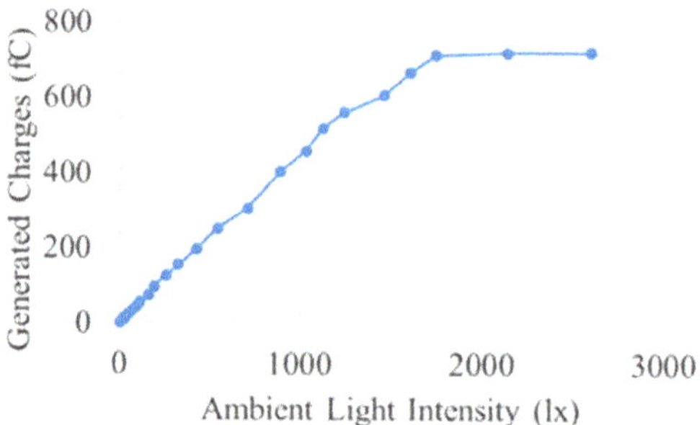

Figure 6. Performance of the PIN sensor in the proposed scheme.

Figure 7a–d show a single-light point image, single-point stitched images (ground truth), unmixing stitched images by a model trained only using L1 loss, and unmixing stitched images by a model trained using L1 loss and edge loss, respectively. It can be seen that the area of a single-point image is obviously small, and the latter three prove no difference in area size. The terminal manufacturer uses false acceptance rate (FAR) and false rejection rate (FRR) to evaluate the pros and cons of the fingerprint acquisition scheme. False acceptance refers to the event when a "wrong" fingerprint is input, but the matching scores is greater than the given threshold. False rejection refer to the event when a "right" fingerprint is input, but the matching scores fall below a given threshold given by the evaluating system. The mathematical formula is as follows:

$$FRR = \frac{NFR}{NGRA} \times 100\% \tag{5}$$

$$FAR = \frac{NFA}{NIRA} \times 100\% \tag{6}$$

where NFR refers to number of false rejections; NGRA refers to number of genuine recognition attempts; NFA refers to number of false acceptances; NIRA refers to number of imposter recognition attempts.

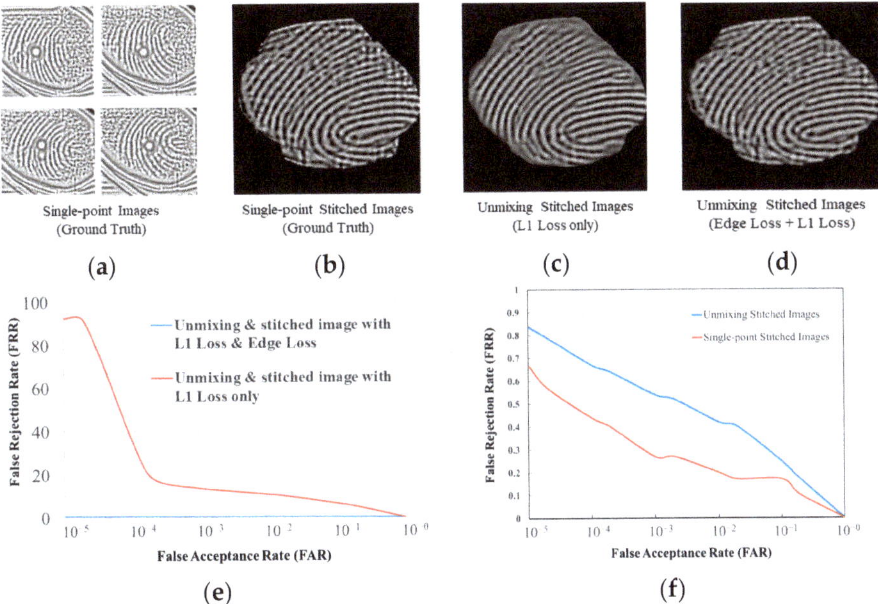

Single-point Images
(Ground Truth)

(a)

Single-point Stitched Images
(Ground Truth)

(b)

Unmixing Stitched Images
(L1 Loss only)

(c)

Unmixing Stitched Images
(Edge Loss + L1 Loss)

(d)

(e)

(f)

Figure 7. Output after processing. (**a**) Image of single-point light source after preprocessing. (**b**) Stitched image of 4 single-point light sources. (**c**) Output image of the proposed unmixing and stitching algorithm trained with L1 loss only. (**d**) Output image of the proposed unmixing and stitching algorithm trained with L1 loss and edge loss. (**e**) Comparison of key indicators with and without edge loss in the proposed unmixing and stitching algorithm. (**f**) Comparison of key indicators between the stitched images of single-point light source and the proposed unmixing and stitching image.

Quantitatively, the general requirements of customer indicators are, in the premise of FAR = 1:50 k, FRR $\leq$ 1%. We took fingerprint data that had not been used for training or tested in the evaluating system. NGRA = 3850, NIRA = 370 k are conducted in this work. Figure 7e demonstrates the influence of edge loss. Edge loss is helpful to improve the accuracy of the network model to restore the concave and convex curves of ridge and valley edges and to reduce the false adhesion and fusion of multiple bifurcated regions. It can be seen that adding edge loss will lead to better performance on FRR. Figure 7f shows the comparison of FRR of ground truth and the unmixing stitched image with edge loss involved. Also, some intermediate evaluation metrics are listed in Table 1 to quantitatively describe the effect of the proposed algorithm. It can be seen in the table and figures that the single-point stitched image and the unmixing stitched image both meet the requirements of customers, although the latter still cannot match the former (FRR = 0.7% versus 0.48%, given FAR = 1:50 k, respectively). Based on our analysis of the recognition result graph, we found that there is a significant FRR difference between the two methods of obtaining fingerprint images when FAR was less than 1%, and thereafter, the FRR curves obtained by the two methods tended to be basically parallel. We believe that this is mainly due to the insufficient number of samples participating in the training, resulting in overfitting of the model during the training process. Overfitting leads to the model mistakenly restoring some fingerprint details to the features of other fingerprints, and there are also very few cases where fingerprint image details cannot be fully restored. To address this issue, we plan to improve model performance by increasing the number of fingerprint samples in the training set. By introducing more training data, we can more comprehensively cover various details of fingerprints and reduce the occurrence of overfitting, thereby improving the accuracy and robustness of fingerprint recovery models.

Table 1. Intermediate evaluation metrics to compare the proposed algorithm and ground truth. In the table, SNR stands for signal-to-noise ratio; DYN for dynamic range; NRSS for no-reference structure similarity; SMD for sum of modulus of gray difference; EAV for edge acutance value; AREA for effective area of the fingerprint image; KEYPOINTS for the number of key points available for recognition.

		SNR	DYN	NRSS	SMD	EAV	AREA	KEYPOINTS
Proposed algorithm	Average value	0.0328	198.22	0.7185	0.0105	202.56	16,586	12.000
	Standard error	0.0026	4.2273	0.0407	0.0015	15.677	1251.6	10.779
Ground Truth	Average value	0.0346	198.75	0.7666	0.0123	217.60	16,161	12.254
	Standard error	0.0026	6.5187	0.0375	0.0016	16.072	1259.2	11.040

The lighting pattern used above is carefully designed. We also compared the image quality generated after the same proposed unmixing and stitching algorithm using different lighting patterns. Specifically, we compared the data of lighting patterns of two points, four points, and six points. From the result of FRR, as can be seen in Figure 8, a four-point pattern is slightly better than six points, which is significantly better than two points. The reasons are as follows: The area of the six-point fingerprint image is larger than that of the four-point fingerprint image, but the most central part of the fingerprint image has more saturated and non-information areas due to the bright spots, as mentioned above in Section 2, which just corresponds to the pressing center of the finger, that is, holes will be generated in the center of the fingerprint image. Although the hole texture can be recovered from the surrounding texture by means of deep learning, for the holes in a large range, there will still be differences between the recovered fingerprint texture and the original fingerprint, and this difference will lead to a difference between the six-point pattern and the four-point one. However, due to the small fingerprint area, the information that can be extracted is less, so the FRR result of the two-point fingerprint image is rather poor.

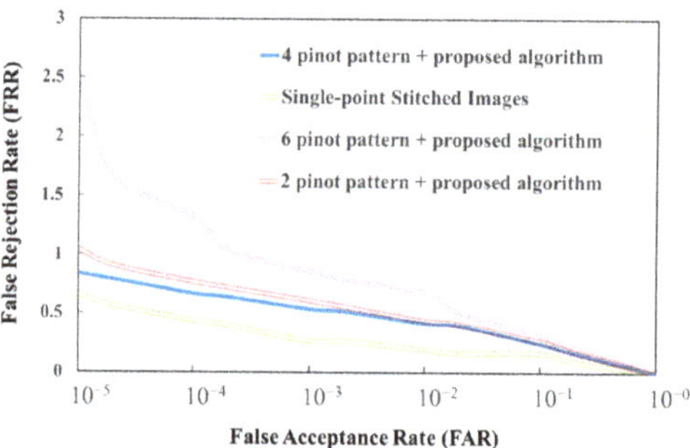

Figure 8. Comparison of lighting pattern.

As far as we know, the in-screen integrated hardware scheme and lighting pattern and the corresponding unmixing and stitching algorithm in this paper are proposed for the first time; thus, there are no identical dataset results to compare with. If we extend the scope to include under-screen schemes, Bae et al. adopt an adhesive hardware scheme, where FRR = 0.73% when FAR= 1:50 k (Figure 6 in ref. [25]). The result of our work is slightly better. In the work of Mathur et al., where a capacitive scheme is adopted, FRR > 0.2% when FAR= 1:100; either the SIFT or VeriFinger algorithm are used for processing, for in-house partial DB (Figure 9 in ref [19]). Our work demonstrates a similar result. Zhang used an under-screen optical hardware scheme, and adopted a fusion feature extraction algorithm to repair the fingerprint. For non-extreme cases, the result obtained is FRR = 1.28% @ FAR = 7.28×10^{-4} (Table 5-4 in [21]), where our work is apparently superior. Raw images captured from the capacitor scheme or the under-screen scheme used in the above three works are with less noise (especially crosstalk noise), better signal-to-noise ratio, and no need to perform light unmixing and stitching processing. Despite the poor quality of raw images and complexity in the algorithm, we still obtained results comparable to peer studies, proving the effectiveness of our scheme.

Therefore, through the design of a four-point lighting pattern and the development of a deep learning algorithm, we realized large-area and high-quality fingerprint image acquisition, which greatly compressed the acquisition time, and solved the problem of too small fingerprint imaging area caused by the single-point light source amplification imaging of the in-screen fingerprint scheme. Hence, one of the main obstacles for the mass production of the in-screen scheme was cleared.

5. Conclusions

By integrating the PIN sensor array into the OLED screen, for the first time, we designed and prepared an OLED in-screen fingerprint recognition system. Compared to current commercial solutions such as adhesive or under-screen schemes, our solution can achieve larger area, better flexible matching, and a significant reduction in cost, turning out to be a better option. In order to solve the problems of insufficient area of fingerprint imaging at one capture and insufficient time of collection at multiple captures in this hardware scheme, we designed an image stitching scheme through four-point light sources and developed an unmixing optical network model to realize optical unmixing and stitching of the one-time-exposure images. According to the matching scores after a large amount of data trained in this model, it is proved that our model, FRR = 0.7% when FAR = 1:50 k, meets the customers' target well. Despite the poor quality of raw images and complexity in the algorithm compared with the above-mentioned schemes, we still obtained results comparable to peer studies, proving the effectiveness of our scheme. The results show the proposed lighting pattern and deep learning algorithm we developed can help obtain high-quality and large-area fingerprint images, which can greatly reduce the time required for fingerprint collection, and thus provide a feasible scheme for the mass production of in-screen integration products.

The work ahead needs to be performed are mainly on three aspects. On the hardware side, PIN performance and integration design need to be optimized to reduce noise interference, especially crosstalk between display and sensing, which will radically reduce the size and difficulty of the algorithm. In terms of lighting pattern, although the mode we adopt in this study can meet the requirements, it still needs to be further optimized according to the hardware scheme and algorithm, so as to obtain enough data in a shorter time and reduce the time required to identify, thus providing a better user experience. In terms of algorithms, the accuracy and robustness of our algorithms will be improved if there are more data samples in the future, or if there are datasets of open access available. In addition, there is still a lot of work needed to be performed in the restoration and reconstruction of fingerprint details.

Author Contributions: Conceptualization, Q.T. and X.D.; methodology, X.C.; software, X.C., Y.L. (Yapeng Li) and R.D.; validation, X.C. and X.F.; formal analysis, X.C., X.F., Y.L. (Yapeng Li) and R.D.; investigation, X.C. and X.F.; resources, X.C., L.W. and M.X.; data curation, X.C., L.W., Y.L. (Yangbing Li) and M.X.; writing—original draft preparation, X.C.; writing—review and editing, X.F., Y.L. (Yapeng Li) and R.D.; visualization, X.C.; supervision, Q.T. and X.D.; project administration, X.F.; funding acquisition, X.D. All authors have read and agreed to the published version of the manuscript.

Funding: This research was funded by R&D center of BOE Technology Group Co., Ltd. under Grant No. 000040003033.

Data Availability Statement: The data presented in this study are available on request from the corresponding author. The data are not publicly available due to privacy issues.

Conflicts of Interest: The authors declare no conflict of interest.

References

1. Chen, H.W.; Lee, J.H.; Lin, B.Y.; Chen, S.; Wu, S.T. Liquid crystal display and organic light-emitting diode display: Present status and future perspectives. *Light. Sci. Appl.* **2017**, *6*, 77–89. [CrossRef] [PubMed]
2. Xu, C.X.; Shu, S.; Lu, J.N.; Yuan, G.C.; Yao, Q.; Wang, L.; Xu, Z.Q.; Sun, Z.Y. 24-4: Foldable AMOLED Display Utilizing Novel COE Structure. *SID Int. Symp. Dig. Technol. Pap.* **2018**, *49*, 310–313. [CrossRef]
3. Ma, H.; Liu, Z.; Heo, S.; Lee, J.; Na, K.; Han, B.J.; Jung, S.; Park, K.; Kim, J.J.; Bien, F. On-display transparent half-diamond pattern capacitive fingerprint sensor compatible with AMOLED display. *IEEE Sens. J.* **2016**, *16*, 8124–8131. [CrossRef]
4. Xu, C.X.; Yao, Q.; Li, X.H.; Shu, S.; He, W.; Xu, Z.Q.; Dong, L.W.; Wang, W.J.; Gao, Z.K.; Yuan, G.C. High transmittance under-display camera structure with COE. *SID Int. Symp. Dig. Technol. Pap.* **2022**, *53*, 48–50. [CrossRef]
5. Lu, Y.; Tang, H.; Fung, S.; Wang, Q.; Tsai, J.M.; Daneman, M.; Boser, B.E.; Horsley, D.A. Ultrasonic fingerprint sensor using a piezoelectric micromachined ultrasonic transducer array integrated with complementary metal oxide semiconductor electronics. *Appl. Phys. Lett.* **2015**, *106*, 226. [CrossRef]
6. Xu, C.; Lu, Y.; Strohmann, J.L.; Panchawagh, H. Large Area Multi-Functional Ultrasound Sensor: Fingerprint, Touch Pressure, Passive Stylus. In Proceedings of the IEEE International Ultrasonics Symposium (IUS), Las Vegas, NV, USA, 7–11 September 2020.
7. Huang, J.D.; Lin, C.H.; Wang, T.Y.; Luo, J.; Yao, I.A.; Lee, T.Y.; Liu, Y.T.; Chang, T.S.; Lin, P.H.; Chang, C.Y. Ultra-Thin Integrations of Optical Array Sensors with Displays and any Transparent Surface for Fingerprint Imaging and Beyond—Toward a Universal Sensor for Display. *SID Int. Symp. Dig. Technol. Pap.* **2023**, *1*, 558. [CrossRef]
8. Jang, K.W.; Kim, K.S.; Jeong, K.H. Ultrathin Contact-Imaging Camera for Fingerprint Imaging Using Microlens Array and Multiple Block Layers. In Proceedings of the International Conference on Optical MEMS and Nanophotonics (OMN), Lausanne, Switzerland, 29 July–2 August 2018.
9. Ko, T.; Krishnan, R. Monitoring and reporting of fingerprint image quality and match accuracy for a large user application. In Proceedings of the 33rd Applied Imagery Pattern Recognition Workshop (AIPR'04), Washington, DC, USA, 13–15 December 2004; pp. 159–164.
10. Lai, C.C.; Wang, W.Y.; Yuan, Y.C.; Hung, H.J.; Li, S.J.; Chen, C.H.; Tai, Y.H.; Liu, H.W.; Tang, Y.Y.; Chang, Y.C.; et al. Designing High-Sensitivity Optical Sensor for In-Cell Fingerprint Sensor with Thick Cover Glass in OLED Display. *SID Int. Symp. Dig. Technol. Pap.* **2023**, *1*, 554. [CrossRef]
11. Zhang, G.; Ye, X.; Shuai, C.; He, R.; Zha, G.; Liu, G.; Yuan, J. Analysis of Large Area Optical Fingerprint Recognition Technology under OLED Screen. *SID Int. Symp. Dig. Technol. Pap.* **2023**, *1*, 565. [CrossRef]
12. Zhou, L.; Qi, C.; Cheng, J.; Wang, R.; Kong, D.; Barbot, A. Organic optical sensor based on a-Si TFT backplane used in fingerprint identification under OLED display. *SID Int. Symp. Dig. Technol. Pap.* **2021**, *1*, 52. [CrossRef]
13. Zeng, Y.; Zhu, S.; Wu, Y.; Yao, Q. Pinhole Matrix Fingerprint on Display Technology for CFOT OLED Display. *SID Int. Symp. Dig. Technol. Pap.* **2023**, *1*, 562. [CrossRef]
14. Tai, Y.H.; Tu, C.C.; Yuan, Y.C.; Chang, Y.J.; Chuang, C.Y. Light-controlled gap-type TFT used for large-area under-screen fingerprint sensor. *IEEE J. Electron Devices Soc.* **2021**, *9*, 517–520. [CrossRef]
15. Helala, A.S.; Muhammad, H.; Hatim, A.; Mansour, A.Z. Cross-sensor fingerprint matching method based on orientation, gradient, and gabor-hog descriptors with score level fusion. *IEEE Access* **2018**, *6*, 2169–3536.
16. Alrashidi, A.; Alotaibi, A.; Hussain, M.; AlShehri, H.; AboAlSamh, H.; Bebis, G. Cross-Sensor Fingerprint Matching Using Siamese Network and Adversarial Learning. *Sensors* **2021**, *21*, 3657. [CrossRef] [PubMed]
17. Wang, L.; Wang, L.; Chen, S. ESA-cycleGAN: Edge feature and self-attention based cycle-consistent generative adversarial network for style transfer. *IET Image Process.* **2022**, *16*, 176–190. [CrossRef]
18. Huang, M.; Zhuang, M.; Zhou, J.; Wu, X. Preprocessing Method Flow of Under-screen Fingerprint Image. In Proceedings of the IEEE 14th International Conference on Anti-counterfeiting, Security, and Identification (ASID), Xiamen, China, 30 October–1 November 2020; pp. 49–53.
19. Mathur, S.; Vjay, A.; Shah, J.; Das, S.; Malla, A. Methodology for partial fingerprint enrollment and authentication on mobile devices. In Proceedings of the International Conference on Biometrics (ICB), Halmstad, Sweden, 13–16 June 2016; pp. 1–8.
20. Wu, Y.Z.; Wu, H.M.; Hsiao, P.Y. Meticulous Alignment for On-Display Fingerprint Sensor on Smartphone. In Proceedings of the IEEE International Conference on Consumer Electronics-Taiwan (ICCE-TW), Penghu, Taiwan, 15–17 September 2021; pp. 1–2.
21. Zhang, R. Research on Image Restoration and Feature Extraction Technology for OLED Screen Fingerprint. Master's Thesis, Huazhong University of Science and Technology, Wuhan, China, 2021.
22. Okamura; Kimura; Shirai; Yamauchi. A 32 × 32 two-dimensional photodetector array using a-Si PIN photodiodes and poly-Si TFTs integrated on a transparent substrate. In Proceedings of the International Technical Digest on Electron Devices Meeting, San Francisco, CA, USA, 13–16 December 1992; pp. 685–688.
23. Chen, F.; Zhou, J. On the influence of fingerprint area in partial fingerprint recognition. In *Lecture Notes in Computer Science*; Springer: Berlin/Heidelberg, Germany, 2012; p. 7701.

24. Loshchilov, I.; Hutter, F. Decoupled Weight Decay Regularization. In Proceedings of the International Conference on Learning Representations, Toulon, France, 24–26 April 2017.
25. Bae, S.; Ling, Y.; Lin, W.; Zhu, H. Optical Fingerprint Sensor Based on a-Si:H TFT Technology. *SID Int. Symp. Dig. Technol. Pap.* **2018**, *1*, 1017–1020. [CrossRef]

MDPI

St. Alban-Anlage 66

4052 Basel

Switzerland

www.mdpi.com

Electronics Editorial Office

E-mail: electronics@mdpi.com

www.mdpi.com/journal/electronics